The Quantum Labyrinth

Fundamental Theories of Physics

An International Book Series on The Fundamental Theories of Physics: Their Clarification, Development and Application

Volume 51

The Quantum Labyrinth

by

Dick J. Hoekzema

KLUWER ACADEMIC PUBLISHERS
DORDRECHT / BOSTON / LONDON

ISBN 0-7923-2066-2

Published by Kluwer Academic Publishers,
P.O. Box 17, 3300 AA Dordrecht, The Netherlands.

Kluwer Academic Publishers incorporates
the publishing programmes of
D. Reidel, Martinus Nijhoff, Dr W. Junk and MTP Press.

Sold and distributed in the U.S.A. and Canada
by Kluwer Academic Publishers,
101 Philip Drive, Norwell, MA 02061, U.S.A.

In all other countries, sold and distributed
by Kluwer Academic Publishers Group,
P.O. Box 322, 3300 AH Dordrecht, The Netherlands.

Printed on acid-free paper

Printed in the Netherlands

To Anneke, Elseline, Tessa, and Renee.

Sometimes the world has a load of questions
Seems like the world knows nothing at all
The world is near but it's out of reach
Some people touch it ... but they can't hold on

David Byrne and Brian Eno
THE GREAT CURVE

Contents

Preface

At the outset of the research leading to this book I held a position somewhere close to 'the standard Copenhagen interpretation' of QM. I was strongly attracted to, in particular, the philosophy of Niels Bohr. However, being aware of some of the problematic sides and ambiguities of his views and of new developments which have taken place in QM after his time, the main challenge would be to develop a more up to date version version of his approach and express it in a philosophically unobjectionable way.

Traces of this original attitude can still be found in views I hold nowadays. For instance, I think that I now know a satisfactory and correct way of dealing with features like 'complementarity', and I still see this as a relevant subject. In many other respects, however, there have been major changes in my position. In fact, during certain stages of my research my views simply started moving and kept on doing so at an irritating pace and for uncomfortably long periods of time. I learned, for example that at least some of the classical ideas about theory structure are much better than I had realized, and cannot just be pushed aside for anything even as impressive as empirical success. I have therefore sought to push QM back into the schemes of classical logic, and various ideas developed in the framework of 'hidden variable theories', the paradigm opponents of the Copenhagen class of interpretations of QM, turned out to be invaluable to this purpose. In particular I have returned to the classical concepts of states and magnitudes, as *subsets of* and as *functions on* a state space, although retaining such features as indeterminism and complementarity. These things are accomplished within a *contextual* approach, in which the usual observables of QM appear as magnitude *classes* rather than magnitudes. This scheme can be worked out in a number of ways, some of which can be tentatively linked to existing interpretations of QM.

Although in some respects the contextual approach remains relatively close to standard ideas about QM, the return to a classical concept of state must be considered as a major piece of surgery into the theory. I do claim, however, that it is more than worthwhile. In the course of my research I have gradually come to the conclusion that standard QM, as it is taught at university, is a conceptual garbage can, where the appearance of coherence rests entirely on some highly debatable analogies with classical theory structure.

The standard formalism of QM can be argued to be malignantly incomplete, resulting in a sizable fracture between the formalism and its 'physical meaning'. From a historical perspective, this may have been beneficial. Holding on to this incomplete formalism has enabled physicists, for a number of decades, to reach a workable consensus by covering up some potent sources of disagreement. This, however, is not what formalisms are meant for. Formalisms are meant to be instruments for expressing, as precisely as possible or desirable, what one *means* to say. Different theories, and ipso facto different interpretations of QM, deserve their own formalism. This will add to the further development of each of the theories itself, and it will even make it easier to form a clear perception of the mutual differences and relative benefits. Fractures between meaning and formalism will be paid for, in the long run. Historically such detours as taken by QM may be unavoidable, and even beneficial for creating 'breathing room' for a new theory, but eventually it will become necessary to redesign the theory and achieve a tighter fit between formalism, meaning, and practical application. The ongoing empirical success of QM is remarkable, but nevertheless there is no way of knowing what damage has been done thus far, as this can be measured only against the perspective of another theory. I will suggest that the contextual approach offers an appropriate, and relatively straightforward, logical and philosophical background for setting this up. It offers, on the one hand, the possibility of accommodating common intuitive ideas to which physicists may wish to hold on, but, on the other hand, also new perspectives on various foundational problems. For instance, the single formula expressing the power of the approach in the most succinct way may be the new probability measure (6.15), which throws a quite new light on the status of the probability concept in QM.

Another development that took place during my research is that I became a sturdier pluralist, in philosophical conviction as well as in matters concerning physics, and in particular QM. There are a number of widely different alternative approaches to the further development of QM. It would not be wise to try to make a collective choice at this moment, and I am not sure that such a moment will ever arrive in the future. Theories are the products of our imagination, subjected to requirements of various types of adequacy, among which empirical adequacy may be regarded as predominant. In a sense a theory represents 'what one is able to maintain of how one would like the world to be'. As there is no general agreement on aims and desires, and as the world seems more liberal than some of us care to admit, uniqueness of scientific development can hardly be expected, except, perhaps, by a lack of imagination. Plurality, however, the supply of, and competition between, many alternative views, may be regarded as a desirable and lasting feature of science, and in fact as a sign of its good health. A theory, a philosophy, a world view, is a work of art; and what is so bad about having plenty of them? For some reason an obsession with absolute truth has, over the ages, become a widespread cultural phenomenon, but there is no need to accede to this. Therefore, part of my research efforts has been put into further developing some aspects of a pluralist philosophy and logic.

I want to thank everybody who has contributed to the completion of this book.

A more extensive list of acknowledgements is included below.

Acknowledgements

I will start by commemorating the late professor Ubbink, who supported my original research proposals and was supervisor of the project when it was granted, until his death in 1986. Further I want to thank Dennis Dieks, especially also for the efforts put in managing the practical arrangements surrounding the project. I owe much to Jan Hilgevoord, and to his willingness to take over the difficult task of trying to supervise the project in its later stages. It was a pleasure working with you Jan, and parts of the text have greatly benefited from it, especially in the chapters 3, 4, 5. I thank also the other members of the department of Foundations of Physics through the years, for their comments, their suggestions, and their patience. This includes in particular Jos Uffink and Dennis Dieks, but also Willem Roos, with whom I worked as a student. Various parts of the present work are derived quite directly from things we were pursuing at that time. I am grateful to all contributors to, what may be rightly called, the 'Hilgevoord seminar'. It is difficult for me to imagine what working on this subject would have been like without these monthly meetings. I may especially mention Roger Cooke, whose influence on this work is greater than he is presumably aware of, Peter Kroes, for commenting on an early draft of some chapters, John Dorling, for his support and his refreshingly nonorthodox approach, and Hans Radder and Hans van den Berg, with whom I had the pleasure to cooperate on some work concerning quantum probability.

An important phase in the research project began when I got the opportunity to work at the department of applied logic in Utrecht, to do research on the semantics of programming languages. Working there added significantly to my modesty, but further provided me with sufficient background to develop with some ease the central ideas of the contextual approach, which I had been grappling with for a considerable period of time. I want to thank Jan Bergstra for arranging this opportunity, and for valuable suggestions and advice about various parts of the text. I thank all members of the logic department for the cordial way in which I was received, and in particular Piet Rodenburg and Albert Visser for many pleasant hours and educative and interesting conversations.

Going back in time, I thank Wim Turkenburg, under whose supervision I worked at the science and society department, previous to the start of my research project on QM. Working there helped to make me see that discussions about topics like 'reality' and 'rationality' are by no means as abstract and academic as they might seem. It further added to my awareness that disagreement between experts may be as common in science as it is elsewhere, and it gave me a better idea of how in practice inextricable mixtures of tradition, keen thinking, and parochial interests are forged into subcultural standards of rationality, and other forms of social ritual which guide decision making in our society[1].

Thanks are due to Hans Maassen for many stimulating discussions on various topics, and to Olivier Costa de Beauregard for a stimulating conversation and correspondence about time symmetry. Paul Feyerabend and Bas van Fraassen are thanked simply because, among contemporary philosophers of science, they have been major sources of inspiration.

I am indebted to Hanneke Pasveer, Riny van 't Schip, and Anneke Kerkhof for typing parts of the text, in the period before I started writing directly into the machine, to Anneke for helping to correct the multitudes of typing errors which appeared since, and to Ruud Hoekzema for his help in setting up my computer system and for his further assistence in getting the text printed. Theo Daselaar and J.H. Kerkhof have been of much help for improving English grammar and spelling.

To Anneke I am indebted most of all for her support, patience, love, and for many conversations which outsiders, if present, might have interpreted as being about art, philosophy, or other silly stuff.

The initial stay at the department of Foundations of Science in Utrecht was supported by the Dutch Organization for the Advancement of Pure Scientific Research (ZWO, nowadays NWO), the later stay was on a contract of the University of Utrecht, and the stay at the department of Applied Logic was made possible by a contract of the N.V. Philips, on an Esprit project concerning the semantics of programming languages.

[1]I do not mean to imply that any of the persons mentioned agrees with any of the views stated in this book.

The Quantum Labyrinth

A Treatise on Quantum Mechanics and Comparative Metaphysics

This is a book about the interpretation of quantum mechanics (QM), but relatively much attention will be paid to other subjects. Surveying the debate on QM several questions may come up, such as: what is an interpretation of QM, how are we to cope with the fact that there are quite a few of them and what methods can we use to study their structure and the relations between them? These questions lead to various topics of a more general logical and philosophical nature, involving issues which are not trivial, not uncontroversial, and not irrelevant. Most of these topics can be understood as relating, either directly or more remotely, to a single theme, which one might regard as the main theme of the entire debate on QM:

> Does QM give us a picture of physical reality and if so, what does this picture look like?

In spite of the fact that it can be phrased in a single sentence, this is not a simple question. As I see it, an attempt to answer it could be a project proceeding through three phases:

> In the first phase one would have to lay down a logical and philosophical environment suitable for handling interpretational problems in QM, and specifically for comparing different interpretations. There is very little reason to believe that standard notions in a shared 'common sense background' are sufficient in this case, so one will have to be more explicit about these matters.
>
> The second phase would then be to fit QM, and to fit different interpretations of QM, into this framework.
>
> The third phase would consist of comparing these interpretations, i.e. to investigate the relations between them and develop standards for their relative appraisal.

In the present book the emphasis will be on setting up phase 1 and 2 of this project. There will be little covering of phase 3, which should be considered as a major project on its own. Before taking a serious start, let us spend some introductory words on each of the three phases.

Phase 1, setting up a framework

Choosing between competing theories, or more generally comparing different theories, involves some severe difficulties on which philosophers and logicians have spent enormous efforts. Unfortunately, this has not resulted in a situation where one has anything like a standard theory for the logic of intertheoretic relations. What luxury it would be to have such a thing on a shelf and to merely have to pick it up and apply it to the problem of comparing different interpretations of QM. Instead, the subject turns out to be a high risk area for the innocent physicist or semi-philosopher who dares to interfere. In the first place, an awful lot has been written that is claimed to be relevant to the issue, so there is no difficulty in reading yourself to exhaustion. In the second place, there is a confusing diversity of views on the matter, so there is no trouble in getting lost or in getting involved in distasteful disagreements.

Nevertheless, the field is of crucial importance to our purposes. In the discussion about QM we are simply confronted with the fact that there are a number of competing views. One may distinguish between three types of reaction to this fact:

i The prevalent attitude among physicists seems to be to think that the differences between the various interpretations of QM are unimportant. By and large we have a mathematical formalism about which there is a high degree of agreement and which one knows how to apply. The rest is merely a matter of philosophy and not really important.

ii Some physicists have made up their mind and have taken a stand for one particular position. Their main purpose in the debate on interpretation is to promote this position and to attack competing views.

iii Other participants of the debate have taken a more neutral position and seek clarification in the first place. In as far as they find it necessary at all to choose between competing views, they want to do so on explicit grounds and after careful consideration.

The present work aims to contribute to the final option. This implies, however, that the task of comparing competing theories must be taken up seriously, which includes paying attention to the nasty details and complications that have surfaced in this area, especially over the past thirty years. Before we make up a strategy and estimate our chances of survival in this battlefield, let us make sure that we really want to go through all this trouble. What are the stakes in this game? Why not settle for a slightly less virtuous but much simpler approach? Let us briefly deliberate upon the alternative options (**i**) and (**ii**) mentioned above.

Unclarity

The history of QM, the entire sequence of success after success, seems to provide evidence for the thesis that the differences between its interpretations can be of at most marginal importance. We have a fantastic theory, which we know how to use, and it apparently progresses largely independently of discussions about interpretation. Then why spend large amounts of energy on an enterprise which seems hardly relevant to the progress of theoretical physics itself? Although, for the present, we do not really have a clearcut answer to this question, we should not fail to notice that the argument is inherently suspect. The bare fact is that we ca not know how relevant or irrelevant the differences between the various interpretations of QM are until we have seriously and systematically investigated what these differences amount to and what consequences they might lead to. For the time being, the situation is comparable to one where people have decided to express their ideas in a specific language, in terms of certain common phrases, even though they are aware of the fact that for each individual in the group the words and sentences may have a different meaning. Such a situation may create a temporary illusion of peacefulness. Diplomats might settle for it if, after fruitless talks, they nevertheless want to save the day by presenting a common resolution, vague enough for each party to be read in its own way.

I would suggest that in theoretical physics one cannot remain satisfied with such a state of affairs. Eventually the unclarities in the interpretation of QM might obstruct the progress of physics in unforeseen ways, and the complacency with which some physicists comment upon the irrelevance of interpretational problems is entirely out of place. If people say the same thing, while meaning different things, then apparently the words do not fit their meaning and should be changed or supplemented. The same holds with respect to the formalism of QM. The present QM-formalism should be regarded as a sort of intersection or common divisor of a class of more extended formalisms, fitting the various interpretations more precisely. If, for instance, Bohr considers QM as being a theory about, what he calls, quantum phenomena, then his ideas are best expressed within a formalism where these phenomena are explicitly represented. In order to achieve this, the architecture of the present quantum formalism must be extended. What can be gained by such an explicit and clear realization of the differences between the various species of QM is that we get more freedom in developing the conceptual apparatus of the theory, or rather the theories, separately within each individual interpretation. Within QM there are a number of problematic concepts, such as locality, causality, determinism, entropy and, of course, reality. It is by no means evident that the natural meaning of those concepts would be 'the same', in whatever sense, within all interpretations. Accordingly, such concepts may be more sharply definable within each individual interpretation than by referring to the common quantum formalism. This might have direct implications for topics like quantum gravitation and quantum cosmology. We cannot be sure, of course, but we are hardly in the position to neglect even the remote possibility that this could be true. Summarizing, I think we have sufficient motives for desiring more clarity about interpretational matters.

The Babylon effect

Let us proceed to discuss the second alternative, option (**ii**). Why not simply allow people to choose their position and let them fight it out. In a debate between intelligent people we may have some confidence that those with the best arguments will win. The problem with such a procedure is that, although it may work well in a situation where problems are relatively clear and straightforward, it may not work at all in a more chaotic situation where problems are ill-defined and conceptual confusion is among the main difficulties to be overcome. For a debate to be decided on good arguments, such arguments must be around. If they are not, we get a quite different type of discussion.

A rather common reaction in this case is that participants in the debate initiate a search for additional arguments. Criteria for theory choice are sharpened until they are 'good' enough to settle the issue. One can have doubts about such methods, in which high priority is given to having issues uniquely settled at all cost. Furthermore, they often fail by lack of agreement about the extra criteria. In the debate on QM we have thus entered a more or less stable stalemate situation in which positions move slowly and arduously. No quick solution seems at hand and in the long run, if victory comes at all, it may simply be won by the side with the most strategic competence and resourcefulness, rather than on the basis of good arguments. Alternatively, if no single party can collect sufficient authority, then physics may share the fate of so many human ventures and fall prey to what can be called the Babylon effect. We may be witnessing the early stages of a scientific schism. Eventually, physics may break up into a number of different traditions, each with its own world view and having little to say to each other, apart from occasional mutual attacks to test the reflexes.

Somehow both alternatives seem a bit pointless. Many of us may have a much brighter picture of science. A picture of a common enterprise which unites people rather than divides them. A picture which offers hope for a better future and a road to peace and good fortune, rather than merely another battleground. Such a picture of science is, no doubt, much too rosy. If we have unreasonable expectations, we are bound to get disappointed and science may end up on the dump heap of salvation tales that have failed to fulfill their promise. On the other hand, in our present case, the above scenario of simply letting people fight it out with whatever means they find appropriate, is built on a *laisser faire* mentality that may be our worst option. We may be able to achieve more satisfactory results by a more careful handling of the situation. Different traditions may arise within physics, but we may be able to achieve that they can live in mutual respect and that they have more to say to each other than: "Me right, you wrong", in various degrees of sophistication. Moreover, if at some stage a single view would become predominant, we might be able to achieve that this event originates from arguments that are clear and authoritative. These perspectives put a bonus on a more careful approach and I think it is worthwhile pursuing it, not only for what is at stake in the present case itself, but also because it can have an exemplary function in a wider perspective. All out competition has its own associated type of logic. Tolerance requires another kind of logic and another kind of background philosophy. Such things do not grow in our backyard. Their development and maintainance demand explicit attention and effort. The interpretation of QM is an area with some nasty pitfalls, but if we do not cope decently with these problems here, then what can we expect in other areas, where comparable problems often appear magnified by orders of magnitude? Clearly, QM can provide a perfect case for testing ideas about these matters. How well will the community of physicists be able to cope with a situation in which, even in the long run, a consensus about the problem of interpreting QM may fail to materialize?

Plan of action

For those who are prepared to conclude that there is sufficient merit in pursuing alternative (**iii**), let us continue and think about how to proceed. We may recall that what we are after, in this phase of the project, is to get ourselves an appropriate framework for the comparison of theories. Accordingly, we turn to philosophy of science and see what it has to offer us. Well, presumably a great deal but inconveniently arranged in a confusing array of different views. How can we manage to collect what is relevant to us and to assemble it into something that is fit to serve our purpose? The following considerations can be called to our aid here. In the first place, after an initial period of distress that comes with entering new territory, we may find that the area is more or less agreeable to be in. People are dealing here with questions that one can learn to find interesting. Moreover, methods are not entirely dissimilar from those in physics. Philosophers study theories in much the same way in which physicists study physical systems. Physicists apply mathematics for describing physical systems, and, in a similar way, philosophers may apply logic to describe the structure of theories. Furthermore, we should keep in mind that for us the whole area is for tresspassing only. We do not want to stay indefinitely, we are there with a purpose and this can lend to our expedition a sense of direction that should make it easier to find our way around.

In the second place, we should take into account a marked cultural difference between physics and philosophy. Physics has a very strong problem solving tradition. In comparison, problems in philosophy seem to be for chewing on, rather than for solving. I do not entirely mean this to sound like criticism. In fact, I think there can be problems that have no solution at all but are still worth spending effort on. Philosophy may have more of such problems than physics. Still, the effect is that physicists are more familiar with the practice of applying Alexandrian solutions to Gordian knot problems. We may need some of this attitude and it is good to be aware of this. It may lend to our expedition a sense of determination that should make it possible to come up with results that might not satisfy every philosopher but which, nevertheless, work sufficiently well to serve our purpose.

Finding a framework for comparing theories makes up an important part of our project. How does this relate to the main theme of this book which was supposed to concern physical reality? We can, in fact, put a rather direct link between these topics. Various forms of scientific realism represent various, more or less sophisticated, forms of the idea that science is a quest for finding out *The Truth* about *The* structure of *Reality*. Much of the logic we use is tuned to this view in the sense that it is about how to deal with truth efficiently. Because competing theories cannot simultaneously be literally true, a strong form of competition is implicitly built into this frame of thought. If we want to compare theories, while taking seriously the idea that they should be treated on equal footing, perhaps forever, it is natural to consider a generalization of our basic ideas about logic. Instead of an absolute truth concept, we may consider truth as a relative concept, i.e., truth as contextual truth, relative to a given context. Accordingly, a logical theory about the concepts of truth and reality can offer an appropriate framework for formulating ideas about how to compare theories.

In our specific approach, phase 1 of our project will comprise the construction of a theory of contextual logic, which we shall use to formulate a theory about the structure and role of the concept of reality. As will be explained more extensively later on, this approach will be built on three corner stones:

the semantic view of theories;

a network approach to modal logic;

a theory of partial interpretations and partial translatability.

Phase 2, application to QM

The logical framework to be erected in phase 1 is meant to enable a more systematic approach to comparing interpretations of QM. In order to achieve this, however, we need sufficiently precise formulations of these interpretations. A serious problem with comparing different theories, especially when they are closely related, is that we cannot take it for granted that common concepts have a common meaning. The debate on QM is so heavily afflicted by difficulties of this type that one may get the feeling of walking on an ice floe. It is often very difficult to assess precisely whether or not the words used in the discussion retain the same meaning, or to find out to what extent the differences between various positions are real or merely a matter of rhetorics. Accordingly, we cannot base our analysis on linguistic similarities and differences between what people say about a subject. We are obliged to look at the meaning of their words. Formal semantics and logic seem to provide the natural framework for dealing with such situations. The relation between QM and logic is a problematic one, however.

QM is a theory with a very peculiar structure, which seems to defy standard logical ideas of what a physical theory should be like. In reaction to this challenge a new type of logic, commonly called quantum logic, has been developed, and it has become a much disputed topic. On the one hand, it can be regarded as a natural generalization of standard logic, but, on the other hand, some very basic intuitions of standard logic are abandoned. There is, at present, little clarity or agreement about the status of quantum logic, nor about its purported claim or purpose. Furthermore, quantum logic has developed into a separate tradition, very much apart from the debate that has taken, and is taking, place between physicists themselves about the interpretation of QM. Physicists, by and large, take little interest in quantum logic and so far quantum logicians have not been able to make clear why this should change.

Quite apart from the eventual fate of quantum logic, formal semantics still seems the natural way to get a better hold on the more widespread discussions that are going on between physicists, and in which quantum logic plays at most a marginal role. There are various alternatives available for standard quantum logic. Bas van Fraassen even speaks[2] of a "labyrinth of quantum logics". In the present approach, however, a main aim will be to work strictly within the framework of classical logic. The contextual logic developed in phase 1 is a classical logic, with some machinery added for the purpose of dealing with multiple points of view. This extra machinery turns out to come in handy for handling the concept of complementarity in QM. Therefore, contextual logic will be used for construing QM as a *contextual theory.* As we shall see, this can be done in a number of ways, and this provides the opportunity of adding to the labyrinth of quantum logics a whole family of classical, contextual, interpretations of QM. Some members of this family closely resemble actual positions taken by physicists in the debate on QM. I shall not claim that one thus obtains historically correct formalizations of these positions. Such claims are very hard to substantiate and probably incorrect. However, the approach certainly provides a solid base for interesting further research.

The second coming of contextual logic, first as a framework for comparing theories, and later as an approach for formalizing QM itself, may seem a bit surprising. It is not as strange as it seems, however. Contextual logic, in the form as presented here, represents an attempt to get more hold on the realism debate. Within the discussion on QM, realism has been one of the major issues almost from the very beginning. Accordingly the subjects are well known to be related. Furthermore, one may observe that there is a certain analogy between, on the one hand, "incommensurability", a term referring to difficulties in the process of comparing theories, and, on the other hand, "incompatibility", a term used in QM to indicate that pairs of physical magnitudes pertaining to a given system do not generally apply simultaneously.

Incommensurability may be described here in general terms as a thesis originating in the nineteensixties, that claimed that the process of comparing theories generally involves difficulties exceeding the scope of logical methods that philosophers, at least at that time, were prepared to employ. Incompatibility satisfies a description in similar wording as referring to the observation that the relations between physical magnitudes in QM cannot be analyzed by means of the logical methods that used to be employed in classical physics. These descriptions provide an admittedly vague but still definite link between the concepts; and I would even go as far as to specu-

[2]Van Fraassen [91].

late that it might not be entirely coincidental that the most persistent proponent of an incommensurability thesis, Paul Feyerabend, had a background of writing about the philosophy of QM[3].

Incommensurability of theories, at least in the general form described above, is something I regard as a truism. It can hardly be denied that, for example, physicists reason in ways that cannot be retraced to textbook logic, that they handle their concepts with a flexibility that standard logical methods cannot account for and that they rely heavily on, what is often called, 'physical intuition' or even 'common sense', even though it is debatable to what extent, and in what sense, sense is common. I think, moreover, that these practices are manifest especially in the way in which physicists deal with varieties of different theories and theoretical models and in the way in which they proceed from one model to another while barely noticing the dramatic conceptual changes involved.

Some of the violent reactions to the whole idea of incommensurability can be traced to the fact that people have very strong intuitions about how science should proceed, about what is good or bad about some theory, about what rationality is, and about what is right or wrong in science. Such intuitions are vital to science, the whole enterprise depends on it. Nevertheless, it remains a fundamental error to think that we can rely on them or that they are a safeguard against any form of error. History provides many fine examples of how intuition can lead us astray. We are therefore obliged to take a serious attitude towards incommensurability, and in my view this entails that:

> in the first place we should try to develop logical systems in which at least part of the flexibility of actual scientific reasoning can be explicitly accounted for. My work on phase 1 represents an attempt to contribute to this;
>
> in the second place, we should use these logical methods to investigate the practice of scientific reasoning in order to get a better view on its structure, uncover hidden premises, and detect limitations and shortcomings.

Work on phase 2 can be regarded as being of this latter type: logical methods are used to get a better view on the structure of QM. This kind of work, besides being useful as a preparation for theory comparison, can have its own merits. We may discover surprising and interesting things about a theory, things that may be useful also for its further development. Let us observe, however, that there is also a dark side to this fact. We are faced with the problem that the theories we study are dynamic and to some extent vague and malleable structures. By studying the logical structure of a theory, and by making a reconstruction within a particular logical framework, we interfere with its structure, and potentially with its further development. If we take a theory, formulated in terms of vague intuitive concepts, and we make a reconstruction of it in terms of a more precise formal framework, then

[3]I recently asked Paul Feyerabend himself about this. He agreed about there being a relation between the two concepts, but he had not been conciously aware of this when he began writing about incommensurability.

in what sense can we say that the formal reconstruction still represents the original theory? The problem is of more than purely academic interest, if only for the fact that a given theory will in general admit a multitude of different formalizations.

A further problem is that, while working out the details of a theory, we never come across a natural boundary with signs reading: "You are now ready". There is generally no fundamental limit to the amount of effort one may invest in the development of a given theory. Problems like these must be provided with some sort of practical solution. In practice, this boils down to taking decisions, such as the decision to accept a certain formalism as a faithful reconstruction of the original theory and to take it as the starting point for further research, for example a comparative study, even though results might depend on the further development of the theory.

Evidently, actual science is full of decisions of this type. Choices are made almost routinely. Daily life is full of situations where similar decisions must be taken, and we are well equipped and well trained to deal with them. The fact that we deal with them easily, however, does not imply that they are, in a logical sense, unproblematic. It merely indicates that we are used to making decisions. More in particular, it does not follow that different people, or different research communities, will independently come to similar results.

These and similar factors also change the character of a comparative study. A comparison of theories becomes a process which turns on a multitude of smaller or larger decisions. It thus becomes a subtle mixture of, on the one hand, finding out the relations between these theories and, on the other hand, deciding about what these relations will be in the future. Accordingly, a comparison between two theories may even influence their further development. There is a sense in which a theory may no longer be the same theory after it has been compared with another one. The heuristics of the two theories are linked by the process. Subtle changes like these often go unnoticed. Once a decision has been made, we take things as they are, in a casual manner, often even without being aware that there was someting to decide about at all, and that history could have taken a different course. Of course we should not let such realizations disturb us, or permit them to interfere with our determination or with the continuation of our work. It is good, however, to be aware of the nature of our actions.

Phase 3

Phase 3, the final phase of the project, includes carrying out a range of comparative studies concerning the various interpretations of QM. We may conceive various types of comparative studies, each with its own structure and purpose, such as:

a survey of the main similarities and differences between a given set of theories;

an inquiry into the meaning of a concept. We may want to know whether, and in what sense, a concept can be said to have the same meaning in different theories;

an evaluative comparison, for instance an argument to the effect that certain theories satisfy a preferably clear and explicit criterium of excellence, whereas some others do not.

Underlying most forms of comparison between theories is a theory about their intertranslatability. The fundamental work of phase 3 will consist of constructing such a theory for the interpretations of QM on the basis of the general theoretical ideas of phase 1 and the more specific analysis of QM of phase 2. The final result, to be presented to the working physicist, should consist of a survey of different quantum theories, in broad outlines, with a summary of their main similarities and differences and their vices and virtues, as seen in the light of some set of prevalent criteria.

As was said earlier, there will be little or no covering of phase 3 in the present book; the emphasis will be on setting up phase 1 and 2. Now with this plan in mind, let us take a deep breath and cross the border towards the territories of philosophy and logic, determined to return home, some day, to the land of physics, and bestowed with honour, trophies and merchandise.

Chapter 1

Realism, empiricism, pluralism

Abstract: The philosophical positions of realism and empiricism are explained, and arguments are given in favour of a third position, metaphysical pluralism.

From the general debate on the interpretation of QM it appears that there is something especially problematic about what QM can be supposed to say about reality. In fact, the question:

> "Does QM give us a picture of physical reality and if so what does this picture look like?"

may be considered as the main theme of the entire debate.

The relation between theory and reality, to some extent condensed in the notion of "truth", is not a trivial issue, and neither is it new. The subject had a significant philosophical tradition long before the advent of QM. It deals with problems which must be both very common and very old. Long ago people must have noticed that it is possible not only to speak the truth or to lie, but that there is a large grey area where all this is not particularly clear. Different people can have different views without it being evident that one of them is right and the others are wrong. This is a problem which, on the one hand, has deep philosophical roots, but, on the other hand, is very directly part of daily life. Thousands, perhaps tens of thousands of years ago, people must have thought about these things.

Let us restrict ourselves to more recent times however. A nice illustration to the theme is provided in Eco's bestselling novel[1] "The Name of the Rose", which is situated in the Middle Ages. The philosophical themes in the book are set against the background of the medieval debate between realism and nominalism, but the main issue is as much to the point now as it was then. It can be summarized by the question:

> "Does reality have a structure?",

or more subtly perhaps:

[1] Eco [68].

> "Is the structure we think to recognize a reflection of a *real* structure of reality?"

In the book a superintelligent priest solves a sequence of murders. The plot he thought to recognize, however, and which provided guidance to his research, turned out to be completely beside the point. The structure he had found was merely a train of coincidences. Even though I shall not argue that this is an entirely common pattern in scientific research it can, on the other hand, not be dismissed as a totally unknown experience. The transition from classical to quantum mechanics is a point in case. It somehow remains beyond comprehension that a seemingly nearly perfect theory like classical mechanics had to be replaced by something as different from it as QM. A transition like this is felt as a shock.

1.1 Realism versus empiricism

In a postscriptum to his book[2], Eco makes a comparison with a labyrinth. He distinguishes three different types of them. The first one is the *classical labyrinth*, which has a simple structure. In order to make it exciting at all, it must contain a minotaur. Secondly, there is the more complex *mannerist maze*, which is sufficiently complicated to make it easy to lose your way. And finally, there is the infinitely complicated *rhizome*, with a structure defying rational comprehension. As an example, one may think of Tarkovski's movie Stalker, where the way through the labyrinth changes all the time, according to unknown rules, or perhaps no rules at all, and is found by throwing a nut with a piece of cloth attached to it.

Nature might be like a rhizome. This cannot be verified or falsified, and it requires a certain amount of faith to believe that it is not the case, and that nature can be understood in terms of a limited number of comprehensible concepts. Philosophical realists are those who are willing to spend this amount of faith. A realist believes that tables and chairs are really there and that atoms and quarks are just as real. The main antagonist of realism in the present century has been empirical positivism which holds that "to believe" is not a scientific activity. To speculate about what does, or does not, really exist, is pure metaphysics; and to ban metaphysics is the core of the positivistic program.

Along with the ongoing discussion, both programs have been forced to compromise on their original views. Indeed, up to date versions of both programs are sophisticated to a degree where it becomes doubtful whether their proponents would be recognizable as such for their predecessors of, e.g., a century ago. Positivism at the turn of the century maintained that direct observation is the only reliable source of knowledge. Theories were thought to merely serve to describe, as economically as possible, the regularities that are found to occur. The idea that direct observations provide a source of certainty, however, was heavily attacked later on. Attempts to divide scientific languages into a theoretical language on the one hand and a non-theoretical observation language on the other, have ended in complete failure. It is

[2] Eco [69].

widely acknowledged nowadays that even the simplest observation is immersed in a theoretical background from which it cannot be separated. This means very serious trouble for the positivist program. As a consequence it is even dubious, whether one can say that positivism has any surviving versions left at all. What may come closest to being a descendant of empirical positivism is Van Fraassen's constructive empiricism[3], which, in very broad outlines, boils down to the view that empirical adequacy is the overriding criterion for theory choice. Any other criteria are of a merely pragmatic nature and are not considered to be central to the scientific enterprise *per se*.

The problems of the realistic program are entirely different. It is hardly interesting to say that nature has a structure, unless you assume at the same time that one can get to know this structure. It is no problem for a realist to acknowledge that some amount of faith is involved here, but it will be important for him to show that his beliefs can be maintained consistently and with some degree of plausibility. Accordingly, a realist should try to make it plausible that in principle it will be possible to include all interesting knowledge within a single theory which gives us the truth and nothing but the truth. He will, therefore, try to show that scientific knowledge is cumulative, and that subsequent theories give better and better approximations of the truth in a monotonically increasing way. Moreover, he will want to make it plausible that these subsequent theories are all about the same unique reality, that there is a sort of stability of meaning, and that reality can be approximated in a more or less decent manner, without too many big surprises. The world should not be completely different every time we find a new theory. Realism has not been shown to be untenable on these points but, on the other hand, it has become apparent that the amount of faith involved is not small. Moreover, anyone who wants to maintain that we are witnessing a gradual approximation to truth must possess a fair amount of intellectual flexibility.

Considerations concerning the plausibility of realism can depend strongly on the way in which science develops in a given period. Let us return for a moment to the metaphor of the labyrinth. Suppose you are walking through a labyrinth and you decide to make a map of it. After quite a while you think you are about ready, but then you find one more obscure little gate. You enter it, and instead of finding an obscure little *cul de sac*, you find the rest of the labyrinth and you soon notice that the original part was only a very small fraction of a much larger whole.

This is roughly what happened to physics in the first decades of this century. The laws of nature were thought to be known up to a few small details. These details, however, blew up to enormous proportions and nowadays relativity theory and especially QM present us with new and totally unexpected pictures of the world. This course of events does not show the untenability of realism, but it certainly does not add to its plausibility. Nevertheless, the labyrinth *might* still be finite, even though as yet its boundaries are not in sight, and its basic structure *might* still be very simple, even though it is beyond our comprehension as yet. It is shocking, however, to see how ideas which seemed indubitable and which belonged to the core

[3]Van Fraassen [93].

of classical physics, do not survive the transition to QM. In as far as QM provides us with a picture of reality at all, it is at least a very strange kind of reality.

Such problems do not bother the empiricist. As long as empirical adequacy is the only significant demand we require from a theory, QM is the most fantastic theory we have ever had. But can we really be satisfied with this? Is it not even a little unethical to mistreat such a fantastic theory with such a rather poor kind of philosophy? Clearly some reflection on these matters can hardly be evaded in thinking about QM. Let us therefore proceed with some more detailed elaborations on the concepts of "realism" and "reality" themselves. What is the meaning of the concept of "reality"?

Ontology versus epistemology

In order to investigate the meaning of a concept one should examine the theories in which it occurs. "Reality", and related concepts like "nature" and "the world", are concepts primarily from the fields of ontology and epistemology. Let us stipulate some conventions with respect to their use in our further discussion. The word "nature" will be used as an ontological term, referring to the totality of everything which exists in an ontological sense, whereas the word "reality" will be used primarily as a concept from the field of epistemology and refers to our theoretical reconstruction of nature. "World" will be used specifically for referring to classes of alternatives, i.e., a world is one of a set of "possible worlds". We may define *naive realism* as the view that there is a very direct relation between these concepts. Good theories are assumed to give a literally true description of nature and to this end they use concepts which are thought to refer to really existing objects, their properties and the relations between them. Within this view the concept of reality can refer either to the 'world *an sich*', i.e., to nature itself, or to an idealized complete and literally true theoretical model of it, which we do not know in detail but which is thought to be possible in principle.

Naive realism is not a particularly sophisticated theory of knowledge but it suffices to provide the concept of reality with a meaning which serves well for many practical purposes, for instance for expressing a demarcation between views originating from phantasy and those intended to be taken literally. Just like other theories it may be applicable only under the appropriate conditons and to the appropriate ends. Using it under the wrong conditions or to the wrong ends may lead to trouble and an indication for this is that, even in daily life, a demarcation between reality and phantasy is not always as clear–cut as naive realism suggests. In order to say some more about the limitations of naive realism let us regard a slightly more complicated view, obtained by splitting the concept of reality into two different concepts.

We shall differentiate between on the one hand internal realities and on the other hand an ultimate reality. Internal reality is a practical concept which refers to a picture of the world as given by a particular theory, used in the appropriate way within its domain of application. Ultimate reality on the other hand is a transcendental concept referring to a perfect picture of the world as given by an imagined perfect ultimate theory which does not exist yet. Let us illustrate the difference by considering the question whether atoms do or do not exist. Suppose we were to agree on the view that they do exist, then what would we really agree upon? What kind of thing is an atom? Is it a little hard sphere, is it like the Rutherford atom, like the Bohr atom, like the atom as described by QM, like ...? As long as we confine ourselves to internal reality answers are quite straightforward. Atoms exist within various useful theories and exactly what it is that exists, depends on the particular theoretical context we choose. With respect to ultimate reality, however, answers disappear like snow in summer. We do not know what ultimate reality is like, and the fact that atoms exist in a sequence of increasingly better theoretical models does not imply that they exist in the ultimate theory. Any subsequent theory in the sequence may either provide a better picture of what an atom is like or, alternatively, give an explanation of why they seem to exist according to our present theories whereas in fact they do not. Furthermore, we should also realize that it is by no means evident that something like an ultimate perfect theory can exist at all. Even if we are able to obtain a sequence of increasingly better theories, by continued scientific research, this does not imply that in the long run the sequence will converge towards a unique final result.

We may conclude from this that there is an intimate relation between ontology and epistemology and that, moreover, the significance of ontology appears to be inversely proportional to the complexity of the associated epistemology. In an unsophisticated epistemology like naive realism the class of significant ontological statements is large. Everything which exists according to those theories which we accept as good is assumed to exist in nature. In a more complex epistemology, however, the relation between theoretical and ontological existence becomes much more problematic.

Realizations of this kind have lead to the development of more sophisticated brands of realism. Among these, *metaphysical realism*, itself in various degrees of sophistication, can be considered as being closest in spirit to the intuitive idea of realism. Applied to science it has been described as the view that[4]

> "Science aims to give us (...) a literally true story of what the world (i.e., in our terms 'nature', djh) is like."

Metaphysical realism can be considered as the philosophy that there is a more or less clear relationship between theoretical and ontological existence and that something interesting can be known about this relationship, or that at least it is beneficial to believe that this is the case. Within the more sophisticated forms of metaphysical realism the relation between the two kinds of existence can be very problematic but ontology can still be regarded as a significant discipline. If one goes to epistemologies beyond metaphysical realism even this becomes questionable. Indeed, if one makes no assumptions at all about the relation between theoretical and ontological existence there remains only a single significant true ontological statement, namely:

[4]Van Fraassen [93] p.8.

"Nature exists". Moreover, this statement can be regarded as analytically true and not extremely interesting.

Among realists, opinions on precisely what parts of science can be considered as ontologically significant differ widely. The spectrum of realists includes for instance, on the one hand, entity realists, like Hacking or Cartwright[5], who hold that entities are real even though our theories about them may be wrong and, on the other hand, structural realists, like Worrall[6], who hold that the mathematical structure of our theories reflects reality, even though the entities occurring in these theories may fail to exist.

Surveying various forms of modern realism, one may try to summarize the situation by saying that scientific realists agree in that they hold scientific theories to be *approximately true*. They may differ, however, in their attitude to the nature of the approximation. We may assess this situation in two rather different ways. On the one hand, one may hold that the various types of realists have something very important in common and there are merely subtle differences of opinion on what they think it means for a theory to be approximately true. On the other hand, one may hold that it is precisely the meaning of "approximately true" which forms the true core of the debate and, accordingly, there is very little reason to place all realists under a common roof. Mutual differences between various forms of realism may be larger than those between some realists and their presumed opponents, such as idealists or empiricists.

Personally, I think that a good look at the debate would practically force us to the second view. It may be noticed, for instance, that it is always possible to include some unintendedly wide meanings of "approximately true" by which realism can be jointlessly fused to either idealism or empiricism. In the latter case, for instance, one may simply define approximate truth as empirical adequacy. A realist may respond that this is not what he means by approximate truth, but in that case he will have to be very precise about what he does intend the concept to mean. As long as we have only vague intuitions about this, we do not have a real issue but only the intuition of an issue, which may as well evaporate completely on closer scrutiny. The example may not be as far-fetched as it seems. Churchland, for instance, appears to maintain an exceptionally liberal form of realism. In Churchland [44], he does not seem to commit himself to any particular form of approximate truth beyond empirical adequacy, but classifies himself as a realist purely because he does not agree with the emphasis that Van Fraassen's constructive empiricism puts on the distinction between observable and nonobservable entities. This leaves open the possiblity that the realist Churchland might in fact hold a more extreme form of nonrealism than Van Fraassen, because one may argue that in Van Fraassen's philosophy the distinction arises from his particular form of empiricism, which includes a realism with respect to observable entities. Even this can be dropped, and again it may not even be as far fetched as it seems. I owe to Jan Hilgevoord the comparison with a rainbow. It may be true that people used to think of it as an object, but now we have

[5] [107], [39].
[6] [224].

theories about raindrops and deflection of sunlight and we know better. Perhaps our theories will someday tell us that all objects we see and touch are rainbow–like 'illusions' in the same vein.

It seems to me that the discussion about the nature of approximate truth is the core of the realism debate and that, moreover, what is interesting about this discussion can be typically rephrased in terms to be introduced below as *comparative metaphysics*. Just like many realists I too have, of course, intuitions about approximate truth, and presumably these arise from the existence of sequences of increasingly better theories. Although we have no way to fix the meaning of "approximate truth" directly by a correspondence to nature, we can investigate a relative version of the concept by viewing the relations between series of subsequent theories. There are may different types of relations, however, and focussing the discussion on merely a concept of approximate truth may confuse rather than simplify matters. Accordingly, I think that comparative metaphysics actually offers a better framework for discussion than the realism debate as such.

Similar remarks can be made with respect to the subject of ontology more generally. Ontology is concerned with existence. In an absolute sense, with respect to what exists in nature, I have difficulties in understanding what this means at all. I have no difficulty, however, in understanding what it means from the perspective of a given theory. Accordingly, I regard *relative ontology* as a coherent subject and issues related to it essentially boil down to questions about theory acceptance and criteria for theory choice. In the same vein I would prefer to rephrase most issues related to the realism debate in terms of questions like: "How do we organize our knowledge and how much freedom do we have in this respect?" In as far as the concept of reality is interesting at all, this is, I think, because it has a cash value with respect to questions like these. Realism offers not only a set of rather questionable assumptions about the relation between theory and nature, but also a logical framework which plays a role in the organization of knowledge. The concept of reality involved here is of course internal reality, reality according to some theory. It is simply unnecessary to make the potential fallacy of assuming a direct relation with the structure of nature itself. From now on we shall disregard the whole idea of a direct reference to nature in an absolute ontological sense. Our attention will be focussed on internal reality, its structure, its construction, and the role it plays in the logical organization of knowledge.

Seen from this angle, the interesting part of the realism debate concerns the question whether the logical framework connected with the notion of reality offers an appropriate environment for managing science. In my view this is only partly the case, and part of the efforts in our further analysis will concern the design of an update.

The decay of empiricism

The modern history of empiricism has been rather more dramatic than the history of realism, owing to the initial success and final demise of its most prominent representative of the present century, empirical positivism.

Empiricism, in a perhaps unusually wide sense, is the view that through observation we can learn about nature. In this wide sense it it opposed to, for instance, a Platonean idealism in which observation rather distracts us from seeing the truth. Within an empiricism of this general type, the next question is how much and what we can learn through observation, and on this issue opinions vary. Metaphysical realism, or at least scientific metaphysical realism, can be regarded as a brand of empiricism, because it regards empirical adequacy as a main criterion for theory choice. It should then be considered as a brand of metaphysical empiricism because it involves assuming particular epistemological theories about the relations between nature and internal theoretical reality. Empiricism in the more usual narrow sense is the view that observation is our only source of knowledge. In this narrow sense empiricism is opposed to realism in holding that science can do without metaphysical assumptions at all. This at the cost of assuming that only those features of a scientific theory which can be directly linked to observation can be considered relevant. This strict empiricism, also called empirical positivism, has been more or less torn apart in the second half of this century, essentially because it became clear that consistent adherence to it would leave us with no relevant science at all.

Constructive empiricism[7] contains some essential improvements on empirical positivism. The basic notion behind it is that theory appraisal is disconnected from truth. Instead, acceptance of a theory involves merely the belief that it is *empirically adequate*:

> "Science aims to give us theories which are empirically adequate; and acceptance of a theory involves as belief only that it is empirically adequate[8]."

Nevertheless, a theory is conceived as something which tells what the world *would* be like if it *were* true. The approach is based on the *semantic view* of theories, in which a theory is defined by its class of intended models. What has been dropped is the idea that there is some direct relation between these theoretical models and nature. One of the consequences of this move is that the defective meaning criterion of empirical positivism can be replaced by a better construction. Entities which are not directly observable can nevertheless be meaningfully discussed, because the theoretical terms referring to such entities derive their meaning from the theoretical models in which these entities occur. Accordingly, talking about such entities can be meaningful even though it does not involve a commitment to assuming their real existence in nature.

What I like about this approach is in the first place that it once again presents a serious alternative to realism, thereby freeing us from having to make ontological commitments which I see as superfluous and potentially fallacious and misleading. Further, relative to empirical positivism, there are major improvements such as a new theory of meaning. At the same time, however, this leads to what I see as the main problem of this philosophy. The defense of positivism against metaphysics

[7]Van Fraassen [93].

[8][93] p.12.

rested upon its strict criterion of meaning. Postpositivism philosophers like Popper[9] and Lakatos[10] have sought for other demarcation criteria between science and metaphysics, but with no conclusive success[11]. I do not see that constructive empiricism comes with decisive answers in this matter. The idea is that empirical adequacy is the only criterion for theory choice which really matters in science, and that additional criteria are pragmatic. I see this as a rather one–sided view on science, which does not do justice to the importance of these additional criteria and to the historical role they have played in science. To say that the choice between empirically equivalent theories is a pragmatic matter is comparable to saying that the choice between a million dollar Michelangelo painting and a million dollar Picasso is a pragmatic matter. From a certain point of view this is true, but it is a point of view which precisely misses the fun in enjoying art.

As I see it, science is inevitably as much an expression of how we want, or are inclined, or are able to see nature as it is a search for empirical adequacy. For the individual scientist empirical adequacy is only one of the factors among many other criteria and circumstances which determine his adherence to a theory. Among the factors which make him interested in doing science at all, it is, moreover, likely to be a relatively minor one. Precisely what it is that drives the individual scientist in doing his work may be hard to tell, but it is certainly crucial for making science tick and I cannot think of this in terms of pragmatic matters, separated from the central aim of science. Evidently, the personal motivations of different scientists may differ greatly, may be hard to analyze, and may not even be important or interesting to investigate. If we are looking for a common goal, which can be clearly formulated and is shared by all scientists, then what we arrive at may not be much more than empirical adequacy. By taking the intersection of what all scientist say to pursue, however, we obtain something which is much too shallow to pass for the central aim of science.

The aim of science

In as far as I would personally want to formulate something like the aim of science at all, I would be inclined to look for something much more abstract, say a search for perfection, where, just like in art, perfection is a vague concept, and hard to grasp, developing along with science itself. Indeed, it may not even be the search as much as the construction of a concept of perfection which is interesting.

Relating a search for perfection to considerations about the structure of nature, we may observe that the degree of success in finding perfect theories is no direct measure for the degree of perfection of nature itself. A painter might make a marvellous portrait of someone who is considered as ugly. In a sense we attempt to do something similar in physics, by trying to formulate beautifully simple laws of nature, while hiding the chaos in for instance boundary conditions or in symmetry

[9] [170], [169].
[10] [140], [141].
[11] See, e.g., Feyerabend [76].

breaking processes or by merely ignoring what does not suit us. Science provides fascinating pictures of those sides of nature that let themselves be portrayed by its theories; but what this implies for the structure of nature itself remains rather unclear. In addition, it is quite evident that to any highly perfect theory one may construct numerous empirically equivalent theories satisfying quite different standards of beauty. Although these might seem very ugly to us, they can, nevertheless, be preferable to someone whith a radically different taste. Accordingly, perfection, in as far as it goes beyond empirical adequacy, is crucially a relative and not a monadic concept.

Human tastes and ideas about perfection are a most appropriate issue for philosophical reflection, and let us make just a few general remarks on the subject, merely to indicate that these matters are not entirely beyond further investigation.

A first thing worth observing is that humans are very keen on having some form of belief. It is not quite clear, however, exactly what it means to believe something. "To believe" can be regarded as a strong form of "to assume", but by concentrating on the cognitive meaning I think an interesting aspect of the concept is lost. There is something inflexible to it that is likely to be rooted in the more instinctive parts of our psychology. Belief may be regarded as belonging to a cluster of more or less related phenomena, such as human appreciation of various forms of art and his involvement in ritual behaviour and religion. It seems very reasonable to assume that the background of such phenomena is rooted in instinct. The way in which people put forward opinions on every conceivable or inconceivable issue, for instance, seems rather similar to the way in which dogs piss on each obstacle they happen to meet, i.e., one of the functions of belief is that it serves as a territorial or social marker.

The study of ritual may be especially interesting because of a structural resemblance between rituals, which represent a form of stylized behaviour, and theories, which represent a form of stylized thought. Both phenomena may have a similar function in organizing our mental life. Furthermore, with respect to other phenomena mentioned above, art is interesting, for instance, because the experience of perfection it can generate is highly similar to the kind of emotion aroused by a good theory. This even includes a feeling of uniqueness. A piece of art may generate the feeling that nothing else could possibly be as beautiful, or as meaningful or something like that, as this particular work. In art, this feeling is refuted time and again by encounters with equally impressive work and this can become one step toward learning to appreciate to some extent the rich variety of possible ways to be involved with the world.

In science, ideology and religion, at least within the history of Western thinking, the idea of uniqueness has been much more profound and enduring. It is up to us, however, to decide whether we want things to remain this way. *Metaphysical pluralism* is a philosophy opposing this.

Alien science

The issue of uniqueness is directly linked with the relation between theory and nature and with the debate on realism. One of the issues discussed in a well-known debate between the scientists Mach and Planck[12] concerned the question whether the science of some alien race of intelligent beings would in the long run produce the same theory, or theories, as human science. Planck, a realist, pleaded yes, whereas Mach, a positivist, pleaded no. With respect to this question the preferences of a metaphysical pluralist can be presumed to be basically on Mach's side. The agreement with Mach is, at least in my view, only superficial, however. A more thorough analysis should point out that a simple yes or no answer cannot be given at all.

To see this let us observe that the issue is complicated and possibly highly ambiguous, if only because it involves translations between the alien theories and ours. Translation is a highly complicated affair, owing to, for instance, the fact that both language and our ideas about language are rather vague. Let us try to imagine some of the practical difficulties involved in our communications with an alien intelligence.

In the beginning there will of course be some difficulties in establishing an elementary form of communication whatsoever. If we assume that these can be conquered, what problems can we expect in later phases? A plausible thing to occur is that some texts are relatively easy to translate and other ones extremely difficult. Moreover, for many texts it may be easy to produce a superficial translation but very hard to make a good one. Furthermore, it is not at all unlikely that different translators will start to disagree on how certain texts should be translated; and possibly different schools of experts will develop, each having their own theories about what the aliens mean with their utterances. Of course one may ask the aliens themselves to provide clarification and, if they are as well-organized as we are, we may expect to receive any variety of answers, some of which may be extremely hard to grasp.

One factor behind the above cluster of problems is what we can call the inherent vagueness of natural language[13]. The meaning of concepts, statements and other fragments of text is often not entirely fixed and unambiguous. Moreover, a text may have various levels of significance, some of which are easier to grasp than others. This may turn the question whether the alien theories are equivalent, or otherwise comparable, to ours into a highly controversial issue. It is quite likely that it will be possible to construct some sort of translation between their theories and ours, but its correctness may remain in doubt. It is also quite likely that translations of their theories will in some respects resemble our own theories. They may be mixed, however, with ideas which seem queer to us, or which escape our comprehension and, on the other hand, some features of our theories may be entirely missing in theirs. Accordingly, the comparison between the theories also hinges on an assessment of

[12] The discussion was reprinted in Toulmin [208].

[13] See, e.g., also Waisman [211] p.120 where, moreover, a further distinction between types of vagueness is made.

how important such differences are.

Besides issues related to the vagueness of language, there is another problem in comparing the alien theories with ours. In due time, after a sufficiently long period of contact and after learning enough about their ways of doing and thinking, we may feel that we understand their language perfectly well and we may be able to produce excellent translations between their texts and ours. Before we get this far, however, we will have learned a lot about, and likely from, them. New concepts will have been introduced into our language and existing concepts may have obtained a different meaning. To what extent then, can we maintain that we have translated their theories into our language? It might be more accurate to say that we have adapted our language to fit their theories. In other words, natural language is not only vague, it is also highly malleable.

Keeping in mind the vagueness and the malleability of language, it is not difficult to imagine how the comparison between the alien theories and ours might forever remain a controversial and unsettled issue, not by a lack, but rather by an excess of answers.

Meaning construction

Let us now try to connect such insights to our considerations about the aim of science. Clearly the idea that the aim of science has to do with a search for an undefinable sparkle of perfection is a very personal view, which may rest upon a romantic streak in my character which I do not care to analyze further. Other people may have quite different things in mind. People with a greater social consciousness may define the aim of science much more in terms of social benefit, and people with a great desire to exercise control over their environment may put their emphasis much more on the goal of gaining control over whatever they are studying. We may draw the conclusion that in fact science is a multifaceted social enterprise which cannot be simply characterized in terms of a central aim. Different people search for different things in science, they exploit it in different ways, they contribute to it in different ways, and together they create an institution which leads its own life, which no one really understands, and of which no one really knows what it will bring in the future, for him personally, for his society and culture, or for mankind in general. In as far as science can be regarded as a unity, this is not as much because its participants share a central aim, but because people have learned to perceive it as a unity, and because it is to some extent institutionalized and organized as a unity. There are communication channels between different fields and disciplines, and between rival groups working in the same area, and these help to create some amount of cultural coherence. This coherence, however, produces a diffuse kind of unity with much internal diversity.

Looking at this diversity from a different angle, suppose that two persons A and B have different opinions about some aspect of science, for instance its aim. Let A be someone in search of some sort of artistic perfection, whereas for B science is directed primarily at social benefit. Is it clear now precisely on which points A

and *B* disagree; is it, for instance, clear how this difference affects their preferences when it comes to theory choice? No, evidently not at all. The behaviour of *A* and *B* in practical situations will depend on what meaning they give to concepts like "perfection" and "social benefit". Both these terms are open to large differences in interpretation, and when it comes to practical decision making they may *a priori* not denote much more than a tendency to arrange arguments in a certain way. This may induce a push in some direction, but to give the concepts a definite meaning requires highly nontrivial argumentation. No book on standard deductive logic will tell us how to do it. Moreover, the result of such a 'meaning construction process' may typically be rather surprising. It may well be, for instance, that at the beginning of the process *A* agrees heartily with C, who also says to be a perfectionist, but in the course of the process comes to agree much more with B, because it turns out that what *B* means by "social benefit" is in some respects unexpectedly close to what A calls "perfection", whereas what *C* calls "perfection" is of unexpectedly bad taste.

Meaning construction is evidently a very delicate process, but communication, and the functioning and development of language crucially depend on it. Moreover, the success of science may originate largely from getting more grip on some phases of the process. Science works by replacing vague intuitive pictures by more precise, often mathematical ones, and the rise of science is to a large extent the history of finding better ways to construct and handle precise concepts. Considering the development of a specific theory from the point of view of meaning construction, we may think of it in terms of Eco's three types of labyrinths. The first phase in the ascent of a new theory is typically rather confusing from a conceptual point of view. New concepts are created, or are 'borrowed' from another context, but they are still vague and lack a precise meaning. During this phase the new theory is comparable to a rhizome. Meaning construction may be applied to get the theory out of this phase. In science this usually involves the use of mathematical models, which serve to stabilize concepts relative to one another. In fact mathematics may be regarded as the science of handling concepts in a precise way. For a given cluster of intuitive notions one may generally construct many different mathematical representations, and accordingly successful meaning construction results in the transformation of a rhizome into a maze. Different mathematical representations of a given intuitive idea are not necessarily equally adequate. Therefore the next step may consist of developing criteria for choosing between the alternatives, possibly resulting in a further simplification of the labyrinth into a classical labyrinth.

Example: An intuitive characterization of the natural numbers is given by "1,2,3, etcetera". Clearly "etcetera" is a very vague term here, and one may try to apply mathematical techniques, involving for instance some mixture of set theory, formal logic, and recursion theory, in order to construct a more precise meaning. It then appears that indeed etcetera is a difficult concept, and in fact on an infinite domain its meaning cannot be precisely specified. This results in a maze of different, nonisomorphic systems of natural numbers. One of these, the smallest, is the "standard model", but its properties cannot be described by a decidable set of axioms (in a language with a comprehensible notion of deducability).

We shall encounter, and apply, a related figure of play in the discussion of QM. It will turn out that the standard formalism of QM is logically incomplete in a rather malignant way. The well–functioning of the theory crucially depends on parts which are not formalized, and this gives the theory a highly rhizomelike quality. This situation may be repaired by continued formalization, but this can be done in a number of quite different ways, which yields us a maze of quantum theories. It is not apparent that conclusive arguments for choosing between these alternatives will become available in the near future.

Note: The above three processes of concept introduction, meaning construction, and selection may each act on parts of a theory during a given period, but it may sometimes be necessary to reverse them and redesign parts of a theory. Although the natural direction in which one tries to work is from rhizome to classical labyrinth, it can also be fruitful to regard new questions and look at problems from a new point of view, and this may involve 'melting' established concepts and turning a well formulated mathematical theory into a rhizome again.

Let us now return to our original topics, concerning the aim of science and the realism–empiricism debate. The main question in the realism–empiricism debate concerns how much nature resembles our theoretical pictures of it. I consider this as an interesting question, but only as long as one has the courage not to try to answer it. More practically interesting aspects of the debate concern the logical structure of scientific theories and the views of different parties on development in science. In my view, the main challenge for philosophy of science in this respect is to come to grips with the astounding variety of more or less related theoretical pictures produced by science, i.e., to find better ways to handle the labyrinthlike character of the enterprise. I do not see that either realism or empiricism is particularly well equipped to meet this challenge, and in this book I shall prefer to work on the basis of a metaphysical pluralism, which I consider much more appropriate.

1.2 Metaphysical pluralism

The particular version of metaphysical pluralism which we shall now proceed to develop will be based on the following, quite minimal epistemological assumption, which, in this respect, will place us quite close to constructive empiricism:

> We have, or at least some among us have, an imagination which enables us to create theories and some of these theories can be empirically interpreted to some extent. The latter means that we can use them to describe our observations and to construct claims about what we should or should not expect to observe. Now what we can learn about nature through observation is that if the empirical claims of a theory are (are not) borne out by observation then apparently nature is such that this theory is empirically adequate (inadequate) with respect to these observations.

Nature is thus seen as an entity which restricts the applicability of our freely invented ideas. It is not necessary to assume that its real structure is thereby revealed to us. If, for instance, we say that we observe something, this should be seen as a theoretical statement which does not include the ontological commitment to the view that this something really exists in an ontological sense as a separate part of nature. Indeed, nature may not even consist of separate parts at all.

Empirical adequacy, with respect to a given set of data, is in itself not a sufficient criterion for theory choice. One of the problems we encounter is the problem of induction. After ten sips of beer from my glass I may have the, up to then empirically adequate, theory that every sip will result in receiving beer in my mouth. The eleventh sip may disappoint me however. More in general, from empirical adequacy with respect to a given set of data we cannot logically conclude empirical adequacy with respect to new sets of data. Moreover, it is always possible to construct theories which agree with reference to the data given, but which disagree outside this domain. Accordingly, we need additional criteria which enable us to proceed from an observed empirical adequacy with respect to given data to an assumed general empirical adequacy with respect to a generalized class of data or possible data. These additional criteria may not always be explicitly clear and, moreover, it may not be clear precisely what the empirical content of a given theory is. Even apart from these facts, however, there is the problem that, if our imagination is rich enough, even such a general empirical adequacy is not sufficient to single out one particular theory. Consequently, if we nevertheless want to select a unique specimen we use even more criteria. All these additional criteria will be regarded by definition as metaphysical. Metaphysical pluralism can now be introduced as the view that there may be more than one interesting set of metaphysical criteria for theory choice.

The choice to regard criteria going beyond empirical adequacy as metaphysical indicates a major difference with constructive empiricism, and it expresses the fact that I cannot, in general, consider them as pragmatic. To choose on pragmatic grounds has, for me, a number of connotations, such as being aware of choosing, knowing what to choose between, a freedom in deciding and perhaps even the possibility to reverse a choice or make a different choice some other time. In many of the commitments that are in fact made by us or for us, however, I think we are about as free as in choosing between being a spider or a human being, and I find it more appropriate to describe our situation in the following terms:

> We may imagine a variety of conceivable world views, which we shall call the metaphysical universe, and ourselves as exploring some small corner of it. The choices and commitments to which we are bound define a shape in this universe. To some extent we are able to change this shape, but it usually requires much effort.

Empiricism seeks to ban metaphysics from science, and in its latest version it tries to do this by considering nonempirical criteria for theory choice as pragmatic. There are conceivable circumstances such that I would agree to this; in particular if the class of alternatives is transparant, in the sense that, in the first place, it is clear what the alternatives are and that, in the second place, it is clear what the consequences of a particular choice are. My choice for a metaphysical pluralism is motivated by the idea that, in a broad perspective, neither of these criteria is satisfied. Usually the class of alternatives is unimaginably large, and the consequences of a particular choice, e.g., with respect to the further course of scientific development or with respect to the fate of mankind, are almost totally unforeseeable.

What makes it possible to choose at all is that we rely heavily on criteria which are undisputed, but by no means self-evident, that the class of alternatives is duly restricted by our own shortsightedness and lack of imagination, and that we simply ignore unknown consequences of our actions. Moreover, we may often not even be aware of making choices. We simply design theories and become committed to the views which are implicit within them. In other words, on a larger scale, the class of alternative world views can be considered as highly opaque, but we conveniently restrict our considerations to those parts which are visible through this fog of ignorance. For the time being, however, we seem to enjoy the good fortune that nature permits us to survive by these methods.

In this view, there is no clear demarcation between science and metaphysics. Science does not offer the right metaphysics, as is the aim of various forms of realism, nor has it been able to reduce metaphysics to pragmatics, as is the aim of empiricism. In as far as science can free us from nasty things like ignorance and superstition this is achieved by increasing our metaphysical agility and flexibility, i.e., by making it easier to change our metaphysical shape and look at the world from different points of view. Realism can be regarded as being based on the assumption that nature is simple enough to be understood, at least in some sort of approximation. Empiricism, on the other hand, is based on a rather high estimate of our own intellectual capabilities, where it comes to being aware of the consequences of our decisions and actions. I regard metaphysical pluralism, however, as the philosophy which best expresses our condition of being tiny creatures in a large and complex universe.

Anything goes

In the so called post–positivism debate the term metaphysical pluralism has become intimately connected with Feyerabend's credo "Anything goes". I am, in fact, willing to defend a moderate version of this slogan, but clearly we have now arrived at a delicate subject, where words are easily misunderstood. The advocates of an objective reality and a rational scientific method, whatever it may be, like to see their position as the only alternative to complete chaos. Proponents of other views are even invited to jump from the topwindow of a 50–story building[14], because they are not supposed to adhere to the truth of scientific theories, such as the law of gravity. This is an excessively cheap way to get rid of your opponents and matters must be slightly more subtle than this. Why is it that any theory of rationality is in the last instance always based on mechanisms that are too vague to really

[14]See Feyerabend [75] p. 221.

comprehend, such as 'peer review', or even 'common sense'[15]? Invariably, it seems, the rules of argumentation in this debate must be stretched beyond the limits where logic can account for them[16]. What do you do, when you find out that without metaphysics you get nowhere at all, but nevertheless you want to avoid making an arbitrary choice for some specific faith? The least you can do is to consider the class of possible, and imaginable world views and investigate an interesting part of it.

From here there are various ways to proceed. One way is to assume that the conceptual apparatus concerned with truth and reality has lost its value and to replace it with, for instance, a consensus directed approach. The quality of a theory is then regarded in the first place as the result of theory appraisal by a community of scientists, i.e., it is turned into a primarily social phenomenon. One may, alternatively, take a much weaker interpretation of "anything goes", which puts more emphasis on the significance of nature in determining what is, or is not, a good theory. Nature, however, may not provide unambiguous answers and this is what generates pluralism. One may furthermore distinguish two levels of pluralism. In the first place, it may occur that the standards of theory appraisal are clear and unambiguous, but that still they are not sufficient to single out a unique theory as the best one. This will be called *theoretical* or *physical pluralism*. In the second place, also the standards of theory appraisal themselves can be ambiguous, resulting in a *metaphysical pluralism*.

Metaphorically speaking, metaphysical pluralism turns metaphysics into a landscape. "Anything goes" now obtains the meaning that there is no *a priori* preferable place or direction in this landscape. You will find something wherever you go and whether what you find suits you can only be determined by knowing what the world is like from that particular spot. Moreover, even if you have in mind some general idea of where you want to go, you may still get further and travel easier by changing directions now and then. In the physical world too, a straight line is in practice seldom the best way to go. The difference between the strong and the weak interpretation of "anything goes" concerns primarily the emphasis on empirical evidence. The weak version that we shall hold here works from the assumption that "nature cannot be fooled[17]" by any social construction whatsoever. What we are talking about is the amount of freedom left *after* empirical adequacy has been taken into account. This, however, still leaves a remaining set of possibilities which may be large and even huge, even though this does not mean that *everything* is possible and that theory choice is *merely* social construction. The kind of idea I have about the structure of the set of adequate theories is that it is something like a large system of small holes through a vast universe of nonsense. Since, however, many of the accessible corridors may lead to a dead end, we had better stay well aware of the fact that there are different possibilities and that, occasionally, it may even be necessary to explore the wasteland of nonsense in order to get anywhere at all.

[15]See, e.g., Lakatos [140], [141] and Feyerabend's [76], where the practical content of Lakatos' rationality theory is brought down to that of a footnote in defense of current social ritual.

[16]Siegel [193] is a point in case.

[17]Quote Feynman, report on the Challenger accident.

From this point of view the conceptual apparatus concerned with truth and reality has not lost its value, but is rather in need of proper amendment. This involves investigating the way in which these concepts are used within science, and to extend them to fit into a pluralist frame of thought. Obviously it will be impossible, even in a pluralistic approach, to maintain the idea that any view is interesting and deserves attention. This would soon be the end of science as an organized social activity. Views can be interesting for many reasons, but clearly some views are definitely more interesting than other. Comparing different views and evaluating the standards by which we compare them are important aspects of scientific work.

World views can be evaluated in the first place with reference to their own internal criteria. Furthermore, they can be compared to each other, but it is not always possible to do this in a unique, strictly logical and rational manner, because there may be no consensus about the criteria to be used. Different world views can be incommensurable. The criteria involved in comparison are then essentially external, and they can leave room for differences in taste or style or any other types of factors considered relevant. The importance of this type of comparative research, however, may greatly exceed that of analyzing differences in taste. One of the important activities for a scientist is the massage of his intuition. Time and again he must try to raise the imagination to consider a problem from yet another point of view. With science proceeding to higher and higher levels of complexity it will become necessary to organize such activities in a more systematic way. To study the relations and the transformations between world views then becomes of crucial importance. The art of transforming world views is comparable here to mathematics in the sense that both are involved in studying the a priori structures which scientists use to represent their knowledge.

Literature: Boyd [34, 33], Cartwright [39], Churchland [44], Cushing [54], Eco [68, 69], Feyerabend [75, 76], Van Fraassen [93], Hacking [107], Kuhn [209], Lakatos [140, 141], Laudan [146, 145], Leplin [148], Popper [170, 169], Putnam [172], Siegel [193], Staal [195], Stein [197], Toulmin [208], Waisman [211], Worrall [224].

Chapter 2

Comparative metaphysics

Abstract: Comparing different theoretical pictures of a complex system may itself be a very complicated matter. Some of these complications are discussed, and some suggestions are made about how one might approach such problems in a more or less systematic way.

From the point of view of metaphysical pluralism, studying the relations between different theories and world views naturally becomes a central issue in philosophy of science, and comparative metaphysics seems an apt name for this kind of activity. Comparative metaphysics may be regarded as an enterprise directed at trying to increase transparancy in the metaphysical universe and at trying to reach, as much as possible, the situation that metaphysical transformations and theory choice can *become* pragmatic matters. Success is not self-evident however. Large scale success may be unattainable at all and even minor success may require major efforts. The general problem of comparing different views is a very difficult one, especially if they fit into different metaphysical backgrounds. One of the problems is how to deal with the vague parts of a view. In labyrinth terminology, a view will generally have rhizomelike parts which are open to differences in interpretation. Even if two views are crystal clear, however, and formulated in very precise mathematical terminology, this does not mean that the relations between them are clear. Comparing different views will generally involve constructing a logical framework in which this comparison can take place. This may itself be a matter which can be conducted in many different ways.

Problems of this type, concerning the *incommensurability* of different views, were extensively discussed especially in the sixties and the seventies. I am not entirely happy, however, with the general perspective from which much of the discussion has taken place, and which included much emphasis on the problem of *rationality* in science. Briefly said, the rationality problem concerns showing the superiority of the scientific method over alternative methods, and the problem is rather difficult. Solving it would require a number of difficult steps. In the first place it requires knowing what kind of structure the universe of possible methods is. Secondly, it requires knowing precisely what the scientific method is and where it is situated in this universe. Thirdly, it requires finding an ordering of which everybody would agree that it would represent the quality of 'being a better' method. And finally, it would require showing that the ordering has a maximal element, being indeed the scientific method which comes out best.

I do not think that on any of these points we have more than vague intuitive ideas to rely on. It is true, of course, that there are restricted rationality problems which are more clearly defined and may have a definite answer. These concern questions like whether a given theory is better than some other given theory, given a sharply defined criterion of success. Many such restricted problems boil down to more or less elementary debugging, which is generally difficult enough and by all means let us continue to put effort into this vital activity. It is not evident, however, that solutions to such problems can ever be combined into a more general notion of scientific rationality, and in fact I do not see the general rationality problem as being decidable at all. One may note, however, that fortunately the success of science does not seem to depend on its success in solving the rationality problem, and we may well be able to do without it. It may further be noted that dreams about cultural superiority occur in many societies, and it is almost reassuring to find that science makes no exception. It might be clarifying, however, to compare the situation of science to that of, say, a car factory. These do not usually have a special department for proving their superiority. They have departments for designing cars, presumably they try to do their best at it, and if they fail and deliver a bad design their competitors will do their best to let them feel the consequences. It may not be entirely unhealthy to regard science simply as a factory for producing world views.

Meanwhile, the rationality issue has resulted in what may be regarded as a distortion of the discussion of incommensurability. Lots of energy have been invested in case studies in history and sociology of science, either in defense or in attack of some tentative concept of rationality. Such studies can be interesting in, for instance, what they say about the structure of theories, about what criteria were considered important in a certain period, or about some restricted rationality problem. However, placing the analysis in the perspective of a general theory of rationality does not add much value to the analysis as such; on the contrary, it rather adds a slightly neurotic twist to the conversation. It might have been more rewarding to focus the attention on practical questions concerning the logical and metaphysical aspects of the incommensurability problem, such as questions concerning what kind of logical framework is in fact available for comparing theories.

2.1 Ambiguity and meta–ambiguity

Theories may be compared in many different ways and for many different reasons. One may look at the general structure of a theory, or at its associated ontology, at any particular details, at the way it combines with other theories, and so on and so forth. Generally, however, it does involve having definite ideas about the structure of these theories and being able to represent them in a format in which they become, to some extent, 'comparable'. In later chapters some general tools connected with this kind of work will be discussed, and be collected into a framework, a toolbox called contextual logic. Because a common motive for wanting to compare theories is that one has a problem of theory choice, theory comparison often includes theory appraisal, especially of theories relative to each other. The topic is related to the rationality issue, and, moreover, realists may like to this discuss it in terms of approximation of the truth. Nevertheless, the subject is also relevant from a wider perspective, and some general equipment for dealing with it may be introduced without imposing spurious restrictions.

Theory appraisal

A particular epistemology may also be viewed as being associated with a particular theory about the structure of nature. Nature might be *simple* in the sense of really possessing a comprehensible structure which we can learn through science and which can be described in a perfect way by some perfect theory. In this case, an ultimate reality does exist. It is the picture of nature as described by this perfect theory and naive realism applied to this particular theory is fully correct. Moreover, naive realism is also partly correct if applied to theories which are not themselves literally true but which approximate the true theory in some well-defined way. Naive realism is then correct with respect to those concepts which retain their meaning with respect to the true theory, i.e., those concepts which can be coherently translated into the concepts of the true theory.

Sequences of increasingly better theories, approximating the truth in a coherent way, may exist irrespective of whether a final true theory can or will ever be found. In this case, nature may be called *semisimple*, and realism can be upheld with respect to those concepts which retain enough of their meaning up the sequence of theories. It is up to the more sophisticated forms of metaphysical realism to explicate what it means for a sequence of theories to approximate the truth and what it means for a concept to retain its meaning.

It is not necessary for nature to be simple or semisimple in the sense defined above. We shall consider two possibilities, or strictly speaking one, because the first is a special case of the second.

The infinitely complex nature:

> For every theory $\mathcal{T}$ *there is a better theory* $\mathcal{T}'$ which *is conceptually quite different.*

The inherently ambiguous nature:

> For every theory $\mathcal{T}$ there is a theory $\mathcal{T}'$ which *is at least as good* as $\mathcal{T}$, but *not conceptually equivalent* to it.

With respect to these possibilities, let us note in the first place that apparently it is possible for nature to be such that realism cannot be fulfilled. The presumed aim of science, to provide a literal description of nature, may be unsatisfiable. Let us then note in the second place, however, that matters are not entirely as clear as this. Both examples are ill–defined as they stand. Everything depends on the meaning of the emphasized criteria:

> What does it mean to say that $\mathcal{T}$' is better or at least as good as $\mathcal{T}$?
>
> What does it mean to say that $\mathcal{T}$' is conceptually quite different or not equivalent to $\mathcal{T}$?

One may try to answer these questions by formulating explicit criteria, say E, for the conceptual equivalence of theories; and by formulating criteria, say Q, for theoretical quality. Criteria for conceptual equivalence concern for instance the degree of translatability between theories. Do the theories use the same concepts, are they about the same entities, etcetera? Criteria for theoretical excellence concern for instance the empirical content and empirical adequacy of the theory, its simplicity and esthetic value, and the compatability with other theories. If one does it well, one would thus expect to obtain an equivalence relation $\simeq$ and an ordering relation $\succeq$, either partial or complete, on the class of theories $\mathbf{T}$. The question is whether it is at all possible to do it that well. The first of various problems is that the class of theories is not a well–defined category. We have no sharp notion of what exactly a theory is. Instead, we have a historically grown concept which is not immune to change and which varies between the various fields of science. The concept is vague, it determines only in a heuristic sense what is or is not a theory. Accordingly, one may encounter problematic specimens for which it is not *a priori* clear whether they should be counted as theories or not. Let us override this particular problem by assuming the existence of sharply defined criteria C which yield a sharply defined set $\mathbf{T}_C$ of theories. Assuming, likewise, that the criteria for conceptual equivalence and for theoretical excellence are sufficiently sharp, one obtains a structure:

$$< \mathbf{T}, \simeq, \succeq > \tag{2.1}$$

where

> $\mathbf{T}$, $= \mathbf{T}_C$, is a well–defined set of theories.
> $\simeq$, $= \simeq_E$, is an equivalence relation on $\mathbf{T}$ expressing conceptual equivalence.
> $\succeq$, $= \succeq_Q$, is an ordering relation on $\mathbf{T}$ corresponding to theoretical excellence.

Returning now to the question whether or not nature is simple, it is clear that the answer, and thereby the tenability of realism, depends heavily on the criteria C, Q and E and, therefore, even the question itself may be highly ambiguous. Furthermore, it may be noted that the question can never be answered without a leap of mere faith. One can never know that a theory is literally true. The point is that nature does not appear to us by revealing its true structure but merely by restricting the applicability of our theories which are themselves the freely invented products of our own imagination. It does tell a lot about nature that a particular theory can be maintained to some degree of satisfaction, but it does not mean that nature is what the theory pictures it to be. The scope of our imagination and our willingness to include or exclude certain types of theories are unavoidably crucial factors with respect to our view of nature.

Comparative metaphysics is concerned with the organization of knowledge rather than directly with its contents. A systematic approach to theory appraisal within the framework of comparative metaphysics may be obtained by studying possible sets of criteria C, E and Q and their resulting triples $<\mathbf{T}, \simeq, \succeq>$. The approach may be illustrated by a, fairly trivial, example.

Example: Consider a measurement of the gravitational constant G, appearing in Newton's theory of gravitation. Such a measurement is relevant to a particular problem of theory choice, as it enables us to choose between alternative versions of Newton's theory with different numerical values of G. Analyzing this problem of theory choice along the lines of comparative metaphysics, one starts by introducing sets of criteria C, E and Q.

C: The universe of all possible theories of gravitation may be cut down to a managable set by restricting the analysis to Newtonian gravitation. The gravitational constant G is a parameter to be measured and is treated as unknown to some extent. Accordingly, $\mathbf{T}$ is a set of theories $\mathcal{T}(G)$ of Newtonian gravitation, with G varying in some range of values:

$$\mathbf{T} = \{\mathcal{T}(G) | \, G \in \text{ some range}\} \tag{2.2}$$

E: There do not appear to be many interesting choices for conceptual equivalence in the present case. The theories $\mathcal{T}(G)$ differ only in the constant G and at a conceptual level this does not seem to be a very interesting difference. Accordingly, all $\mathcal{T}(G)$ will be regarded as conceptually equivalent.

$$\mathcal{T}(G) \simeq \mathcal{T}(G') \text{ for all } G, G'. \tag{2.3}$$

Q: Criteria for theoretical excellence usually blend demands of empirical adequacy with demands of theoretical beauty. As there is little reason, however, to prefer one value of G to another on the grounds of theoretical beauty, only criteria of empirical adequacy remain. These, however, can be chosen in several nonequivalent ways, such as:

$Q_1 : \mathcal{T} \succeq_1 \mathcal{T}'$ iff predictions of $\mathcal{T}$ are usually in better agreement with a given body of evidence than those of $\mathcal{T}'$.

$Q_2 : \mathcal{T} \succeq_2 \mathcal{T}'$ iff predictions of $\mathcal{T}$ are always in better agreement with a given body of evidence than those of $\mathcal{T}'$.

$Q_3 : \mathcal{T} \succeq_3 \mathcal{T}'$ iff predictions of $\mathcal{T}$ are always in perfect agreement with a given body of evidence whereas those of $\mathcal{T}'$ are sometimes not.

$Q_4 : \mathcal{T} \succeq_4 \mathcal{T}'$ iff predictions of $\mathcal{T}$ are always in perfect agreement with any evidence available whereas those of $\mathcal{T}'$ are sometimes not.

Obviously these criteria Q_i are quite vague as they stand and require further specification of what is meant by terms such as "prediction", "usually", "better/perfect agreement" and "given body of evidence". Nevertheless, the general idea is clear: criteria for theoretical excellence can be given here in various grades of restrictiveness.

Turning now to the issue of convergence, it may be noted that the outcome is not necessarily the same for all Q_i. Nature could, for instance, be such that one finds a fluctuating gravitational constant. In this case, criterion Q_1 would be more widely applicable than Q_2. Given an arbitrary body of evidence, the likelihood for finding convergence is large within the triple $< \mathbf{T}, \simeq, \succeq_1 >$ and decreases up the sequence of criteria $Q_1 ... Q_4$. Furthermore, the type of evidence included may be important. Criterion Q_3, for instance, may be well applicable to a whole range of laboratory and astronomical evidence and convergence may be found to the limits of measurement accuracy. However, as soon as the perihelion shift of Mercury, or some other general relativistic effect, is included in the evidence, criterion Q_3 ceases to be interesting because none of the $\mathcal{T}(G)$ can account for it.

Convergence

Clearly, the question whether nature is simple is itself ill–defined and ambiguous. Besides depending on the structure of nature, it also depends on choosing a perspective, determined by some particular triple of criteria $< C, Q, E >$. In practice, especially in debates about realism, realist are likely to prefer criteria which allow them to reconstruct the history of science as a relatively smooth process, producing ever closer approximations to the truth, whereas antirealists are generally delighted to contradict this. Instead of getting stuck in questions like who is right, or who is most right, or whether some people may be approximately right, or whatever, one may approach these matters from a quite different angle.

Might it not be better to *believe* in the comprehensibility of nature? What is the point of doing science unless one has faith in the possibility of success? Some people may argue that there are other forms of success and fullfillment, besides reaching a final goal. Some may even regard faith as a dirty word, a form of self–deception that is strongly poisonous to the true intellectual, but surely there is a real issue here. What is so bad about faith, and can it be avoided at all? Clearly, man and his knowledge are fallible and *all* knowledge has roots in faith to some extent. Faith should be judged on its merits, not on its being a faith. Accordingly, let us dedicate some discussion to the merits of just *assuming* nature to be simple.

Obviously, from a certain point of view it would be extremely convenient if nature had a comprehensible structure. To assume that some of our theories come close to being literally true would explain why they work so well and also why new theories often successfully predict hitherto unobserved phenomena. Moreover, if nature were simple this would allow us to maintain the comfortable picture of science as an enterprise which can accumulate knowledge in a more or less straightforward way. Furthermore, it would assure us that continued research efforts in a given area should lead to a unique final result: the truth. Evidently the view that nature is simple appears to have very attractive features and it seems to be in good agreement with Occam's razor to abstain, as long as possible, from introducing the additional complications associated with another point of view. Accordingly, one may decide to hold on to it as long as it can be maintained with some degree of plausibility.

Although this approach is most powerful in its directness and will serve well in many instances, it is also a bit narrowminded and largely based on wishful thinking. A more sophisticated view may be more complicated and perhaps a little inconvenient for practical purposes, but it may occasionally yield more interesting results and a fuller account of what is actually going on. Convergence of theories or ideas is typically an issue which can serve to illustrate this. Adhering to the simplicity of nature one is easily tempted to interpret the occurrence of convergence of scientifically built opinions as an instance where literal truth is found or at least approximated. Absence of convergence, on the other hand, is seen as an indication of scientific failure. The view that nature may be ambigious invites to a quite different approach. Keeping in mind that there are possibly several nonequivalent theories which can be simultaneously maintainable, disagreement is not necessarily a sign of failure. Furthermore, if convergence of opinions does occur, literal truth is not necessarily the underlying mechanism. In fact one can hardly fail to observe that convergence, either fully or more often partially and temporarily, is a phenomenon which occurs regularly in many fields where truth can hardly be held responsible, such as art, fashion and religion. It would be quite remarkable if the psychological and social factors behind this phenomenon were not at all active with respect to science.

We may conclude that assessing the merits of the view that nature is simple is not a simple thing. It becomes even harder when we realize that the coupling of this view to a specific research strategy is in fact much weaker than is suggested above. Looking at it at a superficial level, the view that nature is simple seems to imply quite another research strategy than the view that nature is ambiguous: you either search for the unique final truth or you try to develop and compare various interesting approaches. The connection is not as direct as this, however. Both strategies clearly have their own merits and shortcomings and can be used independently of considerations about the ultimate complexity of nature. Even if you are in search of a single real truth you may realize that at some stages of your search theoretical pluralism may be the best strategy to find it. On the other hand, even if you aim at developing numerous views it may be wise at some stage to concentrate on a single one and work it out in detail.

Example: The problem of convergence may be viewed from a different angle. There is much debated analogy between scientific development and evolutionary change. Although the merits of using this kind of speculative analogies are indeed debatable, the analogy can be more or less based on the observation that nature does not appear to us by revealing its true structure but by restricting the applicability of our freely invented ideas. Nature is a well-known restricter, not only of ideas but also of the types of animals living in a given area in a given ecological niche.

Few people would maintain that for a given ecological niche a unique 'true' species exists. However, thinking of theories, and recalling the existence of different perspectives defined by different triples $< C, Q, E >$, it is interesting to note that with respect to evolution too one may conceive of a sense in which something like an optimal 'true' animal does exist and that in fact a kind of convergence can and does occur. It is quite unusual for two different species to share exactly the same niche, and species occupying comparable niches tend to develop similarities. The thought can be illustrated by referring to the functional similarities between completely unrelated species occupying a comparable ecological niche, for instance the similarities between the ichtyosaurus, an extinct marine reptile, and the dolphin. Nature has produced two species here of very similar but unrelated animals. It seems that at least at a functional level there can be something like an optimal animal associated with a specific ecological niche and in the above example this optimal animal is approximated by both ichtyosaurus and dolphin.

Apart from the functional level, where ichtyosaurus and dolphin are similar, there are, of course, other levels where it is unlikely that there are any far going similarities at all, for instance the morphological level, the level of protein chemistry, and the genetic level. Accordingly, in this case too, the appearance of similarity depends on choosing a perspective. Thus, the analogy leads back to the realization that the question whether nature is simple is itself ambigious. It all depends on the criteria $< C, Q, E >$: which theories are included in our considerations, when are we prepared to call one theory better than another and when are we prepared to call them conceptually equivalent? There may be various ways to choose these criteria, just as there are various levels that can be chosen to compare the ichtyosaurus and the dolphin. It may occur that nature turns out to be simple on one set of criteria and ambiguous on another. Moreover, a debate on what is the appropriate, 'true', set of criteria may turn out to be futile because each set can be interesting in its own way, just like it is interesting to know both that there are functional similarities between ichtyosaurus and dolphin and that at other levels they are quite unlike each other. A careless belief in nature's comprehensibility could make one miss this point by inducing an excessive preoccupation with finding the one and only real truth.

As we can see now, the belief that ultimately nature is simple in the sense of being fully and unambiguously comprehensible to man entails two things: in the first place that we shall be able to find a triple of criteria $< C, Q, E >$ which stand out as the most natural 'true' ones, in the second place that these criteria will suffice to single out a unique true theory. Whether one likes to maintain a faith like this is primarily a matter of personal psychological inclination, i.e., of taste.

Most likely, philosophers will remain divided on the subject as they have been for centuries. My own taste, however, leads me to reject metaphysical realism, i.e., to reject the view that nature can be unambiguously understood. My own experience is that many interesting problems permit a multitude of views and I do not see this as a feature which can or should be considered to be repairable. To me, ultimately, nature is a mystery. I like it this way and, moreover, I also think that such an attitude can be defended against the onslaughts of scientific progress, and without degrading science as a valuable and exciting enterprise. I just fail to see the benefits of regarding science as a quest for finding the one and only real truth, just as I do not see art as a quest for finding the one and only ultimate painting. I see it as the aim of science to advance human knowledge by developing interesting views on nature. It may happen that, in a specific area of science, combined demands of empirical adequacy and theoretical beauty, together with the limitations of human imagination, give rise to a unique final product. Perhaps such an outcome may even be regarded as convenient, but there is little reason to require it in advance.

Scientific progress

The notion of progress has links with convergence, although not very conclusively. One may have convergence without progress, for example in fashion, as well as progress without convergence. Recalling the analogy with evolution one may expect a very diffuse and complex kind of progress in science, whereas, if one believes in the unity of science, one might expect a much more singular and straightforward concept of progress to be applicable. However, taking the point of view of comparative metaphysics it is clear that the notion needs a context: first a perspective, then a notion of progress. Unless one finds a unique set of truely compelling criteria there will be many such notions.

With respect to evolution there are various complications which make it difficult to define a singular concept representing progress. In the first place, any modification to a species, even if successful, is likely to have drawbacks as well. For example, an adaptation may be successful by making the species better suited for a given ecological niche, but specialization may also have the drawback of preventing the species to move on to another niche. Secondly, the consequences of a modification may be entirely unforeseeable. For example, a gene may be inactive by being situated on a chromosome site which is not expressed. Millions of years later the gene may be transferred to another site and turn out to be of crucial importance. Thirdly, genetic transformations are often 'package deals'. For example, an individual may be successful in a population by having good eyesight, accidentally compensating his bad hearing, caused by a genetic defect which is accidentally coupled to his good eyesight by being situated on a nearby chromosome site. In this way, defects may ride horseback on improvements and become important at a later stage. Fourthly, one may differentiate between different levels of success. For example, success of particular genes, of individuals, of social groups, success of a species, success of an entire eco-system, and so on and so forth.

The above list is not meant to be exhaustive, but it contains some major items of which one may wonder whether similar difficulties might be relevant to defining measures of progress in science. It might seem that, at least in some respects, science is qualitatively different. For instance, improvements achieved in one scientific discipline may find applications in many other disciplines, whereas genetical mutations are restricted to a given species[1]. On the other hand, one may argue that increasing specialization in science is making communication between disciplines increasingly difficult. Science is relatively young and speciation may not have proceeded to the level where interbreeding becomes impossible. An argument for this may be that it seems indeed quite difficult for ideas originating in other cultures to penetrate into the scientific world.

Example: The current rise of chaos theory may provide illustration to some of the difficulties surrounding the notion of progress in science. The ancient Egyptians in 2000 BC had a special deity for chaos and their myths comprised elaborate philosophies about the relations between chaos and order and man's role in finding some sort of balance. Scientific culture, on the other hand, has inherited various traits from the Christian culture where it originated, and which included, in some respects, much more primitive ideas about the structure of the world, such as the strong emphasis on uniqueness, one God one truth, a rather naive separation of good and bad, and ambitious ideas about controlling the world, God controls the creation and we, as his appointed representatives on Earth carry out part of the job. The suggestion of determinism, raised by the early results in Newtonian mechanics, apparently fitted very well into this world of thought. Along with the rise of classical mechanics, and in combination with the expansion of western culture, determinism became a central ingredient of the dominant world view.

Significantly, determinism in classical mechanics turned out to be a misconception from the very start, a point which became apparent only recently by the developments in chaos theory. Sir James Lighthill expressed this very clearly[2] in a lecture to the Royal Society on the three hundredth anniversary of Newton's *Principia*, where he says, speaking "on behalf of the broad global fraternity of practitioners of mechanics":

> "We are all deeply conscious today that the enthusiasm of our forebears for the marvellous achievements of Newtonian mechanics led them to make generalizations in this area of predictability which, indeed, we may have generally tended to believe before 1960, but which we now recognize to be false. We collectively wish to apologize for having misled the general educated public by spreading ideas about determinism of systems satisfying Newton's laws of motion that, after 1960, were proved to be incorrect."

With respect to our present subject there are various interesting aspects to this affair. One may, for instance, regard the idea of determinism as being comparable

[1] Although viruses may introduce new genetic material which might even come from another species.

[2] Lighthill [151], here quoted from Reichl [178].

to a genetic defect, which, by its close association with an extremely successful gene, becomes very widespread. Determinism, although it is now recognized as being defective in mechanics, did have a major influence on the intellectual climate in Europe. It has certainly contributed to a view of nature as being controllable, and scientists have been looking for strict, deterministic laws in many areas of science, including typically chaotic fields like economics. It is very difficult to imagine what the consequences might have been if this flaw in the interpretation of mechanics had been detected at a much earlier time.

The fall of determinism and the rise of chaos theory are currently leading to new branches of science, where more sophisticated ideas about the structure of reality are rapidly being developed. There is little evidence, however, for there being a major impact from ideas from other cultures, such as the ancient Egyptians. Their general world view is too remote for us to easily incorporate their ideas into our own worldview, even though some of these ideas were markedly superior to dogmas dominating western science for various centuries. Apparently, cross fertilization between different cultures is not always as easy as it might seem.

Example: The rapid development and growth of science in the past few centuries may be regarded as being comparable to what may happen when a new gene is introduced into the genepool of a species, or when a new species is introduced into an eco-system. Occasionally, such an introduction can be extremely 'successful' and lead to major changes in the entire population, or the entire system. In science the great changes were triggered most of all by the combination of newly developed mathematical tools, especially differential calculus, and areas where these tools could be applied, which happened to be available as well as extremely fruitful. This can be regarded as being very similar to what happens if a new genetic trait opens up a new ecological niche with unexplored resources to a species.

Analyzing such an event in terms of success and failure, or progress and degeneration, is generally a very difficult matter, because of the complexity and uncertainty of the situation and because of the dependence on choosing a particular perspective. For instance, from the point of view of the successful gene its rapid spreading may be a great victory, but for the species involved it may also be dangerous. The gene may turn out to be accidentally linked to significant defects, its rapid spreading may seriously deplete genetic variation within the species, which may mean a greater vulnerability to disease, and it may supersede other valuable genes.

In other words, rapid change in a complex system may mean great progress from one perspective, while at the same time creating an unbalanced situation, in which almost anything can happen, involving a great vulnerability to random disturbances and possibly great risks. Generally, the system will need much time to recover from such processes. It is quite evident that the rise of science has indeed brought the general system of life on Earth in an extremely unbalanced situation, and that it is questionable whether there will be the time to recover. However, if recovery is possible, what will it be like? For an eco–system adapting to a new species it means various things. There will be genetic pressures, both on the new species itself and on other species in the system. For the new species itself there will be a tendency to specialize into the available niches, generating a process of speciation, i.e., the development of subspecies and in the long run a breakdown of the species into different ones. Among the other species in the system, presumably some will disappear, whereas other ones get new opportunities, for instance by preying on the newcomers. In science similar effects may be expected or are already at work. Many arts, crafts, and areas of traditional knowledge have disappeared or are disappearing, or are adapting themselves, for instance by adopting scientific jargon. Grand scale specialization can be witnessed all over the place, and is likely to result in the generation of separate subsciences and of schisms, comparable to the ones that have occurred in most religions.

Note: In a biological species, the speciation process may to some extent be countered by active interbreeding. Similarly, in science the speciation process may to some extent be countered by retaining effective communication channels. In this sense, philosophy of science might be given some responsibility for guarding a unity of science, however, not by arguing that it *is* in some philosophical sense a unity, but by spending a lot of work in *negotiating* it into some sort of unity. Incidentally, it may be noted that one of the major threats to a unity of science consists of the 'publish or perish' regime to which scientists are submitted nowadays. On the one hand, it makes it impossible for anyone to read more than a few percent of the literature in his own field, whereas, on the other hand, it forces him to publish any turd that leaves his bowels.

Example: One may detect in science a peculiar ambiguity between, on the one hand, modest and extremely careful modes of reasoning, and, on the other hand, giant leaps of thought and almost megalomaniac prospects. This is well illustrated by contrasting two well known quotations from Einstein. In his article *Physics and Reality* he writes[3]:

> "One may say "the eternal mystery of the world is its comprehensibility." [...] In speaking here of "comprehensibility," the expression is used in its most modest sense. It implies: the production of some sort of order among sense impressions, this order being produced by the creation of general concepts, relations between these concepts, and by definite relations of some kind between the concepts and sense experience. It is in this sense that the world of our sense experiences is comprehensible. The fact that it is comprehensible is a miracle."

This careful way of expressing himself may be compared to the sweeping speculations in his Herbert Spencer lecture of 1933[4]:

---

[3]Einstein [70] p. 292.

[4]Einstein [70] p. 274.

> "If, then, it is true that the axiomatic basis of theoretical physics cannot be extracted from experience but must be freely invented, can we ever hope to find the right way? Nay, more, has this right way any existence outside our illusions? Can we hope to be guided safely by experience at all when there exist theories (such as classical mechanics) which to a large extent do justice to experience, without getting to the root of the matter? I answer without hesitation that there is, in my opinion, a right way, and that we are capable of finding it. Our experience hitherto justifies us in believing that nature is the realization of the simplest conceivable mathematical ideas."

The contrast between these fragments seems to indicate a strange lack of coherence in the views of one of the great physicists of this century, who was, at the same time, one of the most inspiring personalities with respect to what he says about methodology. It is almost as if we hear two different persons. However, ambiguity, if it does not degenerate into dishonesty, may be a healthy attitude. People are not monolythic structures, and taking different perspectives can be very fruitful. On the other hand, it does have consequences for our picture of what kind of enterprise science is. One may be reminded of biological processes, which often come in antagonistic pairs, such that, by switching them on and off, a very complex spectrum of outcomes may be achieved.

One may also be reminded of an evolutionary process. If the environmental conditions of a species change, or the species enters a new ecological niche, this may result in rapid genetic changes, inducing an unbalanced situation in which the old animal is still 'present' in the new one, under a layer of modifications. This may lead to all sorts of difficulties such as conflicting modes of behaviour. In science one may detect such an imbalanced situation with respect to its views on complex systems, where developments have been particularly profound during this century. Present century mathematics has, presumably, produced the most sophisticated and advanced ideas ever about infinity and complexity. However, the subject remains difficult, conceptually as well as technically. Basic insights are gradually spreading, but are still easily misinterpreted, especially where they contradict optimistic ambitions of a science long drugged by illusions of determinism, in a society high on power and control.

Thoughts about complexity invite to assuming a basically modest attitude, as our capacities of understanding and controlling complex systems are limited. From Gödel's incompleteness theorem, for instance, it follows that there are undecidable questions in mathematics. There are limits to what may be achieved by representing problems in mathematical terms, and, presumably, to what may be achieved at all. In mechanics, for instance, this may result in the occurrence of noncomputable functions in the dynamics of a system, forcing the transition to a probabilistic description in order to retain a dynamical theory at all. However, Gödel's theorem has been repeatedly misinterpreted into showing a quite different thing, namely that the human intellect is fundamentally superior to what may be achieved by Türing machines or similar automata[5]. It is not the fact that some mathematical theorem is misinterpreted which is significant here, but rather the *type* of error. Indeed, the fact that Gödel's theorem is so widely recognized as an important result is presumably also related to the circumstance that it contradicted ambitions that were widespread, and which, to some extent, still linger on.

One may, at this point, consider the possibility of a relation with the heritage of Christian mythology in science. Properly, a mythology may be helpful in dealing with the conceptual problems of living in a complex world, and, thus, in providing some intuitive basis for the more technical results. Christian mythology is of little help in this particular case, however, because at specifically this point it seems to point in a quite different direction. The Christian picture of the world is relatively orderly and simple. God, who is the only one of his kind, created the place, and basically He runs it. Christians are modest in relation to God, but for the rest they are God's representatives on Earth, which puts them high in the pecking order. They are the appointed masters of the creation, which must feel rather different from being a small part of an infinitely complex system, and basically on a par with other parts of it. Accordingly, already the starting point is rather dubious, and several other cultures have succeeded in producing much more elaborate ideas about complexity and our relation to it.

Example: Hindu mythology has produced the intriguing vision of the universe being one of the infinite sequence of dreams of the sleeping god Vishnu. In the Upanishads there is a story about Indra, god of the heavens, who, after killing a great monster with a thunderbolt, gets a very high opinion of himself and decides to build a magnificent palace. However, his wishes keep on growing and growing, and finally the divine carpenter gets fed up and goes to Brahma to complain. Brahma goes to Vishnu, who, in the form of a little blue boy, brings Indra back to the ground by making him aware of the infinite sequence of universes and the infinite sequence of Indras, all equally conceited. However, Indra overcompensates and decides to leave his throne and spend the rest of his days in meditation. This time his wife complains. With the help of the divine priest some self–esteem is reestablished, and Indra, as a sadder and wiser deity, finds himself a more appropriate place in the universe.

2.2 Describing complex systems

If the world is not simple, measured against our intellectual capacities, this has consequences for our view of what kind of enterprise science is. Indeed, the naive idea of science as building a single coherent picture of the world, comprising, in principle, all meaningful knowledge, may be a red herring. Again, the relatively

[5] Most recently by Roger Penrose, see [167], earlier by J.R. Lucas, see [154]. For criticisms see, e.g., [32, 160, 9].

frugal mythology of Christian civilization offers at this point only limited resources for grounding intuitions about a more sophisticated attitude, but interesting views may be obtained from various sources.

Example: In 1968 Carlos Castaneda published a bestselling book[6] about his encounters with Don Juan, a sorcerer of the Yaqui indians. In subsequent books, more is revealed about the structure of the sorcerer's world, in which he is initiated as an apprentice. An interesting episode is when he is introduced to the "sorcerers' explanation", and Don Juan tells him about the *tonal* and the *nagual* (pronounced as toh–na'hl and na–wa'hl). The tonal is described[7] as "the organizer of the world":

> "[...] on its shoulders rests the task of setting the chaos of the world in order. It is not farfetched to maintain, as sorcerers do, that everything we know and do as men is the work of the tonal."

On subsequent pages there are repeatedly phrases like: "the tonal is everything we are", "the tonal is everything we know", and "the tonal is what makes the world". Then, however, the tonal is described as an island, and the nagual is introduced[8] as what is surrounding the island, "the part of us which we do not deal with at all":

> "The nagual is the part of us for which there is no description – no words, no names, no feelings, no knowledge."

In an occasionally comical scene Castaneda then tries to find a description for the nagual, but every attempt ends up as an item on the island of the tonal, as the nagual "can be witnessed, but it cannot be talked about". Even wittnessing it is not a minor affair, however, because any manifestation of it is suppressed by the tonal. The tonal is also described here as a guardian, which, ideally, should be broad–minded and understanding, but usually changes into a guard, becoming narrow–minded and despotic. In order to obtain "knowledge" the tonal must be induced to become "free and fluid", and, therefore, it must be rearranged by leading an "impeccable" life. In this way one may spread the "wings of perception" and become aware of the "totality of oneself".

Example: Another interesting point in Castaneda's narratives is that part of his training is applied in states of "higher awareness", and ideas that are crystal clear become diffuse or are entirely forgotten when he returns to a normal state of mind. Similar phenomena are described in other cultures. For instance, in the *Mahabarata*, the Hindu epic about the struggle between the Pandavas and the Kurus, there is the point where Arjuna, the war leader of the Pandavas, is about to give the signal for the battle to begin. However, seeing all his friends and

[6] Castaneda [41].
[7] [40] p. 122.
[8] [40] p. 126.

relatives, gathered on both sides, Arjuna is overcome by uncertainty. Krishna, his friend and chariot driver, who is an incarnation of Vishna, then shows him his divine appearance and teaches him about, say, the meaning of life. This part of the Mahabarata is known as the Bhagavad Gita. Arjuna reaches a state of great clarity, giving him the strength to signal the beginning of the battle. However, although Arjuna knows the true nature of Krishna, and has seen his divine appearance, he forgets it, by the divine grace of Vishnu, because no man could bear the strain of the continued association with God.

In Christianity a, perhaps similar, transformation of mental states takes place with the descent of the Holy Ghost. A significant difference between these religious visions and Castaneda's states of higher awareness, however, may be that the religious visions are associated with witnessing eternal wisdom and absolute truth. States of awareness, on the other hand, are coupled to different positions of "the assemblage point", and are basically on a par, although it is very difficult or impossible to integrate phenomena witnessed on different levels.

Ideas such as represented in the above examples may be too remote to be carried into a scientific world view in any direct manner. However, as a source of inspiration, and as an antidote against some of the limitations of a general cultural background, they are most refreshing. Further, I would like to contradict the view that we are merely dealing here with fanciful metaphysical ideas, which do not concern science. These ideas represent, at least, phenomena which should tell us something about the structure of our minds, and possibly our brains, but, furthermore, they may represent valuable adaptations to the complexity of the world, and science may come to need them.

Example: A nice picture to keep in mind in thinking about the description of complex systems is that of a fractal. Consider the prospect of investigating the structure of something like the Mandelbrot fractal[9], but without knowing by which equations it is generated. More specifically, imagine finding a machine which, after being feeded a pair of planar coordinates returns a "yes" or a "no", However, if it takes the machine too long to answer one may push a reset button and try another point in the plane. This picture can serve very well to illustrate some of the points discussed earlier.

After having found the machine we may initially try a number of points more or less at random, but in due time we start to have expectations about what we shall find, and we develop a more systematic search strategy. Suppose that there are homogeneous areas in the plane, entirely filled with either yes or no, and that in other areas, the *edges*, 'yesses' and 'nos' can be found close to each other. Let us further assume that we find that in the homogeneous areas the answers come almost immediately, whereas close to the edge the machine takes much more time. In view of these observations we may adopt a search strategy aimed at making a sequence of maps $\mathcal{T}_0, \mathcal{T}_1, \cdots$, each one dividing the plane into three areas, a

[9] Mandelbrot [157].

yes area, a no area, and an edge. In $\mathcal{T}_0$ the yes and the no area are those areas where the machine answers within a second, in $\mathcal{T}_1$ the machine answers within 10 seconds, in $\mathcal{T}_2$ the machine answers within 100 seconds, etcetera. In each case the edge is defined as the remaining area.

If we now think of the concept "edge", then clearly this is a vague, context dependent concept, which falls apart into a sequence of more precise concepts $edge_0$, $edge_1$, ..., representing "edge" within the sequence of pictures $\mathcal{T}_0, \mathcal{T}_1, \cdots$. Between these more precise concepts there is an evident relation. If the machine answers within one second then clearly it answers within 10 seconds and therefore $edge_0 \supseteq edge_1 \supseteq edge_3 \cdots$. We may now ask whether it makes sense to think of an "ultimate edge", being the limit of $edge_n$ for $n \to \infty$. The existence of this limit is not self–evident and, in fact, the question of whether the limit exists may itself not be decidable, either because answering times diverge or because there is too much finestructure. In this latter case $edge_n$ may not converge to a definite area or line. Instead, there may be areas where yes–points and no–points are chaotically dispersed, such that arbitrarily close to each yes there is a no and *vice versa.* If one, or both, of these possibilities appears to be the case, this may lead us to reflect on the question to what extent we are, in fact, able to understand the structure generated by this machine. A superficial understanding is readily obtained within a few minutes, but a more complete understanding may be very hard to get, may require the introduction of more refined mathematical tools, for instance more advanced topological notions, and may even force us to revise the meaning of concepts which seemed to be quite well–defined at an earlier stage of the research. It might, for instance, turn out that on closer scrutiny we are not even able to locate the precise boundaries of the areas $edge_k$. If these have a sufficient amount of finestructure themselves, then in fact the $edge_k$s may turn out not be areas at all.

Suppose that initially we start with a simple idea of what it means to understand a structure of the above type, namely that it would mean to know of every point in the plane its yes or no–value. Complications of the above type can make it interesting to consider more sophisticated forms of understanding, which take into account the existence of restrictions on what one may actually get to know about a system. Obviously, what we can learn about the system depends on the structure and in particular on the complexity of the system, and, accordingly, our view on what we expect to be able to learn about it rests, either implicitly or explicitly, on assumptions about its simplicity.

There is a variety of assumptions one may introduce at this point. It can, for instance, be assumed that there is a limit to the amount of finestructure, i.e., beneath a certain scale boundaries become sharp. Slightly more sophisticated is the idea that there is some regularity in the finestructure of the boundaries. The Mandelbrot fractal, for instance, contains substructures which can be found over and over again at different scale sizes. Moderately sophisticated, also, is the suggestion that there might be a set of simple mathematical equations from which the entire structure can be generated, even if it would contain infinite amounts of finestructure. Most sophisticated, however, may be the idea that the system *might* not be simple at all, in spite of the apparent possibility to form a superficial picture of it in a few moments' time.

Compositals

Some features of the above example of a fractal may admit a generalization to more lifelike situations concerning the structure of reality more generally. However, the word "fractal" already has obtained a quite specific meaning, referring in particular to geometrical structures with a fractional Hausdorff dimension. Therefore, the word *compositalwill* be coined for referring to a more general class of possibly very complex composite systems; more precise definitions will be given later on. In as far as it is possible at all to make an image of such an infinitely complex system, it cannot be a correct image because in a finite image, comprehensible to ourselves, only an infinitesimal fraction of the infinite complexity of the system can be represented. This raises the problem that, in order to make an image of it, one must first find some way to get rid of nearly all of its structure, retaining only the infinitesimal part that one aims to represent.

Note: At this point there is a connection to earlier discussions about realism. Realists hold that good scientific theories are approximately true. However, with respect to an infinitely complex system it is conceiveble that there might be infinitely many descriptions, all approximately true in some sense, yet without being mutually related in a comprehensible manner, and without it being explicable what "approximately true" means.

Sometimes there are apparently natural methods for deleting excess structure. There may, for instance, be some appropriate scale parameter, making it possible to neglect all structure beneath a certain scale size. Examples are the approximate representation of a real number by means of a decimal fraction of finite length, and the representation of the Mandelbrot fractal by means of a computer print with finitely many pixels. There also are more sophisticated methods, such as a representation of the real number π by an algorithm by which it can be computed, and such as renormalization.

Example: In order to apply renormalization there must be a suitable scale parameter and, moreover, the composital must be composed of substructures which can be found over and over again at all orders of magnitude in the scale parameter. As an example, imagine a coil, made of a thread which is again a coil, made of a thread which is again a coil, made of ... etcetera. This composital consists of structures appearing recurrently, in a similar manner, on every scale size, with a natural scale parameter being provided by the recurrency counter n. If we now make an image of the coil we can choose between two different methods. The first method is to neglect all structure above a given value of the scale parameter n. The second one is more sophisticated.

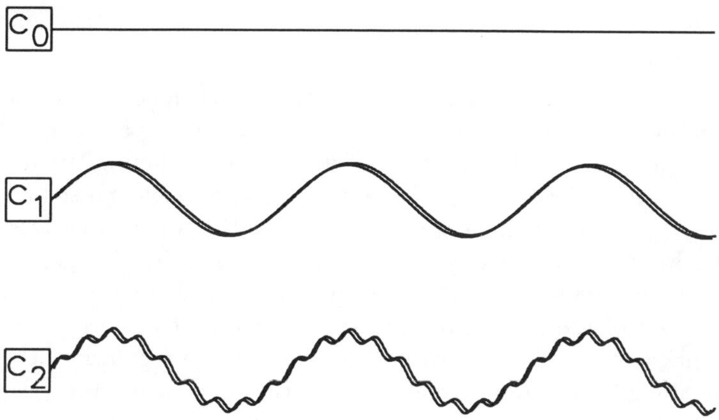

Figure 2.1: Graphical images of C.

Let C be a composite coil of which the radius r_n and the increment s_n are given for every value of the parameter n, i.e., r_n and s_n are known and computable functions of n. One must, of course, acknowledge the fact that a correct image of C cannot be made, at least not in the form of, for instance, an accurate graphic representation. Nevertheless, there is a sense in which this structure is comprehensible to us and we may design a symbolic representation that is, in many respects, almost as good as an accurate one. Let C_k be the coil defined by

$$\begin{aligned} r_n(C_k) &= r_{n+k}(C), \\ s_n(C_k) &= s_{n+k}(C). \end{aligned} \tag{2.4}$$

In a symbolic graphical representation we may now formulate the equality expressed in fig. 2.1.

What we do here is that we regard the coils C_k as being comprehensible to us in spite of the infinite level of coiling that is going on, and this enables us to express useful facts about the relations between these levels. Looking from an abstract perspective, what this amounts to is that we add, to the primitive elements which serve as elementary building blocks of our representations, a class of composite structures which cannot be analyzed in terms of these primitive elements but of which we, nevertheless, claim to have some form of understanding.

We may now compute the relation between the lengths l_n of the coil, as measured on different scale sizes. On the n–th scale, the lenght of C is given by

$$\begin{aligned} l_n &= c_{n+1} l_{n+1} \\ \text{with} \qquad & c_n = \sqrt{1 + (2\pi r_n / s_n)^2} \end{aligned} \tag{2.5}$$

We thus find:

$$l_0 = (\prod_{k=1}^{\infty} c_n) l_\infty. \tag{2.6}$$

l_∞ can now be called the bare length of the coil and l_0 the renormalized, or effective, length at scale $n = 0$.

In quantum electrodynamics comparable things are done with respect to masses and coupling constants. Plain, 'bare', electrodynamics is a theory about charged point particles that exhibit vicious self–interactions, ruining the mathematics of the theory with divergent integral expressions. The problem is solved by introducing 'dressed particles', which are comparable to the coil compositals in the above example. First one makes a dressed theory in which the divergencies of the bare theory are removed by a so–called 'regularization scheme'. One then investigates the behaviour of the regularized theory under scale transformations, i.e., transformations $x \longrightarrow \lambda x$. If, for very small (large) scales, a regularized theory becomes invariant under scale transformations, the theory is called UV (IR) renormalizable. In dressed QED the effective masses and charges of the regularized theory are scale dependent quantities, unequal to the bare masses and charges. The relations between the dressed and the bare values can in principle be found by considering an infinite scale transformation, but in this case the renormalization constants also become infinite, indicating that the bare particles do not really exist in the renormalized theory. In a similar sense there is no bare thread, which is not coiled itself, in a coil composital with diverging renormalization factors.

Abstract compositals

Abstract compositals may be introduced as objects admitting a continued decomposition. Consider an algebra $\mathcal{A}$, consisting of a set of objects **A** and a set of operations **F**. These operations may be unary, binary, or of any other arity, and the set **A** is closed under all them, i.e., if f is an operation in **F** and $\vec{A}$ is a tuple of elements of **A** in the domain of f, then we must have $f(\vec{A}) \in \mathbf{A}$. An element A of **A** will be called *composite* if it can be obtained by the action of some operation on a tuple of elements of **A**, i.e.,

$$\exists f \in \mathbf{F}\ \exists \vec{B} \in \mathrm{Dom}(f)\ \ A = f(\vec{B}). \tag{2.7}$$

The individual components of $\vec{B}$ may again be composite, such that A has a further decomposition into, for instance, $A = f(g(C), h(D, E))$, where C, D, and E may also be composite, and the decomposition may be continued. A will be called a *composital* if the decomposition can be continued forever. The set of compositals **C** is a subset of **A**, and we have that A is a composital, $A \in \mathbf{C}$, iff (if and only if) it has a decomposition of the kind (2.7), where at least one of the components of $\vec{B}$ is a composital, i.e., $B_i \in \mathbf{C}$ for at least one B_i.

Being a composital does not necessarily mean that A is a very complex object. For instance, if **F** contains the identity operation, e, then any element of **A** may be decomposed according to $A = e(A) = e \circ e(A) = e \circ e \circ e(A)$, and so on. Therefore, we must invoke some notion of complexity. To this end we first select a finite set $\mathbf{P} \subseteq \mathbf{A}$ of primitively 'simple' elements of **A**, and a finite set **S** of simple operations. To each element P of **P** we assign a primitive 'complexity', being some finite nonnegative number $c(P)$. Likewise, a complexity is assigned to each operation in **S**. Now assume that A is a composite object, and that it can be decomposed into elements of **P** by means of operations in **S**, for instance, $A = f(B, C)$. To this decomposition of A we may then assign a complexity $c(f(B,C)) = c(f) + c(B) + c(C)$, and to A itself we may assign the complexity of its simplest decomposition. Closing off **P** under the operations in **S** yields a subalgebra $\mathcal{K}$ of $\mathcal{A}$, containing all objects of finite complexity. $\mathcal{K}$ will be called a *comprehensible core* of $\mathcal{A}$.

Example: Consider the case that **A** includes the set of standard natural numbers. A comprehensible core may be defined by taking $\mathbf{P} = \{1\}$, and $\mathbf{S} = \{S\}$, where S is the successor relation. Assigning complexity 1 to both 1 and S, we find that $c(2) = c(S) + c(1) = 2, c(3) = c(S) + c(2) = 3$, etcetera.

With respect to a given comprehensible core, a composital will be called *reducible* if its complexity is finite. Evidently, the complexity of a composital depends on the definition of the comprehensible core and the assignment of primitive complexities. A composital which is irreducible with respect to one comprehensible core may actually have a low complexity relative to another.

Example: A comprehensible core for the set of real numbers is associated with their representation in a decimal expansion. Assume $\mathbf{A} = I\!R$, and define $\mathbf{P} = \{0, 1\}$ and $\mathbf{S} = \{-, f, g, +\}$, where $-$ is a unary operator turning positive numbers into negative ones and *vice versa*, $f(x) = 10 \cdot x$ and $g(x) = 0.1 \cdot x$, and $+$ is the standard binary operator for addition. Assigning a primitive complexity 1 to 0 and 1, as well as to f, g and $+$, a rational number like $1/2 = 0.5$ may be represented by $0.5 = g(1 + (1 + (1 + (1 + 1))))$, obtaining a complexity equal to $c(g) + 4 \cdot c(+) + 5 \cdot c(1) = 10$. The rational number $1/3 = 0.333\cdots$, however, is irreducible with respect to this particular core.

The irreducibility of $1/3$ may be removed by choosing another comprehensible core, or by extending the present one. For instance, if the division operator "/" is added to **S**, the complexity of $1/3$ is brought down to $c(1/3) = c(/) + c(1) + c(3) = c(/) + 6$.

Example: Thinking back of the example of coil compositals, it may be observed that the renormalization approach can be similarly interpreted as an extension of the comprehensible core of an algebra of coils. In this algebra $\mathcal{A}$, the set **A** contains some set of coils, and **F** consists of a set of operations, each coiling thread into a coil of specific radius and increment.

A comprehensible core may be defined by means of the recursive action of these coiling operations on a bare thread. Relative to this core, a coil composital is irreducible. However, the core may be expanded by adding a set of irreducible compositals of which we, nevertheless, claim to have some form of comprehension, for instance because the parameters r_n and s_n are computable functions. A class of coils that were originally irreducible have now become reducible with respect to this new core, and, as we have seen, this enables us to compute all kinds of useful things such as renormalization factors.

Note: Extending the comprehensible core may simply be a matter of choosing a more adequate representation for a set of objects, but it may also correspond to the transition to a weaker notion of comprehensibility. For instance, the real number π is irreducible with respect to a decimal representation of the natural numbers, and remains so if the rational numbers are added to the comprehensible core. However, the core may be further extended by noting that there is a definite algorithm for computing π. All numbers for which such an algorithm exists may be added to the core, for instance with a complexity equal to their associated Kolmogorov complexity, i.e., the minimal length of a program by which they are computed. As a result, π has become reducible, although we still do not know its billionth decimal place. In a sense, extending the core may thus amount to suppressing, or hiding, infinite amounts of structure which, apparently, can be considered irrelevant for some purposes. Accordingly, the definition of a comprehensible core must reflect the purpose which one has in mind.

For a given composital there may be many different ways of suppressing excess structure and obtain a comprehensible, but still interesting picture; for instance, instead of a single scale parameter there may be many. Let us say that in such cases we get pictures taken from different 'angles'. Further, pictures can come in sequences of varying degrees of detail, i.e., from a given angle it can be possible either to take distant pictures or to 'zoom in' and get a close up. One may thus obtain an entire labyrinth of pictures. Such a method of working may have major advantages, because in this way each of the pictures may yield a comprehensible image, whereas it could be extremely difficult to construct and to understand a combined picture. Moreover, individual pictures can now be tailored to the specific desires for which they are wanted. On the other hand, the danger is also clear: one may be left with a chaos of pictures in which the mutual relations are entirely lost. Accordingly, one can afford these methods only if the background logical framework is sufficiently flexible and capable of dealing with the required shifts in perspective.

At an intuitive level people seem to be quite handy with shifting context rather fluently. Within sciencc, however, I would say there is still a considerable amount of work to be done in establishing the theoretical basis for such manoeuvres. In the next chapter I shall try to make a contribution by introducing some elementary logical equipment for handling context shifts.

Literature: Barr [9], Bowie [32], Castaneda [41, 40], Cushing [53], Einstein [71, 72, 70], Feyerabend [75, 76], Van Fraassen [91], Jammer [130], Kuhn [209], Lakatos [141, 140], Laudan [144, 145], Lighthill [151], Lucas [154], Mandelbrot [157], Mc Carthy [160], Penrose [167], Pokorsky [168], Reichl [178], Suppe [201].

Chapter 3

The construction of physical reality

Abstract: Starting with an intuitive picture of what kind of thing a theory is, we work our way towards a more sharply defined, but at the same time more restricted, mathematical notion, which can be investigated with standard tools of mathematical logic. "Reality" is introduced as a concept which plays a role in organizing the relation between sets of different theoretical pictures.

A brief but fitting account of the role of (internal) reality is given in the following quotation from Einstein[1]:

> "The justification of the constructs which represent "reality" for us, lies alone in their quality of making intelligible what is given by the senses."

The present chapter essentially consists of interpretation of, and comments upon, this statement. Let us begin with a brief comment, drawing a relation with our work in the previous chapter.

We live in a complex world about which, by a variety of more or less obscure mechanisms, we manage to gather a multitude of fragmentary pieces of knowledge. We try to make these fragments intelligible by integrating the bits and pieces into a single coherent picture, reality, or by imagining that at least such an integration is possible in principle. At a recent conference[2] Paul Feyerabend said some interesting things which also point in this direction. Asked about his opinion about realism, his answer was that, typically, the concept of reality becomes important at times when there is a tension between the appearances and deeper ideas. With a stable world view, there is no need for the concept. However, at times when views are developing, and one is at the verge of grasping the coherence in newly developed ideas, then the notion of reality becomes important. It is used for expressing the contrast with the older views, up to the time when its importance fades, as the new ideas are fully integrated into the general world view.

[1] Schilpp [187] p.669.

[2] The Erasmus Ascencion Symposium, Oosterbeek, the Netherlands, may 1992.

Reality may thus be understood as a concept which serves to organize our picture of the world. It provides a kind of glue between the fragments. We shall try to develop a more detailed view on this mechanism and also investigate the role of pluralism in this way of organizing our knowledge. One of the main objectives in this analysis will be to get a picture of how the construction of reality proceeds in classical physics and of what goes wrong in QM.

3.1 A model of theories

Views on language also have a bearing on our ideas about theories, which, after all, are usually formulated in language. Therefore, let us make just a few more remarks about language. One of the intriguing things about language is that in spite of its vagueness and malleability, and in spite even of the fact that in our daily use of language basic grammatical rules are often grossly violated, if not totally crushed, it can be used in a very precise way as a highly effective means of communication. In our use of language we have achieved a fascinating combination of, on the one hand, extreme flexibility and very general applicability and, on the other hand, precision and expressive power. Precisely how we do this is not so clear but it can be assumed that the malleability of language plays a key role and that the overall conceptual vagueness of language is necessary for sustaining this malleability.

For language as used in a grammatically strict way, within a sharply defined context, it may be possible to draw a satisfactory picture in terms of a set of expressions, each with a definite meaning. The relations between different expressions and their respective meanings may, furthermore, be analyzable in a more or less formal way in agreement to some extent with standard logical methods as applied to formal languages. For language in general, however, let us drop this approach. Contextually bound languages, where expressions have obtained definite meanings, can in a sense be considered as the final products of our linguistic efforts, whereas language more generally is like a factory where such products are made. Language can still be viewed as a structured reservoir of expressions, but these expressions do not have a definite meaning. Instead, the meaning an expression *might* get in a particular context is constrained by the way in which the expression, or in which isolated concepts or other parts of it, are used on earlier occasions under similar conditions. In other words, instead of a set of expressions with a definite meaning we have a set of expressions of which the future usage is determined in merely a heuristic sense by paradigm examples of their earlier application.

The view on theories which we shall now proceed to consider will be based on the above picture of language. Our linguistic apparatus, however, does not stand on its own. It is an integral part of a mental machinery which in its totality encompasses much more than language alone. We process sense impressions, we form mental images, we remember smells and sounds, we perform actions which influence the outside world, and so on and so forth. Knowledge is represented within the system in a variety of ways, ranging for example from genetically fixed neural structures to standard routines, tacit knowledge, memories and explicit linguistically expressed rules. If all is well, all this together constitutes a more or less coherent body of knowledge. A general study of the structure and functioning of such a body is way beyond the scope of our present approach. For the sake of our analysis here, however, we shall assume that within a body of knowledge as a whole we can isolate separate pieces that behave as structural units and that can be studied more or less apart from the rest. Theories will be regarded as being structural units of this type. In a very general conception of theories, any such unit could be regarded as a theory. Here we shall restrict ourselves, however, to units that are within the reach of conscious scrutiny and that, moreover, are capable of being expressed in language and open to criticism.

In a closer analysis of the concept of a theory we may distinguish between, on the one hand, its external structure, i.e., its relations to other theories and more in general its embedding in the totality of our knowledge and, on the other hand, the internal structure of the theory itself. We shall make some further assumptions with respect to both types of structure. In the first place we shall assume that relative to a particular theory $\mathcal{T}$ the local environment of $\mathcal{T}$ can be divided into three levels:

Three levels:

> The first level is the observational level, or empirical level, containing observation theories which serve to process observational input into facts that are digestible at the level of $\mathcal{T}$.
>
> The second level is the theoretical level, containing $\mathcal{T}$ itself and possibly also alternative and rival theories to $\mathcal{T}$.
>
> The third level is the metatheoretical level, containing anything that is otherwise relevant to the integration of $\mathcal{T}$ in its environment.

Evidently the well–functioning of $\mathcal{T}$ depends crucially on its being properly connected to its environment. In this sense we find agreement here with Quine, who argues that[3]:

> "The unit of empirical significance is the whole of science"

On the other hand, it is clear that the well–functioning of a battery in a car depends on its being properly connected to the electrical circuitry of the car. One may, therefore, argue that the unit of transportational significance is the car as a whole. This is quite true, of course, but nevertheless it is possible, and most convenient, to remove the battery from the car and test its capacities in isolation of the other parts. We shall assume that theories can be treated in much the same way.

We are thus allowed to turn our attention to the internal structure and functioning of a theory. Here we shall avail ourselves of the same, somewhat detached, attitude which we took with respect to language. A theory is not assumed to possess

[3] [175] p.42.

a really sharply defined meaning. Instead a theory is like a factory. It is a data-processing plant, a conceptual apparatus where the incoming data concerning some situation are converted into a coherent picture, a theoretical model of the situation. Such models are the final products of applying a theory and we may conceive them to provide a well-defined meaning to a selected class of descriptive expressions, the language which we use to describe the features of these models. Theories as a whole, however, have a meaning only in the sense that they heuristically constrain the features of the resulting models. In as far as we take a theory to be expressed, or at least to be expressible, in language, the most appropriate way to think of it may be to regard it as a collection of manuals. In the first place a manual for the construction of theoretical models and in the second place a manual for developing the adequate terminology to describe these models. In general there is more to a theory, however. Analyzing a particular situation we may, for instance, be able to choose whether to make a crude or a more detailed model of it. Furthermore, a theory may provide information on how a particular situation will change after a specified type of event or action. More generally, a theory often provides us not merely with a model of a situation but also with a flexibility in changing the features of this model. Let us express this by saying that the theory also provides a manual of admissible transformations on its models. In this case too, the word "manual" is understood to express the heuristic aspect of the situation. A theory may not actually determine a sharply defined set of rules to generate these transformations.

In addition, let us observe that theoretical models generally have features that are considered to be irrelevant, and also that they are usually incomplete in the sense that they cover only some aspects of a given field of phenomena. Accordingly, along with a model, a theory must provide claims about which features of the model are relevant to the description of a given situation. The picture of a theory as it now emerges can be summarized by stating that a theory is assumed to comprise:

A shelfful of manuals:

a manual for the construction of theoretical models of situations,

a manual for developing the language to describe these models,

a manual of admissible transformations on these models,

a manual for constructing relevance claims for each specific model, with respect to a particular type of situations.

Example: As an example let us consider a theory about the movement of the planets in the solar system. To begin with, we distinguish three levels. The observational level is concerned with looking at the sky at clear nights and with measuring angles between bright points. The theoretical level is concerned with our knowledge of the movement of some of these bright points. The metatheoretical level, in this case, contains background theories such as classical mechanics with Newtonian gravity or, possibly, general relativity.

Looking more closely at the theoretical level we find, for instance, several pieces of mathematical calculus taken from classical mechanics, together with some paradigm examples of how this calculus is applied, especially to systems of relatively small masses moving in bound orbits near a relatively large mass. Furthermore, we shall find some general mathematical knowledge about transformations between coordinate systems in $\mathbb{R}^3$, because we need this to establish the correspondence with the observations. Such a cluster of fragments of mathematics and examples of how it is applied constitutes a theory of celestial mechanics. Someone who has acquired some skill in working with this theory will now be able to construct theoretical models of the solar system. A simple model is obtained by assuming the Sun to be a large point mass and the planets to be small point masses and by neglecting the mutual interactions between the planets. We then obtain a class of models whose specific characteristics depend on a number of parameters, such as the Sun's mass and the initial positions and velocities of the planets. A first set of transformations on these solar system models is obtained by varying these parameters. Relevance claims are based on both observational and theoretical grounds. The observations are used to select those parameter values for which the match between model and observations is optimal. Furthermore, the theory is used to make an estimate of the consequences of the various approximations involved. Neglecting the interplanetary interactions, for instance, restricts the precision and the long term relevance of the models.

A second type of transformations is obtained by introducing corrections to the original models, for instance by means of perturbation terms in the dynamical equations. This includes in the first place the interplanetary interactions and secondly things like the effect of the quadrapole mass moment of the Sun and general relativistic terms. Such perturbation terms will make the theory much harder to apply, but this is the price for an increase in precision.

Note: Let us observe that the distinction between different theoretical levels is entirely pragmatic and bears no relation to the complexity of the various levels and the amount of theory needed to organize them; i.e., what is the empirical level in a given situation is not necessarily less 'theoretical' than the theoretical level. Consider for instance a very simple point mass model of the solar system, neglecting all mutual perturbations of the planets on each other etc. As input for calculations with this model we may use empirical data that can be taken from some astronomical handbook. Such data, however, are obtained by using techniques requiring much more sophistication than the simple model that we have placed at the theoretical level. Planetary masses, for instance, may be estimated by elaborate calculations of mutual perturbations of the various solar system objects. In fact, even transformations of observed sky coordinates into solar system coordinates may be more sophisticated and elaborate than the calculations we might perform on what we have chosen to be the theoretical level. Accordingly, this distinction in levels is not meant to suggest the applicability of anything like an ordering of degrees of theoriticity.

Note: In the above it has been left largely unspecified what kind of entity a theoretical model is supposed to be. The terminology more or less suggests a mathematical model, but for all that matters one may also think of mental images, drawings on blackboards, material contraptions, tables, computer programs or whatever. In a general sense we may consider a model to be any entity that carries meaning. Considering more specifically the relation between language and meaning we may say that a model is a structure which makes some statements true and other ones false, i.e., it generates a valuation on at least some of the statements of a language. Within science mathematical models play a very important role because scientific meaning is very often expressed in terms of mathematical structure.

3.1.1 Logical labyrinths

The above considerations give a general, but intuitive picture of what kind of thing a theory is. We shall now try to connect this to the more precise notions which are studied in mathematical logic. First, let us briefly summarize the intuitive notion.

The basic idea was that isolated pieces of knowledge derive their meaning from being immersed in a larger background, which we may call a body of knowledge. Bodies of knowledge as a whole, such as maintained by a person or a scientific community, are complex entities and presumably no one can presently claim to have a detailed understanding of their structure and functioning. In order to obtain a more manageable object for study we focus our attention on structural units within such a body, which we shall call theories. We shall want, however, to study a theory also in relation to other theories. In order to obtain a manageable object for this type of study, we regard a theory as part of a structural unit one order of magnitude larger, which we shall call a theoretical network. This terminology will be used in a scale invariant way, i.e., what is called a theory on one occasion may be called a theoretical network on another. This is tantamount to assuming that we can indeed deal with knowledge in this scale invariant way. I think that, in a number of respects, this is a quite reasonable as well as a relevant assumption.

After we have chosen some appropriate theory and a theoretical network as our objects of study we can proceed to demarcate three levels, i.e., we distinguish between a theoretical, an empirical and a metatheoretical level. Even though these steps should greatly contribute to obtaining a comprehensible field for research, individual theories remain complex entities for which it is not evident that standard methods of formal logic can be made to apply. Therefore, we go one step further. The crucial step by which we shall try to relate the intuitive theory concept to more precisely defined mathematical notions is that instead of trying to analyze a theory in its totality, we shall look at instances where it is applied. The theory itself is regarded as a collection of manuals. We shall be concerned not with the logical structure of these manuals but with the structure of the constructs obtained by applying them, i.e., the language that we shall assume to be accessible to formal investigation is not the language in which the theory itself is formulated, but the language in which the target system is described.

A decision to analyze a particular situation by means of a particular theoretical network initiates a process of model building on various levels. Answers to specific questions are found by studying specific models and are used for the extension and refinement of other ones. What is constructed is a network of models and linguistic structures. We shall call such a network a *logical labyrinth.* Logic can now be viewed as the study of labyrinth structure and also as the study of transportation through a labyrinth. What can be taken from where to where and how?

The relation between a theoretical network and the logical labyrinth which results from applying it can be summarized by saying that the theories provide the blueprints for labyrinth construction. Theories can exist and be communicated in many forms, generally some mixture of explicit rules and paradigmatic examples of their application. The most crucial difference between a theoretical network and a logical labyrinth, in our usage of these terms, is that theories have a heuristics, largely contained in the form of tacit knowledge and intuitive pictures. This gives them an inherent flexibility that is absent in the labyrinths, which we shall consider to have a fixed and definite structure. In the sequel we shall restrict ourselves to the analysis of labyrinth structure. Heuristics will be largely ignored.

Note: By ignoring heuristics I do not mean to suggest that it would be a subject which is beyond logical analysis, or that our use of intuitive pictures and tacit knowledge will forever place us above the logical automata, which presently do not have these capabilities. On the contrary, I have in outlines quite definite ideas about how one might approach the idea of implementing a logical labyrinth on a machine and program at least a primitive kind of heuristics in order to make the machine to some extent self–adapting in a changing environment. Evidently, however, this subject is highly speculative and far outside the scope of our present work.

The elementary building blocks of a labyrinth will be called *contexts.* We shall conceive them to be units in which certain fragments of knowledge can be expressed by means of linguistic expressions that obtain, within the given context, a specific well–defined 'meaning' , of which we shall assume that it can be analyzed in terms of reference to mathematical structures. In this way we obtain a notion which is straightforwardly related to the concept of a theory in the sense as studied in mathematical logic. The standard conception of a theory in mathematical logic is the *syntactic view,* in which the theory is conceived as a deductively closed set of sentences in a formal language. A relatively new development is the *semantic view,* in which the theory is defined by a set of intended models. The 'contextual view' readily accommodates both notions.

Example: Consider for example a word like "Mars". We might want to give the meaning of this word in terms of a reference to something out there in nature, a large clump of matter which is really there, circling around the Sun. Nature itself, however, is only accessible to us through the theories we have about it, and

> through the mental images we form by using these theories. Therefore, in order to speak about the meaning of the word "Mars" in a more concrete manner, we can let it refer to an imagined world, a partial, mental reconstruction of nature, built from the mental images that we associate with Mars. Indeed, communication between people about Mars is based on apparent similarities between the images associated with the word by different people. Nevertheless, the associations one can have with the word "Mars" can be very diverse. One can think of a reddish star, of specks on photographic plates, of a little rotating ball in a planetarium, of blackboard drawings, schoolbook texts, mathematical formulas, or even of a Roman God. From a wide range of possibilities we choose, for a given purpose, those images that seem relevant at the moment, i.e., we choose a temporary meaning from a range of possible meanings, but even in the course of a single conversation this meaning may change. Within a given context, however, we shall take the meaning of an expression to be fixed, so a change of meaning is also regarded as a change of context. Per context one thus has a relatively restricted concept of meaning, that can be extended, however, by analyzing the relations between different contexts and by combining different aspects of an expression within a wider context.

Within science it is often extremely important to fix the meaning of what we say in a very precise manner, and this is achieved by representing mental images of the world by means of mathematical structures. Much of the creative work in science consists of trying to find appropriate formal models for intuitive pictures, which is generally a quite difficult job. Once they have been found, however, they can greatly increase the rigidity of meaning. On the other hand there is also a certain risk involved, because vague intuitive pictures generally admit many different formalizations. Once a seemingly reasonable formal model is found it does not always seem attractive to go on searching for alternatives, and moreover it is not always easy to switch swiftly from one model to another. Accordingly, with a bit of bad luck, one can get stuck with a very bad model without even noticing.

Keeping this in mind, let us go back to the concept of a context. We shall conceive a context to be constituted from a language with a fixed primary semantics in terms of mathematical structures. For instance, within the context of a simple point mass model of the solar system we can think of "Mars" as referring to one of these point masses. This semantics is called 'primary' because it can be further extended by taking into account the relations with other contexts, such as a context in which observations are described concerning a particular reddish speck of light in the sky.

The above set–up provides us with a framework for investigating the logical structure of a theory, or a theoretical network, at least per application. We shall try to further clarify the approach by a more concrete example. First, however, we note that, within such a framework of 'contextual logic', one may distinguish, with respect to a given context, between local logic, which concerns meaning and reasoning within this particular context, and connective logic, which concerns the relations between different contexts. The boundary between local and connective logic is ambiguous because we ourselves are the ones to decide what is to be included in a given context and we may do this at our own convenience. Nevertheless, it will be relevant, for our specific purposes, to make a clear distinction between the two types of logic, and with respect to the subject of "reality", connective logic will be our field of primary interest. Individual contexts are used for storing fragments of knowledge. Reality was conceived above as providing a glue between the fragments; it should generate some form of coherence within the labyrinth. This idea will be further developed in the sequel, and reality, relative to a given set of contexts, will be introduced as a new and special kind of context in which all knowledge contained in the other contexts is combined.

3.2 Contexts

As an example let us look at celestial mechanics and investigate the main lines of how a theoretical model of the solar system relates to a target body of empirical phenomena. Imagine an observer $\mathcal{O}$ who has gathered such a body by regularly looking at the night sky over some period, say a number of years, and writing down the positions of planets visible. This simple example suits well for introducing some basic ideas about the structure of a context, in the sense in which we are going to use the word. Let us associate with this particular body of observations a context called **e**. Such a context is characterized in the first place by a language $\mathcal{L}_{\mathbf{e}}$, in which the observations can be represented and written down. A very simple language can suffice for the job, because, besides dates, hours, and celestial positions, there is little that needs to be represented. Of course there is some freedom in precisely how to construct $\mathcal{L}_{\mathbf{e}}$. The observer is free to use different coordinate systems, different calendars, different notations, numbers, words, figures, etc., as long as it is clear what is the meaning of the expressions he chooses to use. In order to fix this he can define the meaning of his expressions by a reference to mathematical structures.

Semi-interpreted languages

Let us expand further on the details of how mathematical structures are used in science for stabilizing meaning. In classical mechanics, for instance, an important mathematical structure associated with a system is its phase space. A phase space is an example of what is more generally called a state space, and which is a mathematical representation of some classification scheme. Other examples are the configuration space of a system, in which positions but not the momenta are represented, the set of possible phase trajectories, which determines the dynamics of a system, and many types of data structures that are used to represent empirical data and to classify properties of systems. A good example in astronomy is the Hertzsprung-Russell diagram, which serves to classify stars by plotting spectral class against luminosity. Such structures all have a similar role in the logical structure of the theories in which they occur, and this role can be very elegantly analyzed by introducing a notion introduced by Van Fraassen as a semiinterpreted language[4].

A semiinterpreted language is a triple $< \mathcal{L}, \Gamma, \gamma >$, where $\mathcal{L}$ is a language, Γ is a state space and γ is a function $\mathcal{L} \rightarrow \wp(\Gamma)$, where $\wp(\Gamma)$ is the power set of Γ, assigning to sentences in $\mathcal{L}$ a corresponding subset of Γ. Our further terminology will be to call subsets of Γ states, and elements of Γ elementary states. The state $\gamma(\phi)$, corresponding to a sentence $\phi \in \mathcal{L}$, is defined as the set of elementary states in which ϕ is true. The state space Γ, together with the so–called satisfaction function γ, thus provides a partial semantics for the language $\mathcal{L}$, in which the meaning of a sentence ϕ is given by its associated state $\gamma(\phi)$. For example, the meaning of the sentence "the Sun is a main sequence star" is given by a certain region in the Hertzsprung–Russel diagram, and the sentence is true because the actual spectral class and luminosity of the Sun place it within this region.

Note: The triple $< \mathcal{L}, \Gamma, \gamma >$ is called a *semi*–interpreted language because the semantics is partial, in the sense that only whole sentences obtain a meaning. In a more fully interpreted language other expressions, such as the word "Sun" on its own, would also get a meaning. The matter is important, because for many purposes the state space on its own is not sufficient to know what we are talking about. In classical mechanics the phase space of a system of three free point particles is given by $\mathbb{R}^{18}$, but if someone merely gives us a $\mathbb{R}^{18}$, then how should we know that it is supposed to represent three point particles? It can just as well represent some quite different system. Clearly the structure of the system cannot be reconstructed from the structure of the state space alone. This problem can be dealt with by introducing a richer semantics, a subject to which we will come back later, but for the present semiinterpreted languages will serve fine.

Returning now to the context **e**, used by the observer $\mathcal{O}$ for the representation of his observations, let us think about how to construct a suitable data structure to serve as a state space. $\mathcal{O}$'s notes must in the first place contain a reference to the date of each observation, where we shall take the date to include both the day of the year and time of day. We thus obtain a set of dates D, being a subset of all possible dates. For every date d in his notebook there will be some list of observed planets and their positions. The observed position $q(d,p)$ of planet p on date d can be represented, e.g., by angular coordinates on a half sphere, i.e., by points in the plane $[0, 2\pi) \times [0, \pi/2]$. Further, unobserved planets can be given the value NV (not visible) if they are not above the horizon, or the value NO (not observed) if their position has not been measured for some other reason. We thus arrive at the following data structure:

$$\Gamma_{\mathbf{e}} = \{< D, q > \mid D \subseteq T, \quad q : (D \times P) \rightarrow (([0, 2\pi) \times [0, \pi/2]) \cup \{NV, NO\})\}, \tag{3.1}$$

where T is the set of all possible dates and P is the set of planets.

[4][88].

This state space $\Gamma_{\mathbf{e}}$ represents a set of possible notebooks, i.e., a particular notebook corresponds to a point in this set and individual notes in the book are represented by subsets of $\Gamma_{\mathbf{e}}$. Whatever notation observer $\mathcal{O}$ actually decides to use to write down his observations, the meaning of his sentences should be definable in terms of such subsets. Suppose that he keeps a sort of diary, containing the sentence ϕ: "The red planet is now visible in the South". In order for this sentence to be meaningful we must know that "the red planet" is *Mars*, we must know what date d corresponds to "now", and we must know where "the South" is located at this moment, in terms of a region $S(d)$ in the q_1, q_2 plane. If we know all this, the meaning of ϕ is given by

$$\gamma_{\mathrm{e}}(\phi) = \{< D, q > \mid d \in D, q(d, Mars) \in S(d)\}. \tag{3.2}$$

This set is the state corresponding to ϕ.

Using this type of semantics, the context $\mathbf{e}$ is characterized by a language, a state space and a satisfaction function. In fact, we can simply take $\mathbf{e}$ to be defined by this triple, i.e.:

$$\mathbf{e} = < \mathcal{L}_{\mathrm{e}}, \Gamma_{\mathrm{e}}, \gamma_{\mathrm{e}} >. \tag{3.3}$$

Connective contexts

Besides the notebook of observer $\mathcal{O}$ there can be other bodies of empirical data available, say a number of notebooks of a number of observers $\mathcal{O}_1 \cdots \mathcal{O}_N$ from all over the world. Associated with this combined body of data we can conceive a set of contexts, defined as semiinterpreted languages, just like the above context $\mathbf{e}$. We can also imagine a researcher $\mathcal{C}$ whose task will be to correlate the information contained in the files piled up on his desk. This is not a job for an untrained man. The researcher must either have or develop a theory about the relations between all these contexts. Applying this theory results in the construction of a new context $\mathbf{c}$, in which the relations between the other contexts defined earlier can be expressed. A context of this new type will be called a connective context. In analogy to (3.3) we shall define a connective context as a quadruple

$$\mathbf{c} = < \mathcal{L}_{\mathrm{c}}, \Psi_{\mathrm{c}}, \Gamma_{\mathrm{c}}, \gamma_{\mathrm{c}} >, \tag{3.4}$$

where $\Psi_{\mathbf{c}}$ is the set of member contexts of $\mathbf{c}$ and $\mathcal{L}_{\mathrm{c}}$, $\Gamma_{\mathbf{c}}$ and γ_{c} are again a language, a state space and a satisfaction function. Later we shall deal with the construction of Γ and γ in this case.

An example of what $\mathcal{C}$ should be able to do within $\mathbf{c}$ is to calibrate the coordinate frames used by two observers $\mathcal{O}_1$ and $\mathcal{O}_2$ relative to each other. This leads to conclusions such as:

> "if $\mathcal{O}_1$ sees a planet precisely in the South at time t in his local clock time, then $\mathcal{O}_2$ can witness this event at, say, $t + 23$ minutes in his own local clock time."

For statements like this we introduce a special notation. Let $\phi(t)$ be a sentence in $\mathcal{O}_1$'s language, expressing that a specific planet is in the South at time t in his local time, and let $\psi(t)$ be a statement describing the same event in $\mathcal{O}_2$'s language, then the above relation between these events will be formulated as:

"If $[\mathbf{e1} : \phi(t)]$ then $[\mathbf{e2} : \psi(t + 23min.)]$".

Later we shall deal with the problem of how to construe the meaning of such sentences.

3.2.1 Reality

Expressing the relations between contexts in $\Psi_\mathbf{c}$ purely by means of statements of the form "if $[\mathbf{u} : \phi]$ then $[\mathbf{v} : \psi]$" is a relatively primitive way of dealing with the subject. It is generally more efficient to relate all contexts to a single picture that is both general enough and precise enough to contain a faithful representation of all contexts involved.

To this end the researcher $\mathcal{C}$ chooses, or, if necessary, develops, a language containing the terminology needed to cover whatever he wants to include as relevant from the observation reports of the other observers. This language, which we shall call $\mathcal{L}_\mathbf{r}$, may contain sentences in the researcher's native language, say English, such as "The position of Mars at date d was $< \alpha, \delta >$". The meaning of such sentences is determined by the contents of the various notebooks. More specifically, the sentence is presumably true if one of the observers has seen Mars on the indicated position at the specified date. Formally this can be arranged in the following manner. Let ϕ be the above sentence and let us define an associated sentence $[\phi]\mathbf{u}$ in $\mathcal{L}_\mathbf{c}$, meaning that Mars was at the specified position *according to* observer $\mathcal{O}_\mathbf{u}$. A sentence of the form $[\phi]\mathbf{u}$ must therefore have a meaning in terms of a state within the state space $\Gamma_\mathbf{u}$ of the context $\mathbf{u}$, associated with $\mathcal{O}_\mathbf{u}$'s notebook. Let us call this state $\gamma_{\mathbf{u}|\mathbf{c}}(\phi)$. The construction of the satisfaction functions $\gamma_{\mathbf{u}|\mathbf{c}}$, for all $\mathbf{u} \in \Psi_\mathbf{c}$, can be a major job for $\mathcal{C}$ when he constructs his own global language. It requires that he understands the contents of $\mathcal{O}_\mathbf{u}$'s notebook, and this may involve a fair amount of work, including computational work and educated guesswork that may not at all be straightforward.

Having the sentences $[\phi]\mathbf{u}$, and their interpretation by satisfaction functions $\gamma_{\mathbf{u}|\mathbf{c}}$, $\mathcal{C}$ can now turn to the sentences $\phi \in \mathcal{L}_\mathbf{r}$ themselves. For these sentences he desires two things. In the first place they should obtain an interpretation of their own, by means of a satisfaction function $\gamma_\mathbf{r}$, and this requires the construction of a suitable state space $\Gamma_\mathbf{r}$. The data structure (3.1) will do fine in the present case. The data of all observers should fit into it, if the notes of these observers are indeed sufficiently clear.

In the second place he will want to be able to conclude from $[\phi]\mathbf{u}$ to ϕ, i.e., from "the position of Mars at date d was $< \alpha, \delta >$ according to $\mathcal{O}_\mathbf{u}$" he wants to infer that "the position of Mars at d was indeed $< \alpha, \delta >$". This inference contains one complication that one can hardly fail to encounter in practice. Observer $\mathcal{O}_\mathbf{u}$

may be a fallible person with inadequate equipment and false theories and he can occasionally be inaccurate or wrong. It may happen then that the observation reports of two observers $\mathcal{O}_{\mathbf{u}}$ and $\mathcal{O}_{\mathbf{v}}$ contradict each other. In order to avoid that this leads directly to the inconsistency of the entire logical framework, researcher $\mathcal{C}$ can build in a 'safety valve', such as the following one, based on a concept of *relevance.* If what $\mathcal{O}_{\mathbf{u}}$ says about a given sentence ϕ is considered unreliable by $\mathcal{C}$, then let us express this by means of a sentence $\neg R_{\mathbf{u}}(\phi)$, saying that what $\mathcal{O}_{\mathbf{u}}$ says about ϕ is irrelevant to the issue. If the reports of $\mathcal{O}_{\mathbf{u}}$ and $\mathcal{O}_{\mathbf{v}}$ now clash there is a possible way out by questioning the relevance of at least one of these reports. If, on the other hand, $\mathcal{O}_{\mathbf{u}}$'s statements about ϕ are accepted as relevant then the safety valve is closed and there must be an inference from $[\phi]\mathbf{u}$ to ϕ. To this end a law can be postulated, saying that: "If there is a context $\mathbf{u}$ containing information relevant to the question "ϕ", then from $[\phi]\mathbf{u}$ follows ϕ", or in logical notation:

$$\exists \mathbf{u} \in \Psi_{\mathbf{c}} \left([\phi]\mathbf{u} \wedge R_{\mathbf{u}}(\phi) \right) \Rightarrow \phi. \tag{3.5}$$

Let us briefly reflect on what $\mathcal{C}$ has done so far. Starting from a pile of observation reports, formally represented by a set of contexts $\Psi_{\mathbf{c}}$, he has constructed a unified picture in the form of a new context $\mathbf{r}$, consisting of the triple $< \mathcal{L}_{\mathbf{r}}, \Gamma_{\mathbf{r}}, \gamma_{\mathbf{r}} >$. This new context is sufficiently rich in structure to provide an adequate representation of all features of the other contexts that are considered relevant. We shall call such a context a *reality* for $\Psi_{\mathbf{c}}$.

Theoretical realities

The construction of the context $\mathbf{r}$ involved, more or less implicitly, some steps that cannot be accounted for by accepted forms of deductive logic alone. It was implicitly assumed for instance that all observers reported on the same planet Mars. It is not *a priori* evident that it makes sense for observers from different parts of the world to speak about the same planet, because it is logically conceivable that they are observing different objects. These objects can be identified only on the basis of theoretical assumptions about the structure of the world.

There are many more well-accepted theoretical ideas that can be included in a picture of the planetary system. We may note for instance that the data structure (3.1) that was taken for $\Gamma_{\mathbf{r}}$ assigns positions to planets only at times when they are observed, i.e., on the dates occurring in the set of dates D. It is generally assumed that the fact of their observation is irrelevant to whether or not planets have a position and to what their position is. They are even assumed to have a position when they are not visible at all, in the daytime or when they are under the horizon. This leads us to define another state space $\Gamma_{\mathbf{t}}$:

$$\Gamma_{\mathbf{t}} = \{q |\, q : (P \times T) \rightarrow ([0, 2\pi) \times [0, \pi])\}, \tag{3.6}$$

which provides each planet $p \in P$ with a position, in the form of two angles $q_1(p,t), q_2(p,t)$, at all times $t \in T$. Via similar steps one can proceed on and on.

Two major types of moves can be distinguished in this process[5]. In the first place there are 'widenings of scope', extensions of the state space that create room for the representation of new ideas or new empirical phenomena. For instance, the planetary positions are not fully characterized by the above coordinates, because these are merely projections of their position in space on the surface of a sphere. A more complete representation of their position requires extending the state space into a set of functions $r : P \times T \rightarrow \mathbb{R}^3$. In the second place, there are also contractions of the state space that serve to delete states that are considered to be impossible. Admissible functions q are, for instance, assumed to be continuous and satisfy certain dynamical regularities about which one can make theories. Such theories postulate dynamical laws, saying that many of the logically conceivable trajectories in space are in fact impossible and can therefore be removed from the state space.

Making a good reality, i.e., a satisfactory theoretical picture of a given class of empirical phenomena, can be a difficult and highly complicated affair. The general logical structure of the enterprise is more or less transparent, however, at least in classical physics. Observations can be described on empirical data structures. Theories, in various grades of sophistication and completeness, serve to provide a more coherent picture, satisfying demands that originate from more general assumptions about the structure of the world. A complication in this setup is that theoretical pictures are often themselves incomplete, that they are based on approximations that are necessary in order to generate computable results and that they sometimes introduce excess mathematical structure, also serving merely to aid computability. Consequently, theoretical pictures can have features of which we are aware that they are unreliable. This complication can be coped with by invoking the concept of relevance introduced earlier. The presentation of a theoretical picture can be accompanied by a specification of its relevant and irrelevant aspects. For instance, a point mass model of the solar system can be relevant with respect to questions about the planetary positions next year, but it is not relevant with respect to questions about planetary mass densities. A *reality* we shall define to be a theoretical picture without irrelevant features. Further, a *complete reality*, relative to a given class of empirical phenomena, is a reality in which all phenomena in this class can be faithfully represented.

3.2.2 Plurality

For the purpose of providing a general unified picture for a class of phenomena, each described within one of the contexts of a connective context **c**, we created a new context that we defined to be a 'reality' for **c**. This idea constitutes our basic approach towards the concept of reality and towards the issue of realism. The main attractive side of such an approach, with the emphasis on an internal self–constructed concept of reality, is that it gives some sense to the realism debate.

I do not see how a debate about a literal correspondence between theory and nature could ever be decided on reasonable arguments. On the other hand, a discussion

[5]See, e.g., Van Fraassen [96] p.112.

about internal reality is a discussion about practical matters, about whether certain logical constructs work well in organizing our knowledge. On grounds discussed earlier I think that the main contender of realism in this discussion is pluralism. The theoretical realities one builds are always logically stronger than the observational data they aim to represent. In order to provide a coherent picture one has to assume things that are not actually observed, or may not be observable at all. Theories about the solar system, for instance, assume that planets have a well–defined position even on moments that no one observes them. Accordingly there is no standard logical argument leading from the observations to the resulting reality. A particular reality provides a model for the observed data, but not a unique model.

We have no reason, of course, to think that planets flash on and off depending on whether we look at them or not. Accordingly, theories in which this happens can be readily discarded, and very few people will protest. One can then ask whether a similar strategy is always possible. Do we in general have sufficient insight in the structure of the class of models for a given set of observations, and do we in general have decisive reasons to accept one of the models as the standard model, preferable to all the others? I think that for most real life situations both these questions must be answered negatively. There will generally be good reasons to discard many possibilities because they clash with accepted, 'metatheoretical', constraints about what kind of structure a reality should have. Nevertheless I do not think that such constraints can be both sufficiently compelling to be decisive and strong enough to single out a unique final result. As Feyerabend has noted[6], in a slightly different context, such philosophical and methodological matters are themselves cosmological issues. If we happen to live in a simple universe, then a simple methodology can suffice to straightforwardly lead us to absolute truth. In a more complex universe one needs more sophisticated means and even these may not lead to conclusive results. Laudan's thesis that the scientific method itself is not fixed but develops during the course of science[7] can be understood as a corrolary to this view. By learning more about a subject one can also learn more about how to study it. Things being this way, simple criteria for choosing between theories may not always be available, and sets of possible alternative theories may in fact be infinitely complex themselves.

Thinking of our data about the solar system we can note that classical mechanics with Newtonian gravity is able to provide a very good picture of the system. General relativity performs even better, but is also a more complex theory, at least from a computational point of view. Quantum mechanics should be expected to be able to deal with the situation, both in a version with Newtonian gravity and in a version with relativistic corrections. Quantum field theory can hardly fail to be successful in many different versions and a quantized theory of gravitation, if one ever finds a truely satisfactory representative, must be able to score on this problem. Counting these few theories as conceptually different, and some indeed as *very* different, this can be taken to indicate that the structure of the class of *all* possible models for our set of data is in fact unimaginably complex. Choosing between alternatives one does

[6] [75] ch. 16, footnote 66.

[7] See, e.g., Laudan [146].

not know is difficult even if one has intelligible criteria by which to choose. However, available criteria tend to be vague and often contradict each other, and there seems little reason to think that the problem of picking a standard model generally has a unique solution, nor even a sufficiently precise definition.

This being so, a moderate form of pluralism is likely to be the best strategy, because it stimulates creativity and adaptivity to unforeseen conditions. The amount of plurality should be adjusted to, on the one hand, the availability of sufficiently interesting alternatives and, on the other hand, demands of feasibility, i.e., how many different views can we manage to cope with? The latter question also depends on the strength of our logical framework. If this is sufficiently rich and clear and well-organized, we shall be able to manage more different views than if it is poor, confused, and laborious. My main objection against realism is that it is too poor and can thereby lead to missing interesting views, questions, and answers. Finding the right kind of balance in these matters is, however, an extremely difficult job, for which I do not think there are clear-cut rules. We seem to be stuck with the chaotic and highly fallible way in which decisions are usually made in the community of people. I have very little desire to call this rationality, but apart from trying to do our best I see little alternative. Realizing the nature of the game, and our limitations in playing it, might, however, save us from spoiling our character.

Pluralism, within our present approach to the construction of reality, amounts to pluri-realism. From the point of view of what is logically possible, there is no problem at all in admitting the option of building several different realities, instead of confining oneself to constructing at most only one of these things at a time. There are no compelling reasons to postulate the existence of a unique 'true' reality; in fact, truth should be considered to be relative to reality and therefore pluralism should lead to a multiple truth concept. In ch. 10 we shall set up the formal side of such a theory of truth.

3.3 Further elaborations on contextual semantics

The context e was introduced in (3.2) as a semiinterpreted language, along with a note that for many purposes this is highly insufficient. For a richer semantics we can turn to standard model theoretical methods, as applied in predicate logic. In this case a language $\mathcal{L}$ is interpreted by reference to a 'world', containing objects of possibly different types to which expressions in $\mathcal{L}$ can refer. Such a world is often itself represented by a mathematical structure. The link with the state space approach is now given by assuming that the state space represents the set of possible worlds. To each elementary state $\lambda \in \Gamma$ there is a corresponding interpretation $\mathcal{I}_\lambda$, assigning to expressions in $\mathcal{L}$ objects in the world corresponding to λ. If we now take an arbitrary well-formed expression from $\mathcal{L}$, for instance a term like "Mars", then the meaning of this term can be defined by its reference to objects within the worlds $\lambda \in \Gamma$, i.e., the semantics of $\mathcal{L}$ must specify which object, if any, in each world is the referent of the word "Mars". Suppose that Γ is a set of point mass models of the solar system, then "Mars" refers, in each of these models, to one of the point masses. We can formally represent this by introducing a function η:

$$\eta(x) = \{< \lambda, \mathcal{I}_\lambda(x) > \mid \lambda \in \Gamma\}, \tag{3.7}$$

returning for each term $x \in \mathcal{L}$ a set of pairs $< \lambda, \mathcal{I}_\lambda(x) >$, where λ is a world in which x has a referent and $\mathcal{I}_\lambda(x)$ is this referent. Given η, which we shall call the *interpretation function*, the individual interpretations $\mathcal{I}_\lambda$ as well as the satisfaction function γ are straightforwardly recovered. Sentences in $\mathcal{L}$ can be treated in the same way as other expressions. They are standardly interpreted in the set $\{0,1\}$ of truth values, i.e., a sentence ϕ gets the interpretation $\mathcal{I}_\lambda(\phi) = 1$ in those worlds where ϕ is true and $\mathcal{I}_\lambda(\phi) = 0$ in those worlds where it is false. The satisfaction function γ is now given by:

$$\gamma(\phi) = \{\lambda \in \Gamma \mid \mathcal{I}_\lambda(\phi) = 1\}. \tag{3.8}$$

Including the interpretation function η, the structure of a context $\mathbf{u}$ is given by a quintuple $< \mathcal{L}_\mathbf{u}, \Psi_\mathbf{u}, \Gamma_\mathbf{u}, \gamma_\mathbf{u}, \eta_\mathbf{u} >$, where $\Psi_\mathbf{u}$ is empty for a nonconnective context and $\eta_\mathbf{u}$ is trivial for a semiinterpreted language.

Note: It is relevant here to note that we have obtained two levels of structure within a context. On the one hand, there is the structure of the individual states. On the other hand there is the structure of the state space, which may itself be the habitat of all kinds of transformations and symmetries. These two levels are independent. From the structure of the state space alone, regarded as merely a set of points, the structure of the states cannot be reconstructed. *Vice versa*, from the structure of an individual state we cannot tell in what kind of state space it lives. It may be the only state around or it might be just one of a whole set of other possibilities.

Magnitudes and partial interpretations

Whereas the semiinterpreted languages we started with had less structure than we needed, a full interpretation of all terms in a language will generally have more than we want. Within the syntax of a language we often introduce things that are convenient, but arbitrary. Physical magnitudes for instance are quantities we introduce to label states. New ones can be defined freely whenever we like, but we do not want this to affect the structures we aim to describe. For instance, let X be a real valued magnitude such as the x-coordinate of the position of some object. Under a full interpretation $\mathcal{I}$ the symbols X and x as well as the sentence $X = x$ would obtain an interpretation. Real numbers $\mathcal{I}(X)$ and $\mathcal{I}(x)$ would be assigned to X and x, and a truth value $\mathcal{I}(X = x)$, equal to 0 or 1, to the sentence $X = x$, depending on whether these two numbers are equal or not. The worlds into which $\mathcal{I}$ interprets the language must therefore contain the set of real numbers, but we may not want this. In particular, we may want to describe a physical system, and the real numbers are merely convenient artefacts that we do not want to encounter in the physical interpretation of the language. This means that, although we do want to have an interpretation for the whole sentence $X = x$, the terms X and x by themselves can be left uninterpreted. This can be done by considering partial interpretations of the language.

More specifically the natural way to define physical magnitudes is as functions on the state space Γ. Within the semantics of $\mathcal{L}$ we then define a function $X : \Gamma \rightarrow \mathbf{R}(X)$, and within the syntax of $\mathcal{L}$ we define corresponding statements $X = x$, with $x \in \mathbf{R}(X)$, and their meaning will be defined by

$$\gamma(X = x) = \{\lambda \in \Gamma |\, X(\lambda) = x\}. \tag{3.9}$$

This indeed fixes the meaning of the whole sentence without giving interpretations $\mathcal{I}_\lambda(X)$ and $\mathcal{I}_\lambda(x)$ for the individual terms X and x. These terms are now 'hidden'.

Notable about this way of treating magnitudes is that it works well in classical physics, but not in QM. Within QM both magnitudes and states are treated in a way that is quite different from the scheme discussed here, and this touches the core of why QM is a conceptually difficult theory with an unusual logical structure.

Ontology

Hiding, i.e., the possibility of leaving terms without an interpretation, can be applied to any degree we like. Moreover, it can arise very naturally because syntactic notions are often dragged from one context to another, where they acquire a different meaning. This can easily leave some terms without a meaning at all. To admit hiding introduces the complication that the relation between syntax and semantics becomes less direct. Given an arbitrary expression it is no longer evident that this expression has a meaning in the intended semantics. This will now have to be specified separately. The *ontology* associated with a given theory can be defined as that part of the theory aimed at giving such specifications. Within a given context this means that the ontology determines which linguistic expressions are thought to refer to objects, existing in at least some of the worlds represented in the state space.

When it comes to comparing alternative theories, ontology is often a much disputed issue. This is at least partly due to the circumstance that ontological differences between theories are often particularly radical. At the level of the structure of the state space there is usually much more continuity, owing to the fact that the embeddibility of the empirical data must be preserved in all alternatives, in as far as they overlap in their domain of application. Some examples will illustrate this point.

Example: Consider three different contexts describing the same phenomenon, let us say the cooling of a hot object. The first one, **e**, is a context in which the process is described in phenomenological terms. The second one, **f**, is a context in which the phenomenon is explained in terms of a caloric fluid theory, i.e., heat is explained in terms of quantities of a fluid that pervades material bodies. A hot object is thus seen as a container filled with caloric fluid. The third context, **s**, is one which gives an explanation in terms of statistical physics, where heat relates to the amount of energy in a configuration of molecules.

Clearly some terms are not translatable from one context to another. Caloric fluid, for instance, only exists in **f**, and molecules only exist in **s**. Furthermore, in **f** phenomena like heat capacity and heat conductance would be explained in terms of the porosity of a body with respect to caloric fluid and the viscosity of the fluid, whereas in **s** one seeks for explanations in terms of the available number of degrees of freedom for the molecules etc. Evidently the two contexts are entirely different at the ontological level. Nevertheless, at least some whole sentences can be translated without difficulty. The **e**–sentence: "The object cools down", for example, can be translated to **f** as: "The container is leaking", and to **s** as: "This configuration of molecules is losing energy". Translatability of whole sentences indicates the existence of similarities between the respective state spaces, that are bound to exist because the cooling down of the object must have some representation in every one of these contexts.

Example: In a naive picture of a rainbow one might think of it as a real existing physical object. In a more sophisticated theoretical picture the phenomenon is explained in terms of raindrops and deflection of sunlight and so on and so forth. The rainbow as such no longer exists and one can refer to it only in a metaphorical sense, i.e., the term is 'borrowed' from the naive picture but does not possess a literal meaning. Nevertheless, many sentences containing the term do admit a translation from the naive to the sophisticated picture, owing to the fact that both pictures provide at least a partially adequate description of the empirical facts, which implies that there must be similarities between the state spaces of the two contexts. The sentence: "There is a rainbow now" implies that, according to the sophisticated picture, the state of the atmosphere must be such that the phenomenon is visible at this moment.

Example: In Quantum Field Theory, applied to elementary particle physics, the existence of certain particles, such as quarks, is a regularly discussed topic. It is questionable, however, whether one can indeed think of these 'things' as particles in a more or less usual meaning of the word as being definite physical objects. Especially the field theoretical notion of "identical particles" is highly problematic in this respect, and it may be much more straightforward to build the ontology of field theory entirely in terms of fields. "Particles" may be metaphors as much as rainbows. Given such an ontology, a discussion of the existence of, say, the W^-–boson would in fact amount to discussing the structure of the *state space* of the electro–weak field. Clearly, however, there are states of the field that represent phenomena that would be described in a classical context by the presence of a particle. In as far as both the classical and the field theoretical picture are able to represent certain observed phenomena, there must be corresponding similarities between the state spaces.

Example: Ontological differences can concern the existence of certain entities, but also their type, or 'logical form'. The mass of a physical object in classical mechanics, for instance, is a property of the object, i.e., it is represented by a unary predicate. In relativity theory, however, mass comes to depend on the velocity of the object, and is therefore represented by a binary predicate. This makes relativistic "mass" into a term of a different type, and translating the term on its own becomes problematic. Nevertheless, within the context of describing a given phenomenon that can be adequately described also by the nonrelativistic theory, statements about classical mass and relativistic mass are readily translated into one another.

3.3.1 Comparing theories

The present analysis was based on drawing a distinction between on the one hand theories, conceived as vaguely defined objects dressed up with all the frills needed to make them work in a wide range of practical circumstances, and on the other hand logical labyrinths, conceived as more sharply defined objects whose structure can be investigated by, more or less, standard logical methods. Labyrinths are obtained from theories, or theoretical networks, by applying them to, sufficiently sharply defined, phenomena. This process of application works as a sort of 'clutch' between the theory and its practical results. Whereas it can be relatively straightforward to compare contexts or labyrinths, comparing theories is more complicated because, if the theories produce contradicting results, this can also be caused by flaws in applying them. The presence of this clutch protects theories against being dismissed overnight upon the occurrence of limited amounts of counter evidence. The refutation of a theory thus becomes a matter of deciding to stop spending energy in trying to save it, i.e., one cannot hold down the clutch pedal forever but there is some choice in when to release it.

Leaving heuristics out of the picture, comparing theories boils down largely to comparing selected contexts or labyrinths, and releasing the clutch in order to make results conclusive. Comparing contexts can be a complicated affair if the structure of the contexts themselves is complicated. Basically, however, one can proceed along the following lines. There are two levels of structure in a context that can be investigated independently. There is the level of the structure of the state space and there is the level of the structure of the individual worlds which the states represent. Let us first consider comparing the state spaces of two contexts.

Let **u** and **v** be two contexts for describing the same system. The state spaces $\Gamma_\mathbf{u}$ and $\Gamma_\mathbf{v}$ each contain a representation of the possible states of the system. To compare these contexts we can, for instance, pick an elementary state $\lambda_\mathbf{u} \in \Gamma_\mathbf{u}$ and ask which states in $\Gamma_\mathbf{v}$ are compatible with this given $\lambda_\mathbf{u}$. If the answer is "none" then **v** is not *complete* relative to **u**, because **u** contains states that are not represented in **v**. If, on the other hand, there are several different states in $\Gamma_\mathbf{v}$ compatible with the same λ_u, then **u** is not *accurate* relative to **v**, because specifying the state within **u** can still leave different possibilities open in **v**. These and related concepts can be used to present a picture of some of the relative merits of the contexts **u** and **v**. The basic entity that determines the relations between **u** and **v** at this level is a pair of morphisms

$$\mathcal{J}_{\mathbf{v}|\mathbf{u}} \; : \; \wp(\Gamma_{\mathbf{u}}) \rightarrow \wp(\Gamma_{\mathbf{v}})$$
and
$$\mathcal{J}_{\mathbf{u}|\mathbf{v}} \; : \; \wp(\Gamma_{\mathbf{v}}) \rightarrow \wp(\Gamma_{\mathbf{u}})$$

which give us, for each state in **u** or **v** a corresponding state in the other context. In ch. 10 these matters will discussed in more detail.

The other level at which contexts can be compared concerns the structure of the individual elementary states, i.e., the 'worlds' described in these contexts. This subject is intimately related with ontology. Per individual world one can look at which entities of which types are supposed to exist, and whether these entities have matching representations in corresponding worlds in the other contexts. This work may not always be easy. In particular it may not always be clear that an object b in context **v** is indeed a 'matching representation' of an object a in **u**. Even if **u** and **v** both offer empirically adequate pictures of the same phenomenon, it is not necessarily the case that objects from **u** have matching representations in **v**. Ontological similarity is not a prerequisite for empirical equivalence.

Literature: Bechtel [11], Beth [26], Ben–Menahem [21], Cartwright [39], Churchland [42, 44, 43], Cooke [46], Cushing *et al.* [54], Einstein [71, 72], Feyerabend [75, 76, 78, 77], Van Fraassen [92, 89, 88, 93, 96], Fine [79, 80], Fodor [81, 82, 83], Glymour [101], Hacking [107], Hanson [109], Hintikka [117], Kuhn [209], Lakatos [142, 141], Laudan [146, 147], Maxwell [158], Putnam [172, 173], Quine [175], Schilpp [187], Sneed [194], Suppe [201], Suppes *et al.* [203], Watzlawick [213].

Chapter 4

Quantum mechanics

Abstract: A brief introduction concerning the peculiar logical structure of standard QM is followed by an introduction to quantum measurement theory.

In the previous chapter we introduced the notion of a "context", providing a suitable way to analyze the logical structure of a theory of classical physics, and the notion of a "logical labyrinth", a network of contexts, with various interesting applications, such as providing a framework for formulating the empirical interpretation of a theory and for comparing alternative theories. There is another interesting aspect of the notion of a context, namely that it precisely embodies, what I would like to regard as, the core of classical logic: the idea that meaningful sentences have a definite truth value, determined by their interpretation into a world or a set of possible worlds. For this reason it is indeed alarming to find that the standard formalism of QM does not seem to fit into a logic of this type, the heart of the matter being that in standard classical logic the concepts of "state" and "magnitude" have a quite definite meaning and a quite definite relation to each other, which is violated in QM. Briefly said, this will leave us with the following possibilities:

> The standard formalism is malignantly incomplete.
>
> Classical logic is not adequate to deal with the situation and must be replaced by a more general framework.

Our further work will mainly consist of using contextual logic to work out the first alternative, and I am indeed entirely convinced that it is presently the most interesting one to pursue. It is relatively underdeveloped, in spite of the fact that, I would say, it gives the best match to the general attitude among physicists. Bohr's philosophy may serve as a good example. Bohr was a convinced champion of the completeness of QM but not of the quantum *formalism*; quite on the contrary, he always stressed the need for further interpretation. Unfortunately he did not draw the conclusion that it would be wise to modify the formalism into giving a better representation of his views. A similar attitude seems to be prevalent among many physicists. The need for further interpretation of the formalism is widely recognized, but when it comes to the more formal sides of the enterprise the standard formalism is taken as a starting point and only minor variations come into consideration. This has resulted in a quite unnatural fracture between formalism and interpretation, which I think would have been inconceivable in classical physics.

4.1 Logical peculiarities of QM

In order to see what is so strange about the logical structure of QM, let us first recapitulate some of the main features of what was said about the logical structure of classical physics. Imagine some physical system, say the solar system, that we aim to describe, and a theory, say classical mechanics, that helps us to do this. In order to describe the system we need in the first place a language. To this end we usually take a suitable language around, say English, and add terms from a more specifically theoretical origin, such as "energy", "momentum", "Lagrange function", etcetera. For the sake of logical analysis we can also take, instead of a natural language, some suitable formal language, or some formalizable fragment of natural language. A language by itself, the syntax, is merely a set of linguistic expressions, sequences of abstract signs concatenated into larger units such as words and sentences. A semantics must be added to give a meaning to these expressions. Given the syntax there can be various different semantics that are of interest to us, such as, perhaps in the first place, an empirical interpretation in terms of a body of observational data. Such a strictly empirical interpretation is important but generally also partial and fragmentary. It does not, for instance, provide a meaning to statements about the position of Mars at times that the planet is not observed. Therefore, we are also interested in a more complete semantics, and in physics we generally construct it in terms of mathematical structures.

A phase space in classical mechanics is such a mathematical structure and it can be used to define the meaning of sentences by assigning to a sentence the set of phase points for which it is true. The phase space is an example of a state space. In classical mechanics, for some purposes at least, the configuration space of a system can also serve as a state space, and for dynamical systems the set of possible phase trajectories is a suitable state space for giving a meaning to statements about dynamics. In other theories, in and outside physics, similar structures can be identified. We thus arrive at the notion of a 'semiinterpreted language', consisting of a syntax $\mathcal{L}$, a state space Γ, and a satisfaction function γ, assigning to sentences ϕ in $\mathcal{L}$ a state $\gamma(\phi) \subseteq \Gamma$.

Such a triple is called "semiinterpreted" because it still provides only a partial interpretation for $\mathcal{L}$; only whole sentences such as: "The position of Mars at date d was (α, δ)" obtain a meaning. The word "Mars" by itself remains uninterpreted, however. For many purposes semiinterpreted languages are therefore insufficient to provide a satisfactory semantics, but this can be fixed by regarding the state space as representing a set of "worlds", containing objects to which expressions in $\mathcal{L}$, such as "Mars" refer. These worlds can again be represented by mathematical structures, and one then arrives at a picture in which a state space is a set of structures. For the present, however, we shall stick to the semiinterpreted languages because we do not need this further structure for what we are going to discuss now.

Semiinterpreted languages provide an appropriate way to understand the meaning of physical magnitudes. Physical magnitudes, in classical physics, are functions on a state space. Their main function is to label states. Let $X : \Gamma \rightarrow \mathbf{R}(X)$, where $\mathbf{R}(X)$ is the range of X, be a magnitude, then a sentence "$X = x$" determines a subset of Γ. This is, in fact, all we require from a magnitude. The sentence "$X = x$" must have a well defined meaning, but the symbol "X" itself can be left uninterpreted. Which magnitudes we define to describe a system, and what range we choose them to have, is to a very large extent arbitrary, and for this reason one can even prefer *not* to have an interpretation for X. Therefore semiinterpreted languages suffice to understand magnitudes, and the meaning of a sentence "$X = x$" can be constructed simply in the following way:

$$\gamma(X = x) = \{\lambda \in \Gamma |\, X(\lambda) = x\}.$$

The above few paragraphs sketch in broad outlines the logical structure of any theory of classical physics, and apart from lots of practical difficulties that may complicate a concrete analysis, the general scheme is entirely clear cut and transparent. Let us now turn to QM.

A brief introduction to QM

The logical structure of QM differs markedly from that of classical physics, notably where it concerns the definitions of states and physical magnitudes. Whereas in classical physics magnitudes are given as functions on, and states simply as subsets of a state space, in QM things are quite different. The basic postulates underlying QM can be summarized in the following points:

> Classical phase space is replaced by a complex Hilbert space $\mathcal{H}$, generally of infinitely many dimensions.
>
> Physical magnitudes are represented by self–adjoint linear operators on $\mathcal{H}$, and their possible values correspond to the eigenvalues of the associated operators.
>
> Accordingly, a magnitude X is not a function assigning its values $x \in \mathbf{R}(X)$ to elements of $\mathcal{H}$, but it can be defined as a function on the set of eigenspaces of the associated operator X. Values of physical magnitudes do not correspond to arbitrary subsets of $\mathcal{H}$, but to orthogonal subspaces.
>
> The relation between the state of a system and the result of a measurement of a magnitude is probabilistic.

More details about these rules can be added in the form of a measurement postulate, saying how to compute quantum probabilities, and in the form of quantization rules saying how to find a quantum system corresponding to a given classical system.

By the above rules QM can be actually made to run, enabling us to compute comprehensible predictions demonstrating beyond a trace of doubt that, at least empirically, QM is really a marvellous theory. Physicists have managed to develop keen intuitions about how to use the theory and mathematical details with respect to a wide range of applications have been worked out in great detail. In short, the theory is an enormous success. Nevertheless, if we look at the theory from a logical point of view there are some puzzling questions. Classically, a state represents a set of possible worlds and magnitudes serve to label such sets. It is not clear that this approach can be maintained at all within QM. Assuming that a Hilbert space vector represents a possible world, what is the status of the magnitudes? They do not simply label sets of vectors, because in this case they could be represented by functions on the state space, just like in classical physics. In fact, if we consider a sentence of the form "$X = x$", where X is a physical magnitude and x one of its values, then such a sentence is not interpreted within the Hilbert space in the classical sense. There is no set of vectors in $\mathcal{H}$ such that $X = x$ is true for precisely the vectors in this set. Instead, the relation between Hilbert space vectors and values of magnitudes is assumed to be probabilistic, which is highly peculiar from the point of view of classical logic. Even the relation with probability theory seems to be nonclassical however. In the standard formalism of probability probabilities are assigned to events which, in classical physics, can be defined in terms of subsets of a state space. In classical statistical mechanics, for example, one generally works with probability distributions that are measures on a σ–field of subsets of the phase space. The probabilities usually computed in QM, however, are probabilities not assigned to subsets but to subspaces of the Hilbert space, and the standard additivity requirement

$$\sum_i p(A_i) = p(\bigcup_i A_i) \quad \text{for disjoint } A_i$$

is replaced by

$$\sum_i p(A_i) = p(\oplus_i A_i) \quad \text{for orthogonal } A_i.$$

Accordingly, accepting the standard formalism of QM also implies modifications to probability theory.

Our approach will be, however, to take the fact that magnitudes cannot be interpreted in the standard way as an indication that the formalism is incomplete. Indeed, if one sticks to classical logic and to the standard notions of completeness, this will simply turn up as a theorem[1]. Our solution will be to use contextual logic to remove the incompleteness, and this will require extending the state space and splitting the magnitudes. These things will be discussed in ch. 6, but first we shall take a rather extensive look at quantum measurement theory.

The empirical interpretation of the theory proceeds via the assignment of values to magnitudes, i.e., via sentences of the type $X = x$ for which a satisfactory interpretation is missing. Accordingly, there are direct links between measurement theory and the problems concerning the interpretation of QM. Presumably because

[1]See § 7.3 and § 7.4.

of this, measurement theory has suffered badly from the conceptual unclarities in QM. *Vice versa*, measurement theory has a direct relevance for the interpretation of the theory. Different ideas about measurement theory have also lead to different formulations of the measurement postulate, which gives us the expression for computing probabilities in QM, and these differences are closely linked to some of the controversies about interpretation.

4.2 Measurement postulates

The measurement postulate, that gives us the expression for computing probabilities in QM, has formulations in various degrees of sophistication. We shall discuss a number of different versions, starting from the crudest form onwards.

4.2.1 Pure states, maximal magnitudes

The state of a system, manufactured by means of an appropriate state preparation procedure, is represented by a normed vector ψ in the Hilbert space $\mathcal{H}$. Likewise, the possible measurement results, for the measurement of a nondegenerate or *maximal* magnitude X, are represented by means of the complete orthonormal set of eigenvectors $\{\varphi|\ \mathsf{X}\varphi = x\varphi\}$ of the corresponding operator X. The probability of finding the result corresponding to a given φ is equal to:

$$p_\psi(\varphi) = |\langle\varphi|\psi\rangle|^2. \tag{4.1}$$

4.2.2 Projection operators

The measurement of a nonmaximal magnitude, i.e., a magnitude represented by an operator with degenerate eigenvalues, raises the problem that a measurement result is represented by a multidimensional subspace of $\mathcal{H}$, instead of by a single vector. This problem is solved by taking:

$$p_\psi(P) = \langle\psi|P|\psi\rangle, \tag{4.2}$$

where P is the projection operator on the subspace corresponding to the desired measurement result. It is readily verified that (4.2) satisfies the appropriate additivity requirements and that it becomes equal to (4.1) for one–dimensional subspaces.

4.2.3 Proper mixtures

Let us proceed by considering the state preparation. Suppose that, instead of a single state preparation apparatus, we have several, and we also have a random generator, such as a coin, dice, or an urn with coloured balls, to decide which apparatus to use in a given run of the experiment. Let q be the probability measure produced by the random generator, then the probability for an outcome P can be calculated by means of:

$$p_q(P) = \sum_i p_{\psi_i}(P) q_i, \tag{4.3}$$

where q_i is the probability that the i-th preparation apparatus, producing the state ψ_i, will be used. Eq. (4.3) can be now be rewritten as:

$$p_W(P) = \mathrm{Tr}[PW], \tag{4.4}$$

where Tr denotes the operation of taking the trace of the operator enclosed in the brackets, i.e., $\mathrm{Tr}[X] = \sum_i \langle \alpha_i | X | \alpha_i \rangle$, where summation is performed over an arbitrary orthonormal basis of vectors α_i. W is the so–called density operator, here given by:

$$W = \sum_i q_i |\psi_i\rangle\langle\psi_i|. \tag{4.5}$$

One readily verifies that this yields $\mathrm{Tr}[PW] = \sum_i q_i \langle\psi_i|P|\psi_i\rangle$. It should be noted here that the vectors ψ_i are not necessarily orthogonal, as this depends entirely on the used set of preparation apparatus.

More generally, a density operator is any linear operator on $\mathcal{H}$ satisfying:

$$\begin{aligned} W^+ &= W, \\ W &\geq 0 \\ \mathrm{Tr}[W] &= 1. \end{aligned} \tag{4.6}$$

Two things must be noted about the density operators introduced as above. In the first place, the so–called 'mixed state' represented by such a density operator admits, by construction, the interpretation that the system is 'in reality' described by one of the vectors ψ_i and that q_i merely gives the probability for this real state to occur. For this reason mixtures of this type have been named[2] *proper mixtures.*

In the second place, we must note that information is lost if we describe a mixture by its density operator. Only a pure state, with $W = |\psi\rangle\langle\psi|$, has a unique decomposition of the form (4.5). In every other case there are infinitely many different decompositions of the density operator, describing different, but statistically equivalent, mixed states. Accordingly, although proper mixtures described by their density operator do admit an interpretation in terms of true states and our knowledge of it, information about which states are involved is lost.

4.2.4 Improper mixtures

Mixtures in QM arise also in a different way in the description of composite systems. The Hilbert space of a system $\mathcal{S}^{AB}$, composed of two subsystems $\mathcal{S}^A$ and $\mathcal{S}^B$, is given by

$$\mathcal{H}^{AB} = \mathcal{H}^A \otimes \mathcal{H}^B, \tag{4.7}$$

[2] d'Espagnat [73] ch. 7.

where $\otimes$ denotes the tensor product. Let W^{AB} be the state of $\mathcal{S}^{AB}$, and suppose that the two systems have been separated and that we perform a measurement on the system $\mathcal{S}^A$ alone. The results of this measurement are then represented by operators of the form $P^A \otimes 1\!\!1^B$, and the corresponding probabilities are given by

$$p_{W^{AB}}(P^A \otimes 1\!\!1^B) = \mathrm{Tr}[(P^A \otimes 1\!\!1^B)W^{AB}]. \tag{4.8}$$

Using the so–called partial trace, we can define a density operator W^A on the Hilbert space $\mathcal{H}^A$ of $\mathcal{S}^A$ alone by means of

$$W^A = \mathrm{Tr}^B[W^{AB}], \tag{4.9}$$

and one readily computes that

$$\mathrm{Tr}[(P^A \otimes 1\!\!1^B)W^{AB}] = \mathrm{Tr}[P^A W^A]. \tag{4.10}$$

This means that, for all measurements on S^A alone, the density operator W^A correctly gives the statistical properties of this system. Moreover, the same holds for $\mathcal{S}^B$, to which we can assign a density operator $W^B = \mathrm{Tr}^A[W^{AB}]$.

The interpretation of operators obtained by this procedure is more problematic than that of the proper mixtures discussed earlier. It is not possible to interpret them by means of a probability function over vectors ψ_i representing the real state of the system. Suppose this were possible, then W^A and W^B would describe proper mixtures

$$\begin{aligned} W^A &= \sum_i q_i^A |\psi_i^A\rangle\langle\psi_i^A| \\ W^B &= \sum_j q_j^B |\varphi_j^B\rangle\langle\varphi_j^B|, \end{aligned} \tag{4.11}$$

and correlations between $\mathcal{S}^A$ and $\mathcal{S}^B$ would have to be contained in a joint probability q_{ij}^{AB} such that

$$\sum_j q_{ij}^{AB} = q_i^A, \quad \sum_i q_{ij}^{AB} = q_j^B. \tag{4.12}$$

For W^{AB} this means that it must be given by

$$W^{AB} = \sum_i \sum_j q_{ij}^{AB} |\psi_i^A, \varphi_j^B\rangle\langle\psi_i^A, \varphi_j^B|. \tag{4.13}$$

This is not correct, however. If there has been a quantum mechanical interaction between $\mathcal{S}^A$ and $\mathcal{S}^B$ then W^{AB} is generally a so–called *entangled state* that cannot be decomposed in this way, and the difference is clearly observable in correlation experiments if a sufficient set of different magnitudes can be measured. Therefore, the density operators W^A and W^B do not admit an interpretation in terms of vectors in $\mathcal{H}^A$ and $\mathcal{H}^B$ representing the real state of these systems. For this reason they are said to describe *improper mixtures*[3].

---

[3] d'Espagnat [73] ch. 7.

4.2.5 Proper or improper mixtures?

Clearly the difference between proper and improper mixtures relates directly to one of the most fundamental problems in QM, nonseparability. It indicates that the relation between the physics of a composite system and the physics of its subsystems is highly unusual in this theory. Let us elaborate this point a bit further. Imagine the following situation. A state preparation apparatus yields a composite system $\mathcal{S}^{AB}$ in a state ψ^{AB}, which we shall assume to be an entangled state. The two subsystems are then separated and a measurement is performed on system $\mathcal{S}^B$. We measure a maximal magnitude X^B and the outcome x is represented by a vector $|X^B = x\rangle$. The probabilities for further measurements on $\mathcal{S}^A$, say of some arbitrary magnitude Y^A, are obtained from the joint probabilities for a measurement of $Y^A \otimes X^B$ on $\mathcal{S}^{AB}$, conditionalized on the event $X^B = x$. It is readily verified that one gets precisely the same results if one assigns to the remaining system $\mathcal{S}^A$ the state

$$\langle X^B = x|\psi^{AB}\rangle / \|\langle X^B = x|\psi^{AB}\rangle\|, \tag{4.14}$$

which is a normed vector in $\mathcal{H}^A$.

If no measurement is made on $\mathcal{S}^A$ or $\mathcal{S}^B$ then the composite system is described by the entangled state ψ^{AB}, which cannot be decomposed into a product of state of the individual subsystems. But it rather looks as though, by virtue of the measurement on $\mathcal{S}^B$, we can assign the state (4.14) to $\mathcal{S}^A$, i.e., the entangled state is disentangled by this measurement. In fact, we are more or less forced to admit that this is indeed the correct state for $\mathcal{S}^A$, because most actual state preparations are similar in structure to the procedure described above. A quantum system always starts as a system in interaction with other systems within the preparation apparatus. After it is separated from the apparatus we assign a state to it, without holding back because the system is really in an entangled state so it does not have a state on its own. Therefore, if we do not admit the above procedure as a valid state preparation procedure for the state (4.14) then we are in deep trouble because it is questionable whether we have any preparation procedures left at all.

This leaves us with difficult questions, however. For instance, if the above procedure indeed produces the state (4.14) then when does this happen? If it happens instantaneously upon the measurement of X^B then the transition seems to indicate a nonlocal influence of the measurement on $\mathcal{S}^B$ on the state of $\mathcal{S}^A$, because the two systems can be arbitrarily far apart at the time of measurement. We can even go further, however, and construct a variant of Wheeler's 'delayed choice experiment[4]'. Suppose we wait with the measurement on $\mathcal{S}^B$ until some later time t_B, and meanwhile proceed with any arbitrary measurements on $\mathcal{S}^A$, at a time $t_A < t_B$. Further, suppose that we know that $X^B = x$ before we receive any information about the measurement on $\mathcal{S}^A$. In this case, the state (4.14), assigned to $\mathcal{S}^A$ because of the measurement on $\mathcal{S}^B$ at time t_B, still produces the correct expectations for measurements performed on $\mathcal{S}^A$ at the earlier time t_A. At first sight, it may seem that the situation is entirely comparable to a classical correlation experiment, in which an

[4] Wheeler [218].

increase of knowledge about $\mathcal{S}^B$ also leads to an increase of knowledge about $\mathcal{S}^A$. If we want to hold on to the view that the state vector represents the state of the system, rather than our knowledge of the state, then we must realize that it is also the state of the system, and not our knowledge of it, which depends on the condition that $X^B = x$. Therefore, the above situation, like a delayed choice experiment, seems to indicate a form of retrocausality, i.e., causality backwards in time, because the state of $\mathcal{S}^A$ at moment t_A comes to depend on future events happening to $\mathcal{S}^B$ at time t_B, arbitrarily much later.

One has, at this point, different options. A first reaction might be to question once more the idea that the state vector is comparable to a classical state. Although there is certainly good reason for this question, and although it may lead to a different view on the nonseparability problem, I do not think that it leads to a solution in the sense that the problem disappears. In hidden variable theories for instance it reappears as nonlocality, and in Bohr's philosophy in the form of a holism in which one can properly describe a system only in the context of a quantum phenomenon that is determined by the total measurement set–up. There typically does seem to be the choice, however, between nonlocality proper, or instantaneous causality through influence outside the lightcone, and retrocausality[5]. Thus, holding on to the picture that the state vector represents the state of the system, the above experiment can be described in two ways. One way is to accept retrocausality and assign state (4.14) to $\mathcal{S}^A$ irrespective of the time at which the measurement of X^B takes place. The other possibility is to distinguish between the two cases $t_B < t_A$ and $t_B > t_A$ and produce a nonlocal description for each of these cases separately. If $t_B < t_A$ then before t_B one can assign to $\mathcal{S}^A$ the improper mixed state (4.9), which is then reduced to (4.14) at the moment t_B. If, on the other hand, $t_B > t_A$ then the description is reversed and the state of $\mathcal{S}^B$ is nonlocally reduced at the moment t_A when a measurement on $\mathcal{S}^A$ takes place. One readily computes that this form of nonlocality is not sufficient for transmitting signals instantaneously. One can define concepts of locality which are satisfied by QM, which are sufficient to exclude instantaneous transmission of signals, but which are nevertheless markedly weaker than the classical concept of locality[6].

Let us put these problems aside for the moment and take a closer look at differences and similarities between proper and improper mixtures. Assuming that a measurement of X^B produces the state (4.14) of $\mathcal{S}^A$, we now proceed with a slightly different situation. We have the same composite system, in the same state and again the same maximal magnitude X^B is measured. The measurement is not performed by ourselves, however, but by another observer $\mathcal{O}^B$ who does not tell us the result of his measurement. What can we still say about the state of system $\mathcal{S}^A$ in this case? For every individual outcome we know how to assign a state to $\mathcal{S}^A$ and, moreover, QM gives us a probability equal to $\|\langle X^B = x|\psi^{AB}\rangle\|^2$ for each outcome. This

[5] See, e.g., Selleri [191], see also ch. 7.

[6] See, e.g., Redhead [177], Hoekzema [121].

enables us to assign to $\mathcal{S}^A$ a density operator

$$\begin{aligned} W^A &= \sum_x \|\langle X^B = x|\psi^{AB}\rangle\|^2 \frac{\langle X^B = x|\psi^{AB}\rangle\langle\psi^{AB}|X^B = x\rangle}{\|\langle X^B = x|\psi^{AB}\rangle\|^2} \\ &= \sum_x \langle X^B = x|\psi^{AB}\rangle\langle\psi^{AB}|X^B = x\rangle \qquad (4.15) \\ &= \mathrm{Tr}^B[|\psi^{AB}\rangle\langle\psi^{AB}|]. \end{aligned}$$

This W^A is quantitatively equal to the initial density operator (4.9), representing an improper mixture, but let us note that it does not include all information we have about the remaining system $\mathcal{S}^A$. Besides the density operator we also know precisely which mixture it represents, owing to the fact that we know that $\mathcal{O}^B$ measures the magnitude X^B.

Let us, therefore, consider a still slightly different situation, in which $\mathcal{O}^B$ measures some maximal magnitude but he does not tell us which one. It is readily verified that we can still assign to $\mathcal{S}^A$ the same density operator (4.15), which is already clear from the fact that the last line of this equation does not contain the magnitude X^B. Nevertheless, we have less information about the system. We no longer know which mixture we are dealing with. We still know we are dealing with some definite mixture, however. A specific state vector that applies to $\mathcal{S}^A$ does exist, and the other observer $\mathcal{O}^B$ knows which one it is.

Let us, therefore, consider a fourth situation. We send system $\mathcal{S}^B$ in the direction of $\mathcal{O}^B$ but he is now absolutely free to do whatever he likes with it, and he does not tell us what it is he does. He may perform a measurement, he may put it in a deep freezer and let his grandson perform a measurement a hundred years from now, or he may perform any nonmaximal measurement, or no measurement at all. Whatever he does, we can still use the above density operator W^A to describe the state of $\mathcal{S}^A$, because it correctly describes the statistical properties of $\mathcal{S}^A$. It is no longer evident, however, that an interpretation in terms of a definite mixture of pure states can still be given. We are now dealing with an improper mixture in which the entangled state of $\mathcal{S}^{AB}$ has not been disentangled by a measurement of one of the subsystems.

In all these cases there is the choice between a retrocausal and an instantly–causal description of the situation, and this is relevant to the question whether there is a difference between proper and improper mixtures at all. If one chooses the option of instantaneous causality then $\mathcal{S}^A$ is in an improper mixed state up to the moment of $\mathcal{O}^B$'s measurement on $\mathcal{S}^B$, which turns it into a proper mixture. If one chooses the option of retrocausality, on the other hand, then the state of $\mathcal{S}^A$ is described by a proper mixture from the very start, if $\mathcal{O}^B$ performs some measurement at some arbitrary moment.

4.2.6 The projection postulate

If one performs a number of successive measurements is on a quantum system, then the difference with the preceding case is that a measurement is at the same time part of a preparation procedure, preparing the system for the next measurement. We, therefore, need some assumption about how a measurement affects the state of the system.

For simplicity, we start again with the case of pure states and maximal magnitudes. Let $|\psi\rangle$ be the state of the system before the first measurement and let X be the magnitude first measured. The value x is realized with a probability $|\langle X = x|\psi\rangle|^2$, and a so-called ideal measurement is now defined by the condition that this leaves the system in a state $|X = x\rangle$. This approach is readily generalized to the more general case of mixtures and nonmaximal magnitudes and we then arrive at the so-called projection postulate.

Let W be the initial density operator of the system. Further, $\mathbf{M}$ is the set of projection operators representing the possible measurement outcomes of the first measurement, and $P \in \mathbf{M}$ is the projection operator representing the given outcome. For an ideal measurement the resulting post-measurement state is then equal to

$$W' = PWP/\mathrm{Tr}[PWP], \tag{4.16}$$

where the denominator is a normalization constant that ensures $\mathrm{Tr}[W'] = 1$. In case the measurement is performed but the outcome remains unknown we can construct a proper mixture of states (4.16), leading to

$$\begin{aligned} W' &= \sum_{P \in \mathbf{M}} p_W(P) PWP/\mathrm{Tr}[PWP] \\ &= \sum_{P \in \mathbf{M}} PWP. \end{aligned} \tag{4.17}$$

Likewise, if we have partial knowledge of the outcome, represented by a subset $M \subseteq \mathbf{M}$, the corresponding state is

$$W' = \sum_{P \in M} PWP \,/\, \sum_{P \in M} \mathrm{Tr}[PWP]. \tag{4.18}$$

The Von Neumann chain

A much debated problem about these transitions from W to W', which are generally called *state reductions*, concerns the fact that they cannot be the result of an evolution described by the Schrödinger equation. Nevertheless, application of the projection postulate for the system $\mathcal{S}$ can be shown to be consistent within the quantum formalism by regarding the system, at the time of measurement, as an open system in interaction with its environment, notably the measuring apparatus. The procedure is the following:

Before the measurement the apparatus, now also described as a quantum system with its own Hilbert space $\mathcal{H}^A$, is in a state W^A. The evolution of the composite system $\mathcal{S}^C = \mathcal{S}^A + \mathcal{S}$, of apparatus + system, is given by a unitary evolution operator U^C on $\mathcal{H}^C = \mathcal{H}^A \otimes \mathcal{H}$. After the termination of the measurement interaction the pointer position of the apparatus is measured, and the result is represented by a projection operator P^A. As Von Neumann has shown, there are, in principle, triples $< P^A, U^C, W^A >$ such that

$$\mathrm{Tr}^A[(P^A \otimes 1\!\!1)U^C(W^A \otimes W)U^{C+}] = PWP, \tag{4.19}$$

where P is the projection operator representing the measurement outcome in the Hilbert space $\mathcal{H}$ of the system $\mathcal{S}$. This entire procedure, coupling the system to an apparatus, separating them again and reading the pointer position, can now be accepted as a preparation procedure for $\mathcal{S}$, resulting in a new state $W' = PWP/\mathrm{Tr}[PWP]$, in agreement with the projection postulate. It is readily verified that this gives us the right expectations for any subsequent measurements on $\mathcal{S}$, because any probabilities computed with W' are equal to what is obtained by considering the joint probability for a measurement on $\mathcal{S}^C$, conditionalized on the outcome P^A, i.e.:

$$\frac{\mathrm{Tr}[(P^A \otimes P')U^C(W^A \otimes W)U^{C+}]}{\mathrm{Tr}[(P^A \otimes 1\!\!1)U^C(W^A \otimes W)U^{C+}]} = \frac{\mathrm{Tr}[P'PWP]}{\mathrm{Tr}[PWP]}, \tag{4.20}$$

for any projection operator P' on $\mathcal{H}$.

Thus, the problem that W' cannot be obtained from W by means of a Schrödinger equation is solved by noting that the Schrödinger equation applies only to closed systems, and for open systems evolutions leading from W to W' do exist. Although this is correct, however, it solves the problem only by introducing new ones. In the first place it is clear that we are again confronted with the inseparability problem. The new system state depends on the operator P^A representing the pointer position, but it does not matter at all how far system and apparatus are separated when we read the pointer position, nor when we do this. We can decide to wait a year if we like, but W' will nevertheless produce the correct expectations.

There is one more problem, however. The fact that W' explicitly depends on P^A also implies that we have merely shifted our problem concerning the evolution of the system into a problem concerning the evolution of system + apparatus. The measurement on the system is analyzed by regarding it as a subsystem of (system + apparatus), but then the measurement of the pointer position must presumably be analyzed by regarding it as a subsystem of (system + apparatus) + another apparatus, and so on. This situation of infinite regress is referred to as the 'Von Neumann chain', and it marks one of the great controversies in the foundations of QM.

Literature: d'Espagnat [73], Hoekzema [121], Jammer [130], Von Neumann [165], Redhead [177], Selleri [191], Wheeler [218].

Chapter 5

'Recent' developments in measurement theory

Abstract: A more extensive discussion of quantum measurement theory is given, including some more sophisticated forms of the measurement postulate.

The versions of the measurement postulate discussed so far can be classified as relatively rude constructs, applicable only to special cases. Further elaborations on measurement theory lead to new formulations, including, in the first place, the use of *effect valued measures* and *operation valued measures*, and in the second place of the use of the timesymmetric *ABL measure*. Both developments have been around for some time, but for some reason they have not really become common property.

5.1 Effect valued measures

In (4.2) and (4.4) measurement results were represented by means of projection operators. There is a growing awareness of the fact that this should be regarded as an idealization, applicable only in approximation or in very special circumstances. More generally, the set of possible outcomes of an experiment should be represented by, what is called, an *effect valued measure*, or EVM, or alternatively also a *positive operator valued measure*. Speaking with hindsight, it is quite remarkable that Von Neumann did not introduce these things in his 1932 book[1]. It would have been a natural place and, in fact, he had to go into explicit trouble to evade them. He proves that there can exist measurement interactions leading to a representation of measurements in terms of projections, but if he had admitted a more general class of interactions in his analysis, EVMs, and OVMs which we shall discuss in § 5.2, would have arisen automatically. Things being as they are, we had to wait until the seventies[2] for the occurrence of EVMs and OVMs in the literature. Even after this date, however, their diffusion into the physical community has been unnaturally slow, perhaps owing to the fact that Ludwig's axiomatic approach to QM is highly

[1] [165].

[2] Ludwig [155].

abstract. EVMs did gain solid ground in some interesting but more or less remote areas like algebraic QM and the quantum theory of open systems[3]. The realization that they are very well applicable in practical circumstances, however, and that in fact no serious analysis of the measurement process can do without them, is spreading only gradually.

What is an EVM? Consider an experiment. A quantum system, after having been prepared in a state W, is sent into a measurement apparatus that we further envisage as a black box. After swallowing the system the apparatus produces a measurement outcome which can be labelled by an index k, in an *index set* $\mathbf{K}$. It can then be shown that the outcomes can be represented by means of a set of positive self-adjoint operators $R(k)$, called *effects*, such that the probability for k to occur is given by:

$$p_W(k) = \mathrm{Tr}[R(k)W]. \tag{5.1}$$

Assuming that the measurement apparatus can itself be treated as a QM-system, the existence of such a representation can be proved and in principle the operators $R(k)$ can be computed if one has an appropriate QM description of the apparatus and of the measurement interaction.

Example: Spin measurements, say on a beam of spin-1/2 particles, are commonly performed with a Stern-Gerlach apparatus, where the particles are deflected by a magnetic field such that spin-up particles go one way and the spin-down particles the other way. Beam separation is quite good for heavy particles, e.g. silver ions, but very bad or even worse for light particles, such as electrons. This has as a consequence that measurement outcomes for such particles, produced by detection of a particle in one beam or the other, cannot be represented adequately by means of projection operators. An appropriate EVM is readily found however. Taking imaginary figures, let us say that, for a given type of particles and a given apparatus, 80% of the spin-up particles ends up in the up beam and 20% in the down beam, and for spin-down particles *vice versa.* The detection of a particle in one of the beams can now be represented by the following EVM:

$$\begin{aligned} R(+) &= .8\,|+\rangle\langle+| \;+\; .2\,|-\rangle\langle-| \\ R(-) &= .2\,|+\rangle\langle+| \;+\; .8\,|-\rangle\langle-|. \end{aligned} \tag{5.2}$$

Given an apparatus, how do we find the associated EVM? There are two possibilities. Either we determine the necessary figures empirically, for instance by checking the results against a calibrated preparation set-up, or we compute them. For a Stern-Gerlach apparatus this can be done if we know the magnetic field strength gradients. These enable us to calculate the coupling between the spin and the position of the outgoing particles, and the computed correlation between position and spin can then be entered into an EVM of the above form.

It is quite clear that the Stern-Gerlach apparatus is not an isolated case. Similar considerations hold with respect to many instruments, such as particle counters,

[3]See, e.g., Kraus [135], Davies [56], Davies and Lewis [57], Alicki and Lendi [8], Holevo [124].

polarizing filters etc. There may in fact be very few real world instruments for which a fully adequate representation in terms of projection operators does exist.

Formulating things in a more sophisticated way, we define an effect valued measure on $\mathbf{K}$ in the following manner:

Definition: Let $\mathcal{F}$ be a σ–algebra of subsets of $\mathbf{K}$, i.e.

$$\begin{aligned}&\emptyset, \mathbf{K} \in \mathcal{F} \\ &\text{If } K_1 \cdots K_i \cdots \text{ is a countable sequence of} \\ &\text{elements of } \mathcal{F} \text{ then } K_i^-, \bigcup_i K_i, \bigcap_i K_i \in \mathcal{F},\end{aligned} \tag{5.3}$$

where K^- is the complement of K in $\mathbf{K}$. An EVM is now introduced as a function R, assigning to each $K \in \mathcal{F}$ a linear operator $R(K)$ on $\mathcal{H}$ such that:

$$\begin{aligned}&R(K) \text{ is self–adjoint,} \\ &R(K) \geq 0, \\ &R(\bigcup_i K_i) = \sum_i R(K_i) \text{ if all } K_i \text{ are mutually disjoint,} \\ &R(\mathbf{K}) = 1\!\!1 \,.\end{aligned} \tag{5.4}$$

It readily follows that the function

$$p_W(K) = \mathrm{Tr}[R(K)W] \tag{5.5}$$

is a probability measure, for any density operator W on $\mathcal{H}$.

5.2 Operation valued measures

In cases where a number of successive measurements is made on a system, Von Neumann's theory of measurement provides, by means of (4.19):

$$\mathrm{Tr}^A[(P^A \otimes 1\!\!1\,)U^C(W^A \otimes W)U^{C+}] = PWP,$$

some theoretical basis for applying the projection postulate. Besides the infinite regress problem connected with it, however, (4.19) is also deficient in that it only works for a very restricted class of measurement interactions. Only very special triples $< P^A, U^C, W^A >$ will lead to results such that the r.h.s. of (4.19) can indeed be written in the form PWP. What we would like to have is a theory for more general interactions between apparatus and system, i.e., a theory in which a more general class of triples $< P^A, U^C, W^A >$ can be admitted. The theory of open quantum systems, as developed in the past two decades, is very well–suited to deal with this situation. It does not by itself provide a solution to the Von Neumann chain problem but it readily provides a highly elegant generalization of the projection postulate, leading to, what are called, operation valued measures, or OVMs.

Starting with the expression on the l.h.s. of (4.19) we shall now admit the density operator W^A, the projection operator P^A, and the unitary operator U^C to be chosen arbitrarily, and in fact generality can be further increased by replacing the projection P^A by an effect R^A from the range of some EVM. We shall call the resulting expression $\phi(W)$, i.e., :

$$\phi(W) = \mathrm{Tr}^A[(R^A \otimes 1\!\!1\,)U^C(W^A \otimes W)U^{C+}]. \tag{5.6}$$

For a fixed triple $< P^A, U^C, W^A >$ and a variable W, this expression defines a linear mapping ϕ, generally called an operation, on the set of nonnegative traceclass operators on $\mathcal{H}$, and it can be generalized to more general classes of operators. In accordance with the earlier analysis in the Von Neumann case, ϕ assigns to an initial density operator W a new one:

$$W' = \phi(W)/\mathrm{Tr}[\phi(W)]. \tag{5.7}$$

W' again gives us the right expectations for any subsequent measurements on $\mathcal{S}$, because any probabilities computed with W' are equal to what is obtained by considering the joint probability for a measurement on $\mathcal{S}^C$, conditionalized on the outcome R^A, just as in (4.20).

The mappings defined in this way satisfy a condition called complete positivity, and it can be proved that they therefore admit a representation given by the so-called Stinespring decomposition[4]:

$$\phi(W) = \sum_{V \in \mathbf{V}} VWV^+, \tag{5.8}$$

where $\mathbf{V}$ is a set of linear operators on $\mathcal{H}$. This makes (5.7) into a direct generalization of (4.18), with (4.16) as a special case.

Within the context of standard Hilbert space QM, as opposed to the more general algebraic QM, Stinespring decompositions of (5.6) are in fact readily constructed in the following way: find any resolution of R^A and W^A into

$$R^A = \sum_r R^A(r)\,|r\rangle\langle r|, \quad W^A = \sum_w W^A(w)\,|w\rangle\langle w|, \tag{5.9}$$

where the vectors $|r\rangle$ nor the vectors $|w\rangle$ are necessarily mutually orthogonal. By substitution in (5.6) we find

$$\begin{aligned} \phi(W) &= \sum_r \sum_w R^A(r)W^A(w)\,\langle r|U^C|w\rangle W \langle w|U^{C+}|r\rangle \\ &= \sum_r \sum_w V_{rw} W V_{rw}^+, \end{aligned} \tag{5.10}$$

with V_{rw} given by

$$V_{rw} = (R^A(r)W^A(w))^{1/2}\,\langle r|U^C|w\rangle. \tag{5.11}$$

[4]Stinespring [198].

This result proves (5.8), by means of explicit construction.

From the definition of ϕ it follows that the probability of finding the pointer position represented by R^A is given by

$$\begin{aligned} p_W(R^A) &= \mathrm{Tr}[\phi(W)] \\ &= \mathrm{Tr}[\sum_{V\in\mathbf{V}} VWV^+]. \end{aligned} \tag{5.12}$$

Because the trace is invariant under cyclic permutation of the included operators, this can also be written as $\mathrm{Tr}[RW]$, with R defined by

$$R = \sum_{V\in\mathbf{V}} V^+V. \tag{5.13}$$

R is the effect representing the measurement outcome on the Hilbert space $\mathcal{H}$ of $\mathcal{S}$.

Note: It can be noted that each individual term V^+V in (5.13) is by itself an effect, and it could be tempting to interpret them as representing what is really measured. Their sum R would then represent the, possibly coarse–grained, information we receive about the measurement by looking at the pointer position of the apparatus. This interpretation can occasionally be correct. For a simple example let us return to the Von Neumann case (4.18), where each V is a projection operator. If we do not read the pointer position, or the pointer has got stuck by a mechanical defect, then R becomes equal to $1\!\!1$. In spite of this fact the measurement has been performed and the projection postulate must be used in the form (4.18), which is in perfect agreement with using the operation ϕ defined by $\phi(W) = \sum VWV^+$. Nevertheless, one must be extremely careful with using this interpretation of an operation, because the Stinespring decomposition of ϕ is generally not unique.

Note: Let us also note the complication that expression (5.6) is not the most general of its sort. It is based on the crucial assumption that for the composite system $\mathcal{S}^C{=}\mathcal{S}^A{+}\mathcal{S}$ we have

$$W^C = W^A \otimes W \text{ and } R^C = R^A \otimes R. \tag{5.14}$$

In many cases of interest these conditions may hold either only approximately or not at all, and in this case the above conclusions, along with many other results, need to be reconsidered (see also § 8.5).

A main ingredient in the above analysis is the operator R^A, that was assumed to be one of the values of an EVM of operators working on the Hilbert space $\mathcal{H}^A$ of the apparatus. Corresponding to each R^A we found an associated operation ϕ and an operator R, working on the Hilbert space $\mathcal{H}$ of the system. It is readily verified that the latter is one of the elements in the range of an induced EVM of operators on the system Hilbert space $\mathcal{H}$. It is natural now to regard also the operation ϕ as the value of a measure, in this case an operation valued measure (OVM).

Let $\mathbf{K}$ be the index set of pointer positions, $\mathcal{F}$ a σ–algebra of subsets of $\mathbf{K}$, and R^A an EVM on $\mathcal{F}$, i.e., there is an effect $R^A(K)$ for each $K \in \mathcal{F}$. To each $K \in \mathcal{F}$ an operation ϕ_K and a density operator W'_K may now be defined by

$$\begin{aligned} \phi_K(W) &= \mathrm{Tr}^A[(R^A(K) \otimes 1\!\!1\,)U^C(W^A \otimes W)U^{C+}] \\ &= \sum_{V \in \mathbf{V}_K} VWV^+ \\ W'_K &= \phi_K(W)/\mathrm{Tr}[\phi_K(W)], \end{aligned} \tag{5.15}$$

From the first line it immediately follows, using (5.4), that for every W and for $L = \bigcup_K K$, with all K disjunct, we have:

$$\phi_L(W) = \sum_K \phi_K(W). \tag{5.16}$$

Furthermore, each OVM has an associated probability measure

$$p_W(K) = \mathrm{Tr}[\phi_K(W)] = \mathrm{Tr}[R(K)W]. \tag{5.17}$$

equal to that of its associated EVM, defined by (5.13).

For an experiment consisting of a preparation followed by a single measurement, results are clearly entirely equivalent to what we obtain by using an EVM. By representing measurements by OVMs, however, we have gained the possibility to deal also with sequences of measurements. In fact, the set of OVMs is closed under composition, in the sense that the combined action of two subsequent operations, e.g. $\theta_K(\phi_L(W))$, is again an operation.

5.3 Some cases in measurement theory

Even though it solves neither the problem of the Von Neumann infinite regress nor the nonseparability problem, the use of OVMs to represent measurements is a major step forward in quantum measurement theory. We shall illustrate this with a few examples and to this end we consider a system $\mathcal{S}$ interacting at regular intervals with a series of probes $\mathcal{S}^1 \cdots \mathcal{S}^N$, during some period of time $[0,T]$. Each $\mathcal{S}^i$ interacts with $\mathcal{S}$ for a brief period and afterwards some measurement may be performed on it. Given what types of measurements are made on the scattered probes, the total impact of these interactions on the system $\mathcal{S}$ is given by an OVM ϕ. Further, each of the interactions separately is given by an OVM ϕ_i. Therefore, for a K that can be written as $K_1 \times K_2 \times \cdots \times K_N$, we have

$$\begin{aligned} \phi_K(W) &= \phi_{N,K_N}(\cdots \phi_{2,K_2}(\phi_{1,K_1}(W))\cdots) \\ &= \phi_{N,K_N} \circ \cdots \circ \phi_{1,K_1}(W), \end{aligned} \tag{5.18}$$

where "$\circ$" denotes functional composition. Using a Stinespring decomposition for each of the operations this gives

$$\phi_K(W) = \sum_{V \in \mathbf{V}(K)} VWV^+ \tag{5.19}$$

$$= \sum_{V_N \in \mathbf{V}_N(K_N)} \cdots \sum_{V_1 \in \mathbf{V}_1(K_1)} V_N \cdots V_1 W V_1^+ \cdots V_N^+.$$

Calculations on practical cases of these expressions may become quite involved, but some basic intuitions can be rapidly developed by elementary considerations on a number of special cases.

Example: A simple case is the one in which the interactions between the probes and the object vanish. In this case, whatever one chooses to measure on the probes afterwards, all operations $\phi(K)$ become proportional to the unitary evoluton of the system, i.e., $V = a_V U(T)$ for all $V \in \mathbf{V}(K)$ for all $K \subseteq \mathbf{K}$, where U is the evolution operator $exp(-iHT)$ of the system over the period $[0, T]$. An example of an experiment falling into this category is when the system is left undisturbed, but a measurement outcome is produced by a throw of dice. Corresponding to each elementary event $k \in \{1, 2, 3, 4, 5, 6\}$ this gives us a V_k equal to $\sqrt{1/6}\, U(T)$.

Example: If all $V_i \in \mathbf{V}_i(K_i)$ are one–dimensional projectors for all $K_i \subseteq \mathbf{K}_i$ then one gets a complete state reduction at each encounter with one of the probes.

Channels

Between the two extremes pictured in the above examples, there is a whole spectrum of intermediate possibilities, which we shall investigate by introducing some additional concepts. The index set $\mathbf{K}$ can be regarded as a state space of possible histories of the object, in as far as the history of the object can be reconstructed from looking at the scattered probes. $\mathbf{K}$ is a Cartesian product of N smaller index sets $\mathbf{K}_1 \cdots \mathbf{K}_N$, relating to the indiviual probes. Accordingly, an element $k \in \mathbf{K}$ is a tuple $< k_1, \cdots, k_N >$, representing an elementary history of the system, whereas a history more in general is a set of such tuples. Further, we shall define a *channel* to be a history that can be written in the form $C = C_1 \times C_2 \times \cdots \times C_N$, with $C_i \subseteq \mathbf{K}_i$, i.e., a channel has an associated operation that can be written in the form given in eq.(5.18).

We now note that every operator V can be represented by its 'polar decomposition', i.e., it can be written in the form

$$V = U_V R_V^{1/2}, \tag{5.20}$$

where $R_V = V^+V$ is an effect and U_V is unitary. For a given channel C at a given instant i we can find an associated effect $R_i(C_i)$ by adding the corresponding R_V, i.e.,

$$R_i(C_i) = \sum_{V \in \mathbf{V}_i(C_i)} R_V. \tag{5.21}$$

Given that the state of the system at the encounter with the i-th probe is given by a density operator W_i, the expression $\mathrm{Tr}[R_i(C_i)W_i]$ gives the probability of finding the system within the channel C at this time. Let us now in particular investigate the case that for a given channel C all $R_i(C_i)$ are projection operators. In this case we shall call the channel a *projection quality channel.* Each interaction with one of the probes then results in a state reduction, destroying in the density operator all the cross terms between inside and outside of the channel. If the system is located within the channel after such a reduction then its further evolution depends on the operators $U_V, V \in \mathbf{V}_i(C_i)$.

Example: Consider a *stationary channel*, i.e., one for which

$$\mathbf{V}_i(C_i) = \mathbf{V}_j(C_j) := \mathbf{V}_{st}(C) \text{ for all } i, j. \tag{5.22}$$

As a consequence we get

$$R_i(C_i) = R_j(C_j) := R_{st}(C). \tag{5.23}$$

If, furthermore, the condition

$$[U_V, R_{st}(C)] = 0 \tag{5.24}$$

is satisfied for all $V \in \mathbf{V}_i(C_i)$, i.e., the operators U_V map C into itself, then the system is caught within the channel; once it has entered it cannot escape. We shall call such a channel a *stable channel* and note that there are interesting applications. Relative to a stable channel, the successive interactions with the probes constitute a sequence of so-called 'nondemolition measurements'. Nondemolition measurements are used for instance in gravitational wave detectors[5], where the effect to be measured is extremely small, and other types of measurement would introduce a measurement disturbance that would swamp the signal. In our present terminology, the idea is that the measurement apparatus is prepared in a stable channel in which it is frequently monitored by means of nondemolition measurements. The only way for the apparatus to leave the channel is by an external disturbance, in this case a gravitational wave, and this fact is noticed in the next measurement.

Another application of stable channels concerns the construction of memory devices. A memory device is a system in which information can be stored which can then be read arbitrarily often without destroying the contents of the device. A memory device is usually conceived as a macroscopic system, such as a photographic plate, inkdots on paper, etc. We may note, however, that size as such has little to do with it. The crucial factor is that the system must have a stable channel that is 'immune' to the interactions by which the information is read. Macroscopic objects apparently tend to possess such channels in abundance, but microscopic objects can have them as well if the circumstances are arranged sufficiently carefully.

Example: Consider a spin–1/2 system with a free evolution given by $U_0(t) = \mathbb{1}$ for all intervals of time t. The Hilbert space of the system is spanned by states $|+\rangle$ and $|-\rangle$. The state of the system can be measured by means of a probe which

[5]See, e.g., Unruh [210], Braginsky *et al.* [35].

is also a two–state system, with its Hilbert space spanned by vectors $|u_+\rangle$ and $|u_-\rangle$. Moreover, assume that the Hamiltonian of the combined system during the interaction is given by

$$\begin{aligned} H^C &= E_0\{|+,u_+\rangle\langle +,u_+| + |+,u_-\rangle\langle +,u_-|\} \\ &+ E_1\{|-,u_+\rangle\langle -,u_-| + |-,u_-\rangle\langle -,u_+|\}. \end{aligned} \tag{5.25}$$

and that the interaction between the system and the probe takes a duration of time $\tau = \pi/(2E_1)$. The evolution of the combined system during this time is then given by an evolution operator $U^C = exp(-iH^C\tau)$, equal to

$$\begin{aligned} U^C &= e^{-iE_0\tau}\{|+,u_+\rangle\langle +,u_+| + |+,u_-\rangle\langle +,u_-|\} \\ &- i\{|-,u_+\rangle\langle -,u_-| + |-,u_-\rangle\langle -,u_+|\}. \end{aligned} \tag{5.26}$$

The probe is prepared in a state $|u_+\rangle$ and after the interaction a magnitude with eigenvectors $|u_+\rangle$ and $|u_-\rangle$ is measured on it. Corresponding to the two possible outcomes there are two operations ϕ_+ and ϕ_-, with

$$\phi_+(W) = V_+WV_+^+, \quad \phi_-(W) = V_-WV_-^+, \tag{5.27}$$

where

$$\begin{aligned} V_+ &= e^{-iE_0\tau}\langle u_+|U^C|u_+\rangle = e^{-iE_0\tau}|+\rangle\langle +| \\ V_- &= \langle u_-|U^C|u_+\rangle = -i|-\rangle\langle -|. \end{aligned} \tag{5.28}$$

There are, furthermore, two associated effects R_+ and R_-:

$$\begin{aligned} R_+ &= V_+^+V_+ = |+\rangle\langle +| \\ R_- &= V_-^+V_- = |-\rangle\langle -|. \end{aligned} \tag{5.29}$$

The two operations give rise to two stable channels. Once the system has entered one of them, subsequent interactions with identical probes will always yield the same result. This makes the system potentially applicable as a memory device in which one bit of information can be stored.

Let us now proceed to a more general case and see what happens if we drop condition (5.24), i.e., we are still dealing with a stationary projection quality channel but it is no longer stable because the system can escape. There are a number of different cases to be investigated.

Example: Let $\mathbf{C}$ be a set of stationary projection quality channels, and for each channel $C \in \mathbf{C}$ there is a channel $C' \in \mathbf{C}$ such that

$$U_V R_{st}(C) = R_{st}(C')U_V R_{st}(C) \quad \text{for all } V \in \mathbf{V}_i(C_i), \tag{5.30}$$

where $R_{st}(C)$ and $R_{st}(C')$ are the projection operators on the corresponding channels. What this means is that the operators U_V map all states in channel C into

C'. In this case, if we find the system within the channel C, then we know that it will be in channel C' next time and thereafter it will move to C'', C''', etc. Evolution is deterministic, like in the above case of a stable channel. In fact one can define a stable, but comoving nonstationary channel D by taking $D_1 = C_1, D_2 = C_2', D_3 = C_3'', \cdots$.

Example: Imagine, once again, a set of projection quality channels, but we now drop condition (5.30), i.e., the operators U_V spread the state of the system over a number of different channels instead of mapping it into a single one. This makes evolution stochastic. The system can jump from one channel to another in an unpredictable way. This is a case where the indeterminism of QM is very clearly visible. As an example one can think of a spin–1/2 particle on which a sequence of measurements is made of its spin in a number of different directions. For such a sequence of measurements there is no stable channel, stationary nor comoving, except the trivial one given by $R_{st}(C) = 1\!\!1$.

The condition of projection quality of the channels is in some respect very attractive, but we can also expect it to be violated in many cases of interest. Occasionally, projection quality may be approximated good enough to retain much of what was said above, but let us see what happens if the quality is gradually lowered. For simplicity we first consider a set of stationary channels, still satisfying (5.22), but the operators $R_{st}(C)$ are no longer projection operators; their spectra of eigenvalues contain not only the numbers 0 and 1 but also intermediate values. The quality of a channel is lower according as larger parts of its spectrum have values markedly different from 0 or 1. An interesting phenomenon in this case is that a sequence of repeated measurements may be more informative than a single one.

Example: Imagine a system with a three dimensional Hilbert space spanned by states $|-\rangle$, $|0\rangle$, and $|+\rangle$. On this system a measurement is performed with two possible outcomes, labelled a and b, with corresponding operations be given by

$$\begin{aligned} \phi_a(W) &= V_a W V_a^+, \quad \phi_b(W) = V_b W V_b^+ \\ V_a &= |-\rangle\langle-| + \sqrt{1/2}|0\rangle\langle 0| \\ V_b &= |+\rangle\langle+| + \sqrt{1/2}|0\rangle\langle 0|. \end{aligned} \tag{5.31}$$

For a single measurement we find effects

$$R(a) = |-\rangle\langle-| + 1/2|0\rangle\langle 0| \quad R(b) = |+\rangle\langle+| + 1/2|0\rangle\langle 0|, \tag{5.32}$$

but results can be much improved if the measurement can be repeated, say n times. We can then distinguish between three cases, one in which a is found every time, one in which b is found every time and one in which results are mixed. These cases define three corresponding channels that we shall call C_{-n}, C_{+n} and C_{0n}. The operations corresponding to these channels can be computed. For C_{-n} we find an evolution operator $V_{-n} = V_a^n$, for C_{+n}: $V_{+n} = V_b^n$, and for

C_{0n} we need a sum of many terms containing mixed products of V_a and V_b. It is readily computed, however, that all these terms are multiples of $|0\rangle\langle 0|$ and can therefore be combined into a single operator V_{0n}. Straightforward calculation gives, for $n \geq 1$:

$$\begin{aligned} V_{-n} &= |-\rangle\langle -| + 2^{-n/2}|0\rangle\langle 0| \\ V_{0n} &= (1-2^{1-n})^{1/2}|0\rangle\langle 0| \\ V_{+n} &= |+\rangle\langle +| + 2^{-n/2}|0\rangle\langle 0|. \end{aligned} \tag{5.33}$$

This shows that the overall quality of the three channels becomes better according as more measurements are made, and projection quality is approached asymptotically.

The increase in channel quality in the above example hinges on the circumstance that successive measurements measure the same thing over and over again, without introducing an additional measurement disturbance relative to one another. In the example this is due to the fact that the effects R_V all commute and that the unitary operators U_V are all equal to $\mathbb{1}$. More generally, a relevant condition would be that

$$[R_{V_{i+1}}, U_{V_i} R_{V_i} U_{V_i}^+] = 0 \quad \text{for all } i \text{ and all } \; V_{i+1}, V_i. \tag{5.34}$$

This enables one to repeat the same measurement, in a sense, over and over again without being bothered by the indeterminism inherent in QM. In the above case this leads to a reduction of the density operator in a gradual way, by an exponential decay of the cross terms. More generally, if an additional measurement disturbance does exist, this will interfere with the tendency to channel quality improvement and it will depend on the circumstances which effect predominates. A bubble chamber provides an illustration. A single bubble is at best only an, approximate, position measurement, but a complete trace is also an approximate momentum measurement, at least for a charged particle of which its deflection in the magnetic field in the chamber is momentum dependent. The quality of the trace as a momentum measurement depends on its length; a fair number of bubbles is needed to make a good estimate. Therefore, channel improvement clearly predominates in the beginning of the trace, but in the long run the particle will be stopped by the fluid in the chamber, if it has a long enough lifetime. No further information about its initial state can then be obtained by prolonged observation, indicating that stochastic evolution wins in the end.

The gradual state reduction effectuated by the evolution (5.33) is an interesting phenomenon that can also be witnessed in the following example, which gives a simple model of a system approaching thermodynamic equilibrium with its environment.

Example: A system $\mathcal{S}$ has a two dimensional Hilbert space, spanned by energy eigenvectors $|0\rangle$ and $|E\rangle$. The environment consists of a gas of similar systems and every now and then the system interacts with one of the systems in its environment during some typical period of time, say t. During the interaction the Hamilton operator of the combined system is given by

$$\begin{aligned} H^C &= H_0^C + H_I^C \\ H_0^C &= |0,0\rangle\langle 0,0| + E\{|0,E\rangle\langle 0,E| + |E,0\rangle\langle 0,E|\} + 2E|E,E\rangle\langle E,E| \\ H_I^C &= g\{|E,0\rangle\langle 0,E| + |0,E\rangle\langle E,0|\}. \end{aligned} \tag{5.35}$$

The eigenvectors of H^C are readily found and the evolution operator $U^C = e^{-iH^Ct}$ of the combined system over the period t can be calculated with the result

$$\begin{aligned} U^C = & |0,0\rangle\langle 0,0| + e^{-2iEt}|E,E\rangle\langle E,E| \\ & + e^{-iEt}cos(gt)\{|E,0\rangle\langle E,0| + |0,E\rangle\langle 0,E|\} \\ & - ie^{-iEt}\sin(gt)\{|E,0\rangle\langle 0,E| + |0,E\rangle\langle E,0|\}. \end{aligned} \tag{5.36}$$

Let $\mathcal{S}^A$ be one of the systems of the environment, with a density operator W^A be given by

$$\begin{aligned} W^A &= a^2|0\rangle\langle 0| + b^2|E\rangle\langle E|, \quad \text{with} \\ a &= 1/\sqrt{Z}, \quad b = e^{-E/2kT}/\sqrt{Z}, \quad Z = 1 + e^{-E/kT}. \end{aligned} \tag{5.37}$$

No measurement on the system $\mathcal{S}^A$ takes place after its interaction with $\mathcal{S}$, and accordingly the evolution of $\mathcal{S}$ alone over the period t is given by the operation

$$\phi(W) = \mathrm{Tr}^A[(1\!\!1^A \otimes 1\!\!1)U^C(W^A \otimes W)U^{C+}]. \tag{5.38}$$

Using $1\!\!1^A = |0\rangle\langle 0| + |E\rangle\langle E|$ and $W^A = a^2|0\rangle\langle 0| + b^2|E\rangle\langle E|$ this expression splits into four terms that can each be written in the form VWV^+. The operators V are readily computed; one has

$$\begin{aligned} V_I &= a\langle 0|U|0\rangle = a|0\rangle\langle 0| + ae^{-iEt}cos(gt)|E\rangle\langle E| \\ V_{II} &= a\langle E|U|0\rangle = -iae^{-iEt}\sin(gt)|0\rangle\langle E| \\ V_{III} &= b\langle 0|U|E\rangle = -ibe^{-iEt}\sin(gt)|E\rangle\langle 0| \\ V_{IV} &= b\langle E|U|E\rangle = be^{-iEt}cos(gt)|0\rangle\langle 0| + be^{-2iEt}|E\rangle\langle E|. \end{aligned} \tag{5.39}$$

The total operation ϕ can now be calculated by means of

$$\begin{aligned} \phi(W) &= \sum_V VWV^+ \\ &= \{(a^2 + b^2cos^2(gt))\langle 0|W|0\rangle + a^2\sin^2(gt)\langle E|W|E\rangle\} \quad |0\rangle\langle 0| \\ &\quad + \{(a^2cos^2(gt) + b^2)\langle E|W|E\rangle + b^2\sin^2(gt)\langle 0|W|0\rangle\} \quad |E\rangle\langle E| \\ &\quad + e^{iEt}cos(gt)\langle 0|W|E\rangle \quad |0\rangle\langle E| \\ &\quad + e^{-iEt}cos(gt)\langle E|W|0\rangle \quad |E\rangle\langle 0|. \end{aligned} \tag{5.40}$$

If we further compute the change in the various components of W we see that

$$\begin{aligned} \Delta\langle 0|W|0\rangle &= a^2\sin^2(gt)\langle E|W|E\rangle - b^2\sin^2(gt)\langle 0|W|0\rangle \\ \Delta\langle E|W|0\rangle &= (e^{-iEt}cos(gt) - 1)\langle E|W|0\rangle \\ \Delta\langle E|W|E\rangle &= -\Delta\langle 0|W|0\rangle, \quad \Delta\langle 0|W|E\rangle = \Delta\langle E|W|0\rangle^*. \end{aligned} \tag{5.41}$$

Assuming $\sin^2(gt) > 0$, the result of a succession of similar interactions of the system with its environment will be that the cross terms $\langle 0|W|E\rangle$ and $\langle E|W|0\rangle$ vanish exponentially and that the components $\langle 0|W|0\rangle$ and $\langle E|W|E\rangle$ approach an equilibrium given by

$$\langle E|W|E\rangle = (b/a)^2\langle 0|W|0\rangle = e^{-E/kT}\langle 0|W|0\rangle, \tag{5.42}$$

i.e., after some time the density operator of $\mathcal{S}$ will be equal to that of the other systems.

An interesting feature of the above example is that the interaction can be switched on and off gradually. A prompt reduction of the density operator in a single step is obtained for $\sin(gt) = 1$, but for $\sin(gt) = 0$ no reduction takes place at all. V_{II} and V_{III} then vanish, whereas V_I and V_{IV} become proportional and can be combined into a single unitary evolution operator.

Protochannels

Another interesting point concerns the nonuniqueness of the Stinespring decomposition and the relation of this with the concept of channels. For a number of successive interactions between the system and the environment in the example, it is tempting to define channels C_I, C_{II}, C_{III}, and C_{IV}, corresponding to the different evolution operators occurring in the Stinespring decompositions of the operations describing these interactions. It will be noted, however, that the original definition of a channel is not applicable here. Channels were originally introduced by means of trajectories through an index set $\mathbf{K}$. There is no such index set here, however, because no measurements have been performed on the scattered probes. In relation to this it can be noted that the particular Stinespring decomposition we used in the example is not unique. It was based on a particular resolution of the unity operator $1\!\!1^A$ as $|0\rangle\langle 0| + |E\rangle\langle E|$ and on the representation of W^A as $a^2|0\rangle\langle 0| + b^2|E\rangle\langle E|$. Any other resolution of either $1\!\!1^A$ or W^A would have given a different but equivalent set of evolution operators. Accordingly, although it is technically unproblematic to define channels C_I, C_{II}, etc., corresponding to the evolution operators of a given Stinespring decomposition, it is not clear that they have a physical meaning comparable to that of the proper channels defined earlier. Let us therefore call them *protochannels.* Further, a protochannel will be called *elementary* if it contains only a single evolution operator V_i at every instant i. Accordingly, every channel or protochannel can be decomposed into a set of elementary protochannels, but not necessarily in a unique way.

The problem concerning the possibility to attach physical meaning to a particular set of protochannels is intimately linked to a discussion earlier in this chapter about proper and improper mixtures. A central issue there concerned the question whether density operators have a preferred resolution, and intimately related to this is the question whether this also holds for effects. The existence of preferred resolutions of $1\!\!1^A$ and W^A is precisely what determines the existence of a preferred Stinespring decomposition of the above operation ϕ_1, and thereby it also determines whether physical meaning can be unambiguously attached to the protochannels based on such a decomposition[6].

Poor quality channels

Finally, let us pay attention to the case that all channels are of very poor quality. In particular we focus on the extreme case that $R_V \sim 1\!\!1$ for all V. This applies when there is no interaction at all between a system and its environment, but there are other cases because it may still occur that there are a number of different operators U_V.

> **Example:** A charged particle enters a chamber with a magnetic field. The field strength is set with a knob in an adjacent room which is controlled by two drunken sailors playing dice for it. These people categorically refuse to communicate the settings to the chief scientist who is commanding the experiment. Corresponding to each possible field strength x there is a unitary evolution operator U_x, and in the game of dice x has a probability $p(x)$ to occur. Under these circumstances the final density operator W' can be expressed in terms of the initial W by means of
>
> $$W' = \sum_x p(x) U_x W U_x^+ = \sum_x V_x W_0 V_x^+, \tag{5.43}$$
>
> with V_x equal to $\sqrt{p(x)}\, U_x$.

Summary

Surveying these examples, it is clear that the theory of operations offers a powerful and widely applicable approach to measurement theory. Many interesting things become comprehensible, at least in outlines, even without detailed knowledge of the mathematical developments that have taken place in this area, and the approach is clearly superior to more conventional work on measurement theory, both in practical scope and in theoretical insight. It leads to the suggestion that the basic notion in investigating the relation between quantum physics and classical physics is that of an 'open system', and it thereby opens the perspective of replacing heuristic principles such as Bohr's 'correspondence principle' and such as various considerations about 'macroscopic systems' by something more tangible.

On the other hand it is also clear that the theory of operations gives no further clues to some of the basic problems of measurement theory, such as in particular those of the Von Neumann chain and of the dependence of present operations on future or far away measurements. Finally, the examples discussed above add some further doubts to whether the Hilbert space is conceptually comparable to a classical state space. If one looks for analogues of classical state spaces, there are candidates that, with respect to logical structure, seem more appropriate, such as the index

[6]See also § 6.4.1.

set **K**, or such as the set of all elementary protochannels in a given Stinespring decomposition of the operations.

5.4 ABL measures

The standard QM probability measures, of the form $p = \mathrm{Tr}[PW]$, have the obvious shortcoming that nonideal measurements and sequences of successive measurements cannot be represented by projection operators. This problem is overcome by applying EVMs and OVMs, leading to measures of the form

$$p(L, K) = \mathrm{Tr}[R(L)\phi_K(W)], \tag{5.44}$$

where R is an EVM and ϕ is an OVM. This remedy is by itself not sufficient, however, for overcoming a second obvious shortcoming, which concerns conditionalization.

The set of probability measures admitted by a well–behaved probabilistic theory should be closed under conditionalization at least on events of positive measure; i.e., the measure obtained from a theoretically possible measure by conditionalization on a given event should again be a possible measure. Accordingly, let p be a possible measure and let A be an event such that $p(A) > 0$, then the measure p_A defined by

$$p_A(B) = p(A \cap B)/p(A) \tag{5.45}$$

must also be admitted. In our present situation this implies that the set of measures of the form (5.44) is not yet sufficiently general. Measures obtained from (5.44) by conditionalization must be included. If we consider the case of a sequence of ideal Von Neumann measurements, then the elementary events in our probability space are possible system histories, represented by sequences of projectors. Including the conditional measures now amounts to admitting measures of a form introduced by Aharonov, Bergmann, and Lebowitz in 1964[7]. Accordingly, including the conditional measures obtained from (5.44) amounts to accepting generalized ABL measures, slightly updated in order to cope with the use of OVMs.

Pre–, retro–, and pretrodiction

Use of the ABL measures is crucial in some discussions, and in particular where it concerns issues like time reversal and time–symmetry. The view it gives on these issues is quite different from what is obtained by holding on to measures of the form $\mathrm{Tr}[PW]$. Starting from the measure (5.44), let us for instance conditionalize on some fixed value of L. This results in a measure

$$p(K) = \frac{\mathrm{Tr}[R(L)\phi_K(W)]}{\sum_{K \in \mathbf{K}} \mathrm{Tr}[R(L)\phi_K(W)]} = C\mathrm{Tr}[R(L)\phi_K(W)]. \tag{5.46}$$

This expression contains a symmetry which is not present in measures of the form $\mathrm{Tr}[PW]$, and which is brought out more clearly by a change of notation. Taking

[7] Aharonov *et al.* [6].

$W_{\text{in}} := W$, $W_{\text{out}} := R(L)$, and writing the operation ϕ_K in a Stinespring decomposition we get

$$\begin{aligned} p(K) &= C \sum_{V \in \mathbf{V}(K)} \text{Tr}[W_{\text{out}} V W_{\text{in}} V^+] \\ &= C \sum_{V \in \mathbf{V}(K)} \text{Tr}[W_{\text{in}} V^+ W_{\text{out}} V]. \end{aligned} \tag{5.47}$$

Expressions of this type allow us to make statistical *predictions* about the outcome of a measurement, on the basis of a known W_{in}, but they also allow making statistical *retrodictions*, on the basis of a known W_{out}. We may further coin the term *pretrodiction* as a concept covering both prediction and retrodiction, in this case on the basis of a known W_{in} and W_{out}, and it may be noted that pretrodiction is a very common affair in physics. A theory of celestial mechanics, for instance, will let us predict Jupiter's position a hundred years ago as well as to retrodict its position hundred years into the future, and knowing the position of a planet at some succession of moments will enable us to make pretrodictions about its position at intermediate times. Likewise, within classical probabilistic theories in physics there is no problem in making inferences in both directions of time. Given a set of possible probability measures on the set of possible histories of some physical process in a classical stochastic theory, one may impose boundary conditions on the process at arbitrary moments in time and use conditionalization to obtain the possible probability measures for the remaining possible histories. It is highly satisfying to find a formulation of QM in which this same procedure can be unproblematically applied.

Time–symmetry and asymmetry

Let us note that time–symmetric conditionalization as such has nothing to do with time–symmetry of the dynamics of a theory. Conditionalization on events in arbitrary time order should be possible in any probabilistic dynamical theory, simply because probability theory has no preferred time direction. However, if one mutilates a probabilistic theory by restricting conditionalization, then this may interfere with the time symmetry that may be inherent in other parts of the theory. With respect to time–symmetry in QM, the situation can be summarized by saying that, except the so–called 'state reduction', the dynamical laws of the theory seem to be essentially time symmetric, or more specifically CPT–invariant.

For the factual time–asymmetry observed in nature two different types of explanations can be nominated. One possibility is that it originates from asymmetric features in the dynamics, perhaps the state reduction process. The other possibility is that there are special boundary conditions, such as a low entropy initial state of the universe. It may also be relevant in this respect to note that one can have different ideas about the physical meaning of the operators W_{out} appearing in the above probability measures. One option is to hold on to the picture of a single density operator $W(t)$, describing a system evolving in time, and interpret W_{in} and W_{out} as manifestations of $W(t)$ at different times t_{in} and t_{out}. An alternative option is to assume that W_{in} and W_{out} are independent quantities, and that the occurrence of W_{out} represents genuine new degrees of freedom in physics. This may lead to a quite different approach to time–symmetry (see also § 7.6.1).

Observations on unknown states

There can be situations in which we have insufficient knowledge to apply the ABL measure, nor any of the other measures discussed. This situation occurs if W_{in} or W_{out} is unknown, if we do not know all magnitudes which have been measured, or if, more generally, there have been unknown interactions with the environment. Instead of having a single given probability measure we are then confronted with a set of possible probability measures. Besides the usual problem of pretrodicting unknown outcomes from a known set on the basis of a known probability, this gives rise to the problem of trying to distinguish between the alternative probability measures, which is a quite different subject. Astronomical observation provides an example of a situation where one tries to perform the latter. In this case one observes the structure of an unknown W_{in}, prepared by nature, instead of by ourselves in the laboratory. On the basis of empirical data we then try to pretrodict something about the conditions prevailing in some part of the universe some time ago, but not in the sense of assigning probabilities to the outcomes of measurements of magnitudes known to have been measured at the time.

We shall not presently pursue the subject any further, but evidently the distinction is highly relevant. We finally note that the subject is directly related to the distinction between, what Hilgevoord and Uffink call "uncertainties of the first and of the second kind", in their recent work about measures of uncertainty[8].

Literature: Aharonov *et al.* [6], Alicki and Lendi [8], Braginsky *et al.* [35], Davies [56], Davies and Lewis [57], Hannabus [108], Hilgevoord and Uffink [116], Holevo [124], Kraus [135], Lindblad [152], Ludwig [155], Von Neumann [165], Stinespring [198], Unruh [210], Wheeler [219].

[8]Hilgevoord and Uffink [116]

Chapter 6

Contextual QM

Abstract: Contextual logic is used to overcome the logical incompleteness of the standard formalism and open the way for further formalization. It will turn out that there are major conceptual differences between different ways to proceed.

The most urgent of several problematic sides of QM that have been identified in the previous chapters concerns its logical structure. The Hilbert space, often considered as the analogue of the classical phase space, is not a proper state space in the classical sense. Any considerations about other peculiarities of QM will remain adrift if we lack a properly developed view on its logical structure, because this is what determines the perspective for discussing anything else.

What else is there to be discussed? Let us briefly summarize what we have encountered so far. Besides the logical matter, there is the problem of inseparability, and the resulting choice between nonlocality and retrocausality. In the third place, the measurement process leads to the problem of the Von Neumann infinite regress. Fourthly, the indeterminism of QM is regarded as a problem by some physicists. Fifthly, nonideal measurement procedures lead to the use of effect– and operation valued measures, which, however, raises questions with respect to the definition of observables in QM. And finally, it was noted that the standard versions of the probability measures of QM are not closed under conditionalization, a difficulty which is overcome by introduction of the ABL measures, but this also has further consequences for various topics. In the coming chapters we shall further discuss these problems, trying to define them more clearly, exhibit links between them, make suggestions about possible solutions, and bring some clarity in the choices we face.

The logical problem, however, is the most pressing one. The core of it is that the standard formalism of QM is semantically incomplete, in the sense that the Hilbert space cannot offer a satisfactory meaning, in a classical sense, to statements in which values are assigned to physical magnitudes, e.g. like "$X = x$". Indeed, the notion of *completeness* can be regarded as the pivot around which several problems of QM turn. The completeness or incompleteness of QM has been a major issue in the foundations of physics for a long time. However, if one holds on, as we want to do, to classical logic and classical probability theory, then it turns out that the incompleteness of the standard formalism appears straightforwardly as a corollary to Gleason's theorem[1]. Therefore, the next major issue will be the construction of a more complete theory. Hidden variable theories aim to achieve this, but generally have additional objectives, such as determinism. Our aim will be to use contextual logic for attaining completeness while staying much closer to standard QM.

It turns out that there are a number of different ways in which this can be done, leading to a maze of different contextual quantum theories. It will be our aim to present an introductory survey of some of these views, and to sketch some of the relations with positions that have been taken by physicists in the actual debate on QM thus far. However, in view of the amount of work involved in a more detailed exploration of the landscape that will enfold before us, it will be impossible here to develop this in much detail. Accordingly, many questions will be left unanswered, or are not even raised.

Contextual quantum formalisms, a summary

In the development of contextual QM some discrete steps can be distinguished, which will also be represented in the structure of the present chapter. After a fair amount of preliminary research, a first definite result[2] was, what we shall further call, the *initial contextual quantum formalism.* If we want a state space in QM, then the Hilbert space or the set of density operators merely by themselves turn out to be insufficient; something must be added. Van Fraassen arrived at a similar conclusion in a slightly different way, and in his 'modal interpretation' of QM he makes a distinction between two kinds of propositions[3], 'state attributions' and 'value attributions'. Essentially the same is done in contextual QM, but in a different way, and in particular with a different modal logic, namely contextual logic. In the initial formalism, the problem of semantic incompleteness is solved in a straightforward way: the description of a quantum system by means of a density operator is retained, but considered incomplete. It is complemented with the specification of a *contextual state.* In this way, a larger state space is created in which, using Van Fraassen's terminology, both state attributions and value attributions receive an interpretation.

The initial formalism achieves in many respects a fair agreement with the standard density operator formalism, but its dynamics contains some simplifications which, although corresponding to assumptions customary made in standard QM, are evidently unacceptable. Therefore, the next step involved generalizing this theory by applying a more sophisticated dynamics, based in particular on OVMs and on the ABL measure[4]. Finding such a generalization was less straightforward than expected, however, as, it turned out that modifying the dynamics had profound implications for the conceptual structure of the theory, in particular owing to the fact

[1]See § 7.3 and § 7.4.

[2]Published in Hoekzema [122], most fragments reprinted in the present chapter.

[3][95] p.100.

[4]See ch. 5.

that the use of OVMs leads to quite different ideas about the definition of magnitudes in QM. After a few months of struggling the result[5] was a *theory of quantum processes.*

The conceptual structure of process theory is quite different from that of the initial formalism, and in some respects also much simpler. Its logical structure is classical, and quite transparant, and the empirical results of standard QM, including the QM of open systems, are readily recovered. There is, nevertheless, one point of concern tagged to it: the notion of a "quantum process" is contained as a primitive concept, whereas one would like to have it explained in more basic notions, such as the history of a particle evolving in time. At this point one arrives at a major crossroad, with a choice between two main options. One possibility is to find an interpretation of process theory where, indeed, a process represents the history of a quantum particle. One way of achieving this is outlined below, in § 6.3, but there are different ways of elaborating on this approach, and there are links with views expressed in the actual historical debate about QM. The alternative is to forget about particles, and find an interpretation of processes in terms of 'events'. This approach is outlined in § 6.4, and a further elaboration into a relativistic event field theory[6] is reprinted in ch. 9. The choice between a particle approach and an event approach here appears as correlated with choosing between an instantly–causal or a retrocausal view[7] in process theory. This may not be logically necessary, but I did encounter some problems in considering the prospects of a retrocausal particle theory, and similarly with those of an instantly–causal event theory.

Surveying the different possibilities, it seems likely that, although some of them may not survive closer scrutiny, many will be left that can be consistently maintained. What to do? As a convinced pluralist I think there is little objection against developing a number of views simultaneously, as is actual practice anyway, but, evidently, there is a limit to the efforts one may wish to devote to this. Moreover, there is, of course, a difference between the choices made by individual scientists and the choices one would like a society to make in this respect. In the present case my own attitude is that I am quite confident that sufficiently many physicists will prefer some version of particle theory to event theory, simply because it is more like what they are used to. My own intuitive feeling, however, is that this class of views is a swamp. If one does not mind inconveniences like wet feet or drowning one may happily stay there indefinitely, but for other people like myself, if they do not mind getting used to some unusual features, I recommend event theory as a much more agreeable ticket to physics in the next century. Therefore, if I were to continue doing physical research myself, I would feel sympathy for people dwelling the moors, but I would not be willing to consider working on any other approach.

[5]Published in Hoekzema [120], most of it reprinted in the present chapter.

[6]Published in Hoekzema [123]

[7]See ch. 4.

6.1 The initial contextual formulation of QM

An important difference between QM and classical physics is the existence of mutually incompatible magnitudes in QM. Such magnitudes cannot be simultaneously measured, and are generally not even conceived to be simultaneously sharply defined. According to Bohr's philosophy they belong to different, complementary, descriptions of the system, and such complementary descriptions are mutually exclusive in the sense that they cannot apply simultaneously. Which magnitudes are sharply defined at which moment in time is determined by the total measurement set–up, that defines a so–called 'quantum phenomenon'. The meaning of a description of a system is always related to such a quantum phenomenon, which is for Bohr the smallest independent unit of physical meaning. Although the whole idea of complementarity has in practice turned out to be conceptually difficult and confusing, at least some aspects of are readily covered within contextual logic, where the concept can be interpreted by assigning separate contexts to different phenomena. The details of a contextual QM can be elaborated in many different ways, and some of the differences relate to different versions of the measurement postulate. The initial formalism of contextual QM, which we shall now proceed to introduce, is based on a relatively simple model, matching the standard density operator formalism with the probability measure (4.4): $p_W(P) = \mathrm{Tr}[PW]$.

In this formulation of QM the description of a physical system has a layered structure. One layer concerns the description of the system by means of its density operator, as in the standard formalism. To this layer, however, we add a second and a third layer, connected with describing the system within a specific *context*, or quantum phenomenon.

In the first place, one will have to specify what the structure of a context is and which context applies at any given moment. A context $\mathbf{u}$ is conceived here as a unit in which a specific set of physical magnitudes, also called observables, is well defined. There are assumed to be many such contexts, with a different set of magnitudes defined in each of them, e.g., in one context position is well–defined, and in another one momentum. All contexts together are collected in a set Ψ. Within a context $\mathbf{u} \in \Psi$ the state of a system can be specified by the values of the magnitudes in $\mathbf{u}$. We shall take each state space $\Gamma_\mathbf{u}$ to be generated by a specific resolution of the unity operator $\mathbb{1}$ on the Hilbert space $\mathcal{H}$ of the system, i.e., each $\Gamma_\mathbf{u}$ is a set of projection operators P such that

$$\sum_{P\in\Gamma_\mathbf{u}} P = \mathbb{1} \,. \tag{6.1}$$

Different contexts are assumed to be mutually exclusive, in agreement with Bohr's concept of complementarity, and the second layer of description of a system concerns the specification of which context applies. In fact, ordering the states in this way amounts to taking the disjoint union ($\uplus_i A_i = \cup_i\{< A_i, a > \mid a \in A_i\}$) of all possible contexts.

Further, the third layer in the description of a quantum system concerns the state of the system within the specified context, i.e., the assignment of values to the magnitudes which are well–defined in the particular context which applies at the given moment. Because we are working within a logic of the classical type, we assume that the magnitudes which are well defined *have in fact* a sharp value, independent of whether we *know* this value.

The three layers together are surveyed within a 'quantum context' **qm**, for which we construct the state space

$$\begin{aligned} \Gamma_{\mathbf{qm}} &= \{< \Gamma_{\mathbf{u}}, P, W > \mid \mathbf{u} \in \Psi, P \in \Gamma_{\mathbf{u}}, W \in \mathcal{D}\} \\ &= \bigcup_{\mathbf{u}\in\Psi} (\{\Gamma_{\mathbf{u}}\} \times \Gamma_{\mathbf{u}}) \times \mathcal{D}, \end{aligned} \tag{6.2}$$

where $\mathcal{D}$ is the set of density operators on $\mathcal{H}$. This set $\Gamma_{\mathbf{qm}}$ can properly serve as a state space for QM, in a classical sense. Accordingly, an elementary quantum state is a triple $Q =< \Gamma_{\mathbf{u}}, P, W >$, consisting of the specification of a context, a contextual state, and a density operator, and at any moment a system is assumed to be in a state of this form.

Note: The density operator W is conceived here as a fundamental state variable, and not as a statistical mixture. The reason for using density operators rather than Hilbert space vectors here is that interactions between different systems may generate improper mixtures[8], which cannot be interpreted as statistical mixtures in a classical sense. We shall later see, however, that the creation of improper mixtures can be avoided in a theory in which retrocausality is admitted, and indeed this will make it possible to express states in terms of Hilbert space vectors.

6.1.1 Dynamics

The dynamics of the theory should provide a theory about the possible trajectories in $\Gamma_{\mathbf{qm}}$ in much the same way in which classical mechanics provides a theory about the possible trajectories in phase space. In order to formulate such a theory we can use in the first place the dynamical equations of standard QM, notably the Schrödinger equation, which apply to the evolution of the density operator at times the system is dynamically isolated. We also need further assumptions, however, about the evolution of the contextual part of the quantum state. We shall here assume that the context applying to a system at all times when it is dynamically isolated, i.e., when it does not interact with other systems, is the trivial context $\mathbf{u}_{1\!\!1}$ with

$$\Gamma_{1\!\!1} = \{1\!\!1\}. \tag{6.3}$$

This assumption is a minimal one. As we shall see later, there is much freedom here to enter alternative ideas. The motivation for choosing the above assumption in the present case comes from the idea that the contextual properties of a system represent interactions of the system with other systems. At times when the system is dynamically isolated there is no such interaction. Assuming the Schrödinger picture,

[8]See ch. 4.

the evolution of the quantum state $Q(t)$ during the time of dynamical isolation is given by

$$\begin{aligned} Q(t) &= < \{1\!\!1\}, 1\!\!1, W(t) > \\ \dot{W}(t) &= -i[H, W(t)]. \end{aligned} \tag{6.4}$$

Now let us proceed by considering the evolution of the system at times when it is not isolated.

With the present formalism we have two objectives. In the first place, we want a theory with a classical structure, with systems having, in a classical sense, whatever properties they are supposed to have at a given time. In the second place, we want to stay as closely as possible to the standard density operator formalism. In order to achieve this we shall make two simplifications with respect to the dynamics of the theory at times when it is not dynamically isolated. In order to make sure that contextual properties can indeed be represented by projection operators we restrict the applicability of the formalism to the case of ideal measurements, and assume that every interaction of the system with another system is an ideal Von Neumann measurement, in agreement with (4.19). Further, in order to be able to assign a definite quantum state to a system at all times, we assume that all interactions take place instantaneously, i.e., they are of infinitesimal duration. In this way we evade the question of what state to assign to the system during the interaction process.

With a specific interaction we now associate a nontrivial context $\mathbf{u}$ with $\Gamma_{\mathbf{u}} = \{P_1, \cdots, P_N\}$, where the operators P_i are the projection operators corresponding to the possible measurement outcomes. Corresponding to this nontrivial context $\mathbf{u}$ there is a bifurcation in the evolution of the quantum state Q into N different possible trajectories, i.e., at the time of measurement t_m the quantum state bifurcates, according to

$$\begin{aligned} Q(t_m - \epsilon) &= < \{1\!\!1\}, 1\!\!1, W > \\ &\rightarrow \left\{ \begin{array}{rcl} Q_1(t_m) &=& < \Gamma_{\mathbf{u}}, P_1, P_1 W P_1 / \mathrm{Tr}[P_1 W] > \\ &\vdots& \\ Q_N(t_m) &=& < \Gamma_{\mathbf{u}}, P_N, P_N W P_N / \mathrm{Tr}[P_N W] > \end{array} \right. \end{aligned} \tag{6.5}$$

and after the measurement, when the system is again dynamically isolated, we get one of the states

$$Q_k(t_m + \epsilon) =< \{1\!\!1\}, 1\!\!1, P_k W P_k / \mathrm{Tr}[P_k W] > . \tag{6.6}$$

Such a bifurcation expresses a dynamical nondeterminism: after the measurement one of the states Q_k will in fact occur but we cannot say which one. QM does assign probabilities to the branches, however. Given the density operator W and given the context $\mathbf{u}$ the probability for Q_k to be realized is equal to

$$p_{\mathbf{u},W}(P_k) = \mathrm{Tr}[P_k W], \quad \text{if } P_k \in \Gamma_{\mathbf{u}}. \tag{6.7}$$

We thus obtain a theory which matches the standard density operator formalism in the sense that the density operator is part of the state description of a system, values of observables are represented by projection operators, and the probabilities are equal to the standard ones. Further, the theory has a classical logical structure in the sense that the state description in terms of elements of $\Gamma_{\mathbf{qm}}$ can be assumed to be exhaustive, and some state of this form applies to the system at all times, regardless of what we do or do not *know* about the system.

6.1.2 Deficiencies of the initial approach

The initial formalism is based on two quite unacceptable simplifications, and, therefore, it must be replaced by something more general. OVMs can be applied to achieve this, but this will turn out to have major consequences for the conceptual structure of the theory. Moreover, from our analysis in ch. 5 it follows that the probability measure of the standard formalism is not generally adequate and must be replaced by the ABL measure. This, too, will have consequences for the conceptual structure of contextual QM.

The introduction of OVMs leads to two kinds of difficulties, related to dropping the two simplifications on which the initial approach was based. If we drop the restriction to ideal measurements in order to accommodate more general interactions of the system with its environment, this raises the problem that the contextual states can no longer be represented by projections. A conceptually important consequence of this will be that the standard representation of physical observables by means of Hermitian operators is no longer generally adequate[9]. Further, dropping the assumption of instantaneous interactions, admitting interactions of finite duration, leads to the question of which state to assign to a system during an interaction. It then appears that the contextual states in the initial formalism can best be interpreted as referring to a *process*, rather than to the state of a system at any particular moment. The assumption of instantaneous interactions enabled us to suppress this problem, but this is difficult to maintain in a more definite theory.

6.2 Quantum process theory

Quantum processes, like the quantum systems in more customary views, are described by assigning a state to them. Because a process is an object in spacetime, rather than an object in space evolving in time, this state does not change with time. States are used for distinguishing processes; they are not dynamical entities in the usual sense.

The central part in the design of quantum process theory will be the construction of its state space. A state description is essentially nothing more than a list of values of parameters needed to characterize a process. In our case, we shall need *operator valued* parameters and *operation valued* parameters, and more specifically a process state will be assumed to be described by a triple

$$\Phi = < W_{\text{out}}, \phi, W_{\text{in}} >, \tag{6.8}$$

[9]See also § 7.3.

where $W_{\rm in}$ and $W_{\rm out}$ are positive operators and ϕ is an operation. For each triple of this kind we further define a real number $L(\Phi)$:

$$L(\Phi) = \mathrm{Tr}[W_{\rm out}\phi(W_{\rm in})], \tag{6.9}$$

and we require $L(\Phi)$ to be finite. We thus arrive at a state space of processes given by

$$\Gamma_{\mathbf{proc}} = \{\Phi |\ L(\Phi) < \infty\}. \tag{6.10}$$

QM will be conceived to be a statistical theory about the occurrence, under sufficiently well-specified conditions, of such processes in nature. In order to obtain such a theory we proceed in two steps. In the first place, we shall think of processes as being arranged in patterns. For a specific experimental set-up there is a corresponding set of possible alternative processes which may be realized in the experiment. Such a set of alternatives will be taken to define a *context* $\mathbf{u}$, with a contextual state space $\Gamma_{\mathbf{u}} \subseteq \Gamma_{\mathbf{proc}}$ containing precisely this set of alternatives. For simplicity we shall here assume $\Gamma_{\mathbf{u}}$ to be countable. We may now define an elementary *quantum state* for a process $\mathcal{P}$ as a pair $Q =< \Gamma_{\mathbf{u}}, \Phi >$, with $\Phi \in \Gamma_{\mathbf{u}}$, and the quantum state space $\Gamma_{\mathbf{qm}}$ as the set of all pairs considered to be physically possible. These quantum states are conceptually comparable to te quantum states of the initial formalism.

> **Note:** This approach leads to a theory in which the operators $W_{\rm in}$ and $W_{\rm out}$ are included in the definition of the contextual states. This makes the link with the initial formalism less direct, but it is more convenient for our further analysis. From the point of view of quantum process theory, the difference is marginal. What matters is merely that all parameters relevant to the description of a process are represented in the quantum state. Precisely how they are arranged is irrelevant at this level.

6.2.1 Probability measures

The second step we take is that, in order to obtain QM as a statistical theory, we assume that within a given context a process Φ has a probability proportional to $L(\Phi)$ of being realized. In view of the experience with standard QM, this seems a natural assumption, and it leads in the first instance to a probability measure

$$p_{\mathbf{u}}(\Phi) = C_{\mathbf{u}} L(\Phi), \tag{6.11}$$

where $C_{\mathbf{u}}$ is a normalization constant. This measure produces a quite straightforward relation between process theory and the empirical content of standard QM.

> **Example:** Imagine a preparation apparatus producing a state $W_{\rm in}$, let ϕ be a unitary evolution, and let $\Gamma_{\mathbf{u}}^{\rm out}$ be a set of out-states representing the possible measurement results of some given apparatus. Combined into a context, this gives
>
> $$\Gamma_{\mathbf{u}} = \Gamma_{\mathbf{u}}^{\rm out} \times \{\phi\} \times \{W_{\rm in}\}, \tag{6.12}$$

and we find probabilities given by

$$p_{\mathbf{u}}(W_{\text{out}}, \phi, W_{\text{in}}) = p_{\mathbf{u}}(W_{\text{out}}) = C_{\mathbf{u}}\text{Tr}[W_{\text{out}}\phi(W_{\text{in}})], \tag{6.13}$$

where $C_{\mathbf{u}}$ may be included in a normalization of W_{in}. This is precisely the result given by standard QM.

A further statistical weight factor w may now be added to the above measure for representing manipulations with the probability space, such as by selecting events in fixed fractions. For instance, one may use a measurement apparatus producing either of two states W_{in} or W'_{in} in fixed ratios. For dealing with a case like this we must introduce a method of mixing quantum probabilities and probabilities of a different origin. There are two options. The first possibility is to distinguish two different contexts $\mathbf{u}$ and $\mathbf{v}$, and compute probabilities

$$\begin{aligned} p_{\mathbf{u}\text{ or }\mathbf{v}}(\Phi) &= p_{\mathbf{u}}(\Phi)p(\mathbf{u}) + p_{\mathbf{v}}(\Phi)p(\mathbf{v}) \\ &= C_{\mathbf{u}}L(\Phi)\chi_{\mathbf{u}}(\Phi)p(\mathbf{u}) + C_{\mathbf{v}}L(\Phi)\chi_{\mathbf{v}}(\Phi)p(\mathbf{v}), \end{aligned} \tag{6.14}$$

where $\chi_{\mathbf{u}}(\Phi) = 1$ if $\Phi \in \Gamma_{\mathbf{u}}$ or 0 otherwise.

The equivalent, but rather more convenient alternative is to retain a single context, introduce a statistical weight function w, and let probabilities be defined by the probability measure:

$$p_{\mathbf{u},w}(\Phi) = C_{\mathbf{u},w}\, L(\Phi)w(\Phi). \tag{6.15}$$

The above two states W_{in} and W'_{in} may now be given weights $w(W_{\text{in}})$ and $w(W'_{\text{in}})$ in order to manipulate the probabilities $p_{\mathbf{u},w}(W_{\text{in}})$ and $p_{\mathbf{u},w}(W'_{\text{in}})$ into the correct ratio.

Example: Suppose we know for certain the final outcome W_{out} of an experiment, suppose that there are two possible intermediate measurement result, represented by operations ϕ_1 and ϕ_2, and let there be two possible preparation states $W_{\text{in},1}$ and $W_{\text{in},2}$. This yields

$$\begin{aligned} \Gamma_{\mathbf{u}} = \{ &< W_{\text{out}}, \phi_1, W_{\text{in},1} >, < W_{\text{out}}, \phi_2, W_{\text{in},1} >, \\ &< W_{\text{out}}, \phi_1, W_{\text{in},2} >, < W_{\text{out}}, \phi_2, W_{\text{in},2} > \}. \end{aligned} \tag{6.16}$$

Suppose, furthermore, that we are 80% sure that $W_{\text{in},1}$ is the correct in-state. The statistical weights w may then be chosen in such a way that $p_{\mathbf{u},w}(W_{\text{in},1}) = .8$ and $p_{\mathbf{u},w}(W_{\text{in},2}) = .2$. We take

$$\begin{aligned} &w(W_{\text{out}}, \phi_i, W_{\text{in},j}) = w(W_{\text{in},j}) \\ &w(W_{\text{in},1}) = .8/C_{\mathbf{u},w}\Big(L(W_{\text{out}}, \phi_1, W_{\text{in},1}) + L(W_{\text{out}}, \phi_2, W_{\text{in},1})\Big). \\ &w(W_{\text{in},2}) = .2/C_{\mathbf{u},w}\Big(L(W_{\text{out}}, \phi_1, W_{\text{in},1}) + L(W_{\text{out}}, \phi_2, W_{\text{in},1})\Big). \end{aligned} \tag{6.17}$$

For $p_{\mathbf{u},w}(\phi_i)$ this gives

$$\begin{aligned} p_{\mathbf{u},w}(\phi_1) &= p_{\mathbf{u},w}(\phi_1, W_{\text{in},1}) + p_{\mathbf{u},w}(\phi_1, W_{\text{in},2}) & (6.18)\\ &= .8 \frac{\text{Tr}[(W_{\text{out}}\phi_1(W_{\text{in},1})]}{\text{Tr}[W_{\text{out}}\phi_1(W_{\text{in},1}) + W_{\text{out}}\phi_2(W_{\text{in},1})]} \\ &\quad + .2 \frac{\text{Tr}[(W_{\text{out}}\phi_1(W_{\text{in},1})]}{\text{Tr}[W_{\text{out}}\phi_1(W_{\text{in},1}) + W_{\text{out}}\phi_2(W_{\text{in},1})]} & (6.19)\end{aligned}$$

and $p_{\mathbf{u},w}(\phi_2) = 1 - p_{\mathbf{u},w}(\phi_1)$.

Example: A simple experiment in QM may consist of a preparation, represented by a density operator W, immediately followed by a measurement, represented by an EVM R on an index set $\mathbf{K}$, and the quantum probability of finding the value $k \in \mathbf{K}$ is given by $\text{Tr}[R(k)W]$. In terms of the above process theory this is expressed by taking a set of states

$$\Phi_k =< W_{\text{out},k}, \phi, W_{\text{in}} >,$$

with $W_{\text{in}} = W$, $\phi = \phi_{1\!\!1}$, $W_{\text{out},k} = R(k)$, producing a state space

$$\Gamma_{\mathbf{u}} = \{< R(k), \phi_{1\!\!1}, W > \mid k \in \mathbf{K}\}. \tag{6.20}$$

Lacking further information about the outcome k, w is assumed constant, resulting in a probability measure

$$p_{\mathbf{u},w}(\Phi_k) = C_{\mathbf{u},w} L(\Phi_k) = \text{Tr}[R(k)W], \tag{6.21}$$

as usual.

Example: In the preceding example, consider the case that there is the choice between two different modes of the preparation apparatus, represented by the two density operators W_1 and W_2, but we do not know which one was used in a particular experiment. Still assuming w constant now leads to

$$p_{\mathbf{u},w}(\Phi_{ki}) = C_{\mathbf{u},w} L(\Phi_{ki}) = C_{\mathbf{u},w}\text{Tr}[R(k)W_i], \tag{6.22}$$

with $\Phi_{ki} =< R(k), \phi_{1\!\!1}, W_i >$ and

$$C_{\mathbf{u},w}^{-1} = \sum_{k,i} L(\Phi_{ki}) = \sum_i \text{Tr}[W_i] = 2,$$

for unit trace density operators.

Example: In the preceding example, suppose there is a further choice between two different modes of the measurement apparatus, represented by two EVMs R_1 on $\mathbf{K}_1$ and R_2 on $\mathbf{K}_2$. This gives a further expansion of the contextual state space into

$$\Gamma_{\mathbf{u}} = \{\Phi_{kji} \mid i, j \in \{1, 2\}, k \in \mathbf{K}_j\}. \tag{6.23}$$

For constant w we simply get $p_{\mathrm{u},w}(\Phi_{kji}) = \frac{1}{4}L(\Phi_{kji})$, but let us also look at a more complicated case. Suppose that we have the information that in 80% of the runs of the experiment in which preparation mode 1 is used, the measurement apparatus was also operating in mode 1. In this case the correlation between the apparatus modes can be expressed by taking

$$w(\Phi_{kji}) = .4\ \delta_{1j}\delta_{1i} + .1\ \delta_{2j}\delta_{1i} + .25\ \delta_{1j}\delta_{2i} + .25\ \delta_{2j}\delta_{2i}. \tag{6.24}$$

6.2.2 Dynamics

From the point of view of process theory, dynamics becomes a matter of breaking long processes into a chain of shorter parts, or *vice versa* of combining such parts into complexes. We shall call these parts the *phases* of a process, i.e., we have a *preparation phase* and a *retroparation phase*, both represented by positive operators, and an *evolution phase*, represented by an operation. We may think of these phases as building blocks which can be fitted to one another. For instance, let $W_{\mathrm{in}}(t_1)$ be a preparation and let $\phi_{(t_2,t_1)}$ be an evolution, then the pair $< \phi_{(t_2,t_1)}, W_{\mathrm{in}}(t_1) >$ can be taken to represent another preparation, with the associated positive operator

$$W_{\mathrm{in}}(t_2) = \phi_{(t_2,t_1)}(W_{\mathrm{in}}(t_1)). \tag{6.25}$$

Likewise, a pair $< W_{\mathrm{out}}(t_2), \phi_{(t_2,t_1)} >$ represents another retroparation, with the associated positive operator

$$W_{\mathrm{out}}(t_1) = \phi^{+}_{(t_2,t_1)}(W_{\mathrm{out}}(t_2)). \tag{6.26}$$

Further, a pair $< W_{\mathrm{out}}(t_1), W_{\mathrm{in}}(t_1) >$, being the combination of only a preparation and a retroparation, can be considered as being equivalent to the triple $< W_{\mathrm{out}}(t_1), \phi_{\mathit{1\!l}}, W_{\mathrm{in}}(t_1) >$, representing a complete process, with the associated real number

$$L(W_{\mathrm{out}}(t_1), \phi_{\mathit{1\!l}}, W_{\mathrm{in}}(t_1)) = \mathrm{Tr}[W_{\mathrm{out}}(t_1)W_{\mathrm{in}}(t_1)]. \tag{6.27}$$

We can now proceed to build complexes. Combining two preparations $W^A_{\mathrm{in}}(t_1)$ and $W^B_{\mathrm{in}}(t_1)$ gives a composite preparation $< W^A_{\mathrm{in}}(t_1), W^B_{\mathrm{in}}(t_1) >$ with an associated positive operator

$$W^{AB}_{\mathrm{in}}(t_1) = W^A_{\mathrm{in}}(t_1) \otimes W^B_{\mathrm{in}}(t_1). \tag{6.28}$$

Likewise the pair $< W^A_{\mathrm{out}}(t_1), W^B_{\mathrm{out}}(t_1) >$ has an associated operator

$$W^{AB}_{\mathrm{out}}(t_1) = W^A_{\mathrm{out}}(t_1) \otimes W^B_{\mathrm{out}}(t_1). \tag{6.29}$$

Further, the combination of two evolutions into a pair $< \phi_{(t_3,t_2)}, \phi_{(t_2,t_1)} >$ is again an evolution, with an associated operation

$$\phi_{(t_3,t_1)} = \phi_{(t_3,t_2)} \circ \phi_{(t_2,t_1)}. \tag{6.30}$$

Using these rules, one can fit elementary phases together to build processes of arbitrary degrees of complexity.

Example: The tuple

$$\Phi = \; << W^A_{\rm out}(t_3), W^B_{\rm out}(t_3) >, \phi^{AB}_{(t_3,t_2)}, << \phi^A_{(t_2,t_1)}, W^A_{\rm in}(t_1) >, W^B(t_2) >>$$

represents a complete process, with an associated L–value of

$$L(\Phi) = \mathrm{Tr}\Big[\{W^A_{\rm out}(t_3) \otimes W^B_{\rm out}(t_4)\}\phi^{AB}_{(t_3,t_2)}(\phi^A_{(t_2,t_1)}(W^A_{\rm in}(t_1)) \otimes W^B(t_2))\Big].$$

Instead of building processes, let us conversely consider the idea of breaking a process into ever smaller pieces. A question of considerable interest is whether it is always possible to split a given evolution into phases of arbitrary short duration. In the special case of an *isolated process* this question is easily answered with *yes*. We define an isolated process as a process in which the evolution is represented by an operation with a unique Stinespring decomposition of the form

$$\phi(X) = e^{-iHt} X e^{iHt},$$

where H is the Hamilton operator associated with the process. Evolutions of this type can always be split into different phases, because the operators $U(t)$ form a one–parameter group, satisfying $U(t+t') = U(t)U(t')$. Therefore, one also has

$$\phi_{(t_2,t_1)} = \phi_{(t_2,t)} \circ \phi_{(t,t_1)}, \tag{6.31}$$

for arbitrary $t_1 \leq t \leq t_2$.

For a nonisolated process, however, the question is considerably more difficult to answer. non–isolated processes are subprocesses of composite processes. Let $\mathcal{P}^C$ be a composite process with a Hilbert space $\mathcal{H}^C = \mathcal{H}^A \otimes \mathcal{H}$. A tuple $< W^A_{\rm out}(t_2), \phi^C_{(t_2,t_1)}, W^A_{\rm in}(t_1) >$ now represents an evolution for a process $\mathcal{P}$ on $\mathcal{H}$ alone, with its associated operation given by

$$\phi_{(t_2,t_1)}(X) = \mathrm{Tr}^A[(W^A_{\rm out}(t_2) \otimes 1\!\!1\,)\phi^C_{(t_2,t_1)}(W^A_{\rm in}(t_1) \otimes X)]. \tag{6.32}$$

For an operation of this type it is by no means evident that for all $t_1 \leq t \leq t_2$ a plausible decomposition into successive phases would exist. The problem is important because it relates to the question of what kind of thing a process is. If we want to interpret a process as the history of a system, then presumably a composite process represents the history of several systems. This raises the question what the state of each system is, before and after, but also *during* their interaction. If the evolution of the presumed system $\mathcal{S}$ cannot be decomposed into phases, however, then assigning a state to it becomes quite problematic.

There are two quite different approaches to this problem. One approach is to find a way to assign states to systems at all times. This can be accomplished, but at the cost of various complications, such as a dynamics which is asymmetric under time reversal and in which the 'reduction of the wave packet' appears as a physically real process. The other approach is to make a theory without systems evolving in time. This idea will be elaborated in the form of a 'quantum event theory'.

Time–symmetry

With respect to time–symmetry in process theory, it may be noted that one may formulate two different versions of its dynamics. In agreement with the distinction made in § 4.2.5, one may choose between an instantly–causal, time–asymmetric, and a retrocausal, time–symmetric view. Considering the operation $\phi_{(t_2,t_1)}$ defined by (6.32), it is clear that it depends explicitly on $W^A_{\rm out}(t_2)$. Now if one takes

$$W^A_{\rm out}(t_2) = U^{A+}(t_3 - t_2)W^A_{\rm out}(t_3)U^A(t_3 - t_2), \tag{6.33}$$

then this dependence represents a direct form of retrocausality, because it makes the evolution at time t_2 of the process $\mathcal{P}$ dependent on what happens in the process $\mathcal{P}^A$ at a, possibly much later time t_3. However, this expression for the $W^A_{\rm out}(t_2)$ is based on a time–symmetric dynamics in which the operators $W_{\rm out}(t)$ evolve backward in time similar to the way in which the operators $W_{\rm in}(t)$ evolve forward in time.

Alternatively, one may adopt an asymmmetric view on the dynamics, in which $W_{\rm in}$ and $W_{\rm out}$ both evolve forward in time via a dynamics which includes state reduction phases. In this case, prior to a measurement being made in the process $\mathcal{P}^A$, the operator $W^A_{\rm out}(t_2)$ is given by

$$W^A_{\rm out}(t_2) = 1\!\!1\ , \tag{6.34}$$

rather than by (6.33), and it can obtain nontrivial value only by virtue of a state reduction, following a measurement being performed.

Whereas, in the time–symmetric view, $W_{\rm out}$ represents independent degrees of freedom, in the asymmetric case it is natural to regard it as an artefact which may be defined in other terms. After all, the reason for introducing it originated in the introduction of the ABL measures. This had nothing to do with the dynamics of QM, however, but only with the absence of an *a priori* arrow of time in probability theory. Nothing compels us to regard $W_{\rm out}$ as an independent state variable or to hold on to a time–symmetric dynamics within QM. Therefore, in the asymmetric view it is natural to define the fundamental process states as pairs

$$\Phi(t_2, t_1) \;= < \phi_{(t_2,t_1)}, W(t_1) >, \tag{6.35}$$

rather than the as the triples (6.8). For such a pair an associated L–function may be defined by

$$L(\phi_{(t_2,t_1)}, W(t_1)) = {\rm Tr}[\phi_{(t_2,t_1)}(W(t_1))], \tag{6.36}$$

and in terms of the earlier definition (6.9), this value of L is obtained by taking $W_{\rm out} = 1\!\!1$. We now note that for arbitrary $t_1 \leq t \leq t_2$ one may write

$$L(1\!\!1\ , \phi_{(t_2,t_1)}, W(t_1)) = L(W_{\rm out}(t), \phi_{(t,t_1)}, W(t_1)), \tag{6.37}$$

if one defines

$$W_{\rm out}(t) = \phi^+_{(t_2,t)}(1\!\!1\ ). \tag{6.38}$$

This, indeed, yields the possibility of defining $W_{\rm out}$ as an artefact, instead of an independent state variable, while retaining equivalence with earlier results.

6.2.3 Statistical mixtures

Within the present formalism triples of the form $\Phi = < W_{\rm out}, \phi, W_{\rm in} >$ are assumed to provide a fundamental state description, where $W_{\rm in}$, $W_{\rm out}$, and ϕ are not to be interpreted as statistical mixtures. Statistics enters the scene only by picking a context and a weight function and by interpreting $p_{\mathbf{u},w}(\Phi) = C_{\mathbf{u},w}L(\Phi)w(\Phi)$ as a probability. The probability measure $p_{\mathbf{u},w}$ defines a statistical mixture in the proper sense, comparable for instance to a classical ensemble. In order to get a better view on the relation of process theory with the initial formalism, and via the initial formalism with standard QM, we shall proceed by looking for a representation of statistical mixtures in terms of *statistically mixed process states.* The basic idea is similar to that of representing mixed quantum states by density operators in standard QM, but the details are somewhat different.

Whatever representation we choose for representing statistical mixtures, an addition and a scalar multiplication must be defined for the resulting statistical states, because we must be able to express that mixtures can be convex sums of other mixtures. We shall discuss two different ways of representing mixtures. The first one is based on a representation of a statistical state $\bar{\Phi}$ as a triple $< \bar{W}_{\rm out}, \bar{\phi}, \bar{W}_{\rm in} >$. This approach will serve well for discussing the relation between process theory and the initial formalism, but for the rest it is rather limited in scope because a convex sum of such triples is usually not again a triple of this kind. The second approach is based on an operator representation of mixed states.

In the first approach, a statistical state $\bar{\Phi}$ is defined by a triple $< \bar{W}_{\rm out}, \bar{\phi}, \bar{W}_{\rm in} >$. On the set of such triples we define a positive real scalar multiplication, which is further defined to be trilinear in the sense that

$$\begin{aligned} r < \bar{W}_{\rm out}, \bar{\phi}, \bar{W}_{\rm in} > &= < r\bar{W}_{\rm out}, \bar{\phi}, \bar{W}_{\rm in} > \\ &= < \bar{W}_{\rm out}, r\bar{\phi}, \bar{W}_{\rm in} > \\ &= < \bar{W}_{\rm out}, \bar{\phi}, r\bar{W}_{\rm in} > . \end{aligned} \tag{6.39}$$

As a consequence of this trilinearity the L-function is linear under the scalar multiplication. We have

$$L(r\bar{\Phi}) = \mathrm{Tr}[r\bar{W}_{\rm out}\bar{\phi}(\bar{W}_{\rm in})] = rL(\bar{\Phi}). \tag{6.40}$$

Let $\Gamma_{\mathbf{u}}$ be a set of processes and w a statistical weight function on these processes. To every $\Phi \in \Gamma_{\mathbf{u}}$ we may now construct its associated *pure* statistical state $\bar{\Phi}_{\mathbf{u},w}(\Phi)$ given by

$$\bar{\Phi}_{\mathbf{u},w}(\Phi) = C_{\mathbf{u},w}w(\Phi)\Phi. \tag{6.41}$$

In order to obtain *mixed* states we need additivity. Within the present approach this is not entirely trivial, but some result can be achieved by making the restriction to the special case that $\Gamma_{\mathbf{u}}$ and w factorize according to

$$\begin{aligned} \Gamma_{\mathbf{u}} &= \Gamma_{\mathbf{u},\rm out} \times \Gamma_{\mathbf{u},\rm ev} \times \Gamma_{\mathbf{u},\rm in} \\ w(W_{\rm out}, \phi, W_{\rm in}) &= w_{\rm out}(W_{\rm out})w_{\rm ev}(\phi)w_{\rm in}(W_{\rm in}). \end{aligned} \tag{6.42}$$

Let $C \subseteq \Gamma_{\mathbf{u}}$ be a set of states which factorizes in the same way into $C = C_{\text{out}} \times C_{\text{ev}} \times C_{\text{in}}$; in accordance with our terminology in ch. 5 it will be called a channel. For such a state one may define an associated statistically mixed state $\bar{\Phi}_{\mathbf{u},w}(C) = C_{\mathbf{u},w} < \bar{W}_{\text{out}}, \bar{\phi}, \bar{W}_{\text{in}} >$, with

$$\begin{aligned} \bar{W}_{\text{out}} &= \sum_{W_{\text{out}} \in C_{\text{out}}} w_{\text{out}}(W_{\text{out}}) W_{\text{out}} \\ \bar{\phi} &= \sum_{\phi \in C_{\text{ev}}} w_{\text{ev}}(\phi)\phi \\ \bar{W}_{\text{in}} &= \sum_{W_{\text{in}} \in C_{\text{in}}} w_{\text{in}}(W_{\text{in}}) W_{\text{in}}. \end{aligned} \tag{6.43}$$

The probability measure can now be written in the following form:

$$p_{\mathbf{u},w}(C) = L(\bar{\Phi}_{\mathbf{u},w}(C)). \tag{6.44}$$

Now suppose that C is the union of a number of channels C_i. If $\bar{\Phi}_{\mathbf{u},w}(C)$ were the sum of the states $\bar{\Phi}_{\mathbf{u},w}(C_i)$, the probabilities would add up correctly. This can be arranged by defining a summation for mixed process states in the following manner:

$$\begin{aligned} \bar{W}_{\text{out}}(C) &= \sum_{\bar{W}_{\text{out}} \in \cup_i \{\bar{W}_{\text{out}}(C_i)\}} \bar{W}_{\text{out}} \\ \bar{\phi}(C) &= \sum_{\bar{\phi} \in \cup_i \{\bar{\phi}(C_i)\}} \bar{\phi} \\ \bar{W}_{\text{in}}(C) &= \sum_{\bar{W}_{\text{in}} \in \cup_i \{\bar{W}_{\text{in}}(C_i)\}} \bar{W}_{\text{in}}. \end{aligned} \tag{6.45}$$

It will turn out that the above representation of mixed states is quite convenient for establishing a relation with the initial formalism. For more general purposes, however, it has serious disadvantages. In the first place, it only works if (6.42) is satisfied, and, further, the above addition only works well for channels. The set of channels is not closed under the set theoretic operations of union and intersection, however, i.e., the union of two sets $A \times B$ and $K \times L$ is not necessarily factorizable itself into $P \times Q$. Therefore, additivity holds only on a domain which does not have the algebraic properties one would like it to have.

State operators and process valued measures

A better representation of mixed states can be obtained by representing processes by operators. To this end we note that not only W_{in} and W_{out}, but also the operation ϕ can be represented as an operator. Being linear operators on $\mathcal{H}$, or equivalently on $\mathcal{H}^+$, W_{in} and W_{out} can also be represented as *vectors* on $\mathcal{H} \otimes \mathcal{H}^+$ or $\mathcal{H}^+ \otimes \mathcal{H}$. An operation may now be regarded as a linear operator

$$\phi : \mathcal{H} \otimes \mathcal{H}^+ \to \mathcal{H} \otimes \mathcal{H}^+, \tag{6.46}$$

which is itself a vector in $\mathcal{H}^+ \otimes \mathcal{H} \otimes \mathcal{H} \otimes \mathcal{H}^+$. Using a Stinespring decomposition of ϕ one may, in fact, represent the number $L(W_{\text{out}}, \phi, W_{\text{in}})$ in a slightly unconventional notation as

$$L(W_{\text{out}}, \phi, W_{\text{in}}) = (W_{\text{out}}| \sum_V V \otimes V^+ |W_{\text{in}}), \tag{6.47}$$

where $|W_{\text{in}})$ and $(W_{\text{out}}|$ are vectors on $\mathcal{H} \otimes \mathcal{H}^+$ and $\mathcal{H}^+ \otimes \mathcal{H}$ respectively. Using the vector or operator representations of W_{out}, ϕ and W_{in}, the process given by the triple $\Phi =< W_{\text{out}}, \phi, W_{\text{in}} >$ can as a whole be represented either by a vector on $(\mathcal{H} \otimes \mathcal{H}^+)^4$, or by a linear operator

$$\bar{\Phi} : \mathcal{H}^+ \otimes \mathcal{H} \otimes \mathcal{H}^+ \otimes \mathcal{H} \to \mathcal{H}^+ \otimes \mathcal{H} \otimes \mathcal{H}^+ \otimes \mathcal{H}. \tag{6.48}$$

For either representation one has the standard addition and scalar multiplication, and it is easily verified that this results in a correct representation of statistical mixtures.

For an arbitrary subset of $K \in \Gamma_{\mathbf{u}}$ one may now define its associated statistical state $\bar{\Phi}_{\mathbf{u},w}(K)$, given by

$$\bar{\Phi}_{\mathbf{u},w}(K) = C_{\mathbf{u},w} \sum_{\Phi \in K} w(\Phi) \bar{\Phi}. \tag{6.49}$$

The function $\bar{\Phi}_{\mathbf{u},w}$ is a *process valued measure*, or PVM, on the subsets of $\Gamma_{\mathbf{u}}$, and the associated real function $L \circ \bar{\Phi}_{\mathbf{u},w}$ is a probability measure. In principle, PVMs may also be used for generalizing the theory for uncountable $\Gamma_{\mathbf{u}}$. For a given $\Gamma_{\mathbf{u}}$ this would require the construction of an algebra of measurable subsets $\mathcal{F}_{\mathbf{u}}$, and of a suitable real valued weight measure w on $\Gamma_{\mathbf{u}}$ such that the above definition (6.49) can be replaced by

$$\bar{\Phi}_{\mathbf{u},w}(K) = C_{\mathbf{u},w} \int_{\Phi \in K} \bar{\Phi} dw(\Phi). \tag{6.50}$$

6.2.4 The relation with the initial approach

Isolated processes can be split into phases at arbitrary moments in time. Therefore, we may also divide such a process into a sequence of very thin slices, bounded by moments t_i and $t_i + \Delta t$, and to a sequence of successive slices we can then assign a sequence of process states

$$\begin{aligned} \Phi(t_i + \Delta t, t_i) &= < W_{\text{out}}(t_i + \Delta t), \phi_{\Delta t}, W_{\text{in}}(t_i) > \qquad (6.51) \\ &\text{with} \\ \phi_{\Delta t}(X) &= U(\Delta t) X U^+(\Delta t) \\ W_{\text{in}}(t_{i+1}) &= \phi_{\Delta t}(W_{\text{in}}(t_i)) \\ W_{\text{out}}(t_i) &= \phi^+_{\Delta t}(W_{\text{out}}(t_{i+1})). \end{aligned}$$

These slices will provide the basis for establishing a relation with the initial formalism.

In order to accommodate nonisolated processes too we here invoke the two simplifications on which the initial formalism was based. In the first place, we assume that all interactions fit into the scheme of an ideal Von Neumann measurement, i.e., we only admit operations which can be represented, via their Stinespring decomposition, by a projection operator. In the second place, we assume that nonisolated evolutions take place entirely within a single slice. As we shall later take the limit for $\Delta t \to 0$, this implies that measurement interactions take place instantaneously. By virtue of these assumptions (6.51) can be maintained also for nonisolated evolutions, with the only modification that now

$$\phi_{\Delta t}(X) = PXP. \tag{6.52}$$

The next step concerns defining contexts and introducing statistical assumptions. Consider an arbitrary slice $[t+\Delta t, t]$. For the process $\mathcal{P}(t)$ associated with this slice we construct a contextual state space $\Gamma_{\mathbf{u}}(t+\Delta t, t)$ and a statistical weight function w. We thereby make the restriction to cases where the factorizability conditions (6.42) are satisfied, and therefore the restricted definition of statistical mixtures becomes applicable, i.e., corresponding to each channel $C = C_{\text{out}} \times C_{\text{ev}} \times C_{\text{in}} \subseteq \Gamma_{\mathbf{u}}$ there is a mixed state $\bar{\Phi}_{\mathbf{u},w}(C)$ given by (6.43).

We proceed by considering channels, and sets of channels, of a very specific type. In the first place, let C_{in} contain only a single operator W_{in}. Suppose, furthermore, that we have no information at all about W_{out}. We can then use the empirical observation that having no information about W_{out} is very well described by taking[10] $\bar{W}_{\text{out}} = 1\!\!1$. Further, let $\Gamma_{\mathbf{u},\text{ev}}$ be given by

$$\Gamma_{\mathbf{u},\text{ev}} = \{\phi_k |\, k \in \mathbf{K}\}, \tag{6.53}$$

where we shall assume that

$$\sum_{k\in\mathbf{K}} \phi_k^+(1\!\!1) = 1\!\!1 , \tag{6.54}$$

because we do not only want to have $\bar{W}_{\text{out}}(t+\Delta t) = 1\!\!1$, but also

$$\bar{W}_{\text{out}}(t) = \sum_{k\in\mathbf{K}} \phi_k^+(\bar{W}_{\text{out}}) = 1\!\!1 . \tag{6.55}$$

We have now obtained a set of channels $C(k)$, of the form

$$C(k) = \Gamma_{\mathbf{u},\text{out}} \times \{\phi_k\} \times \{W_{\text{in}}\}, \tag{6.56}$$

and for each given $C(k)$ there is a mixed state

$$\bar{\Phi}_{\mathbf{u},w}(k) = C_{\mathbf{u}} < 1\!\!1 , \phi_k, W_{\text{in}} >, \tag{6.57}$$

where $C_{\mathbf{u}}$ can be absorbed in W_{in}, to give $C_{\mathbf{u}} W_{\text{in}} = W$. If $\mathcal{P}(t)$ happens to be a nonisolated process then, by assumption, ϕ_k can be represented by a projection

[10]See § 7.6.1.

operator P_k. Otherwise, if $\mathcal{P}(t)$ is isolated, it can be represented by a unitary evolution operator $U(\Delta t) = e^{-iH\Delta t}$, and taking $\Delta t \to 0$ then gives $U(\Delta t) = \mathbb{1}$.

If we finally define

$$\bar{\Gamma}_{\mathbf{u}}(t) = \{\bar{\Phi}_{\mathbf{u},w}(k) | \, k \in \mathbf{K}\}, \tag{6.58}$$

then a pair $\bar{Q}_k =< \bar{\Gamma}_{\mathbf{u}}(t), \bar{\Phi}_{\mathbf{u},w}(k) >$ can be taken to correspond to a quantum state of the initial formalism, and indeed the probability measure of the initial formalism is precisely equal to $p_{\mathbf{u},w}(C_k) = L(\bar{\Phi}_{\mathbf{u},w}(C_k))$.

We have thus managed to find an embedding of the states of the initial formalism into a specific type of statistical mixtures in process theory. This embedding cannot be extended into an isomorphism, but evidently this can hardly be expected. Although the initial formalism solves some of the problems of standard QM, it still inherits a few serious lacunas. Besides the fact that one would like to have a better understanding of the concept of a 'quantum process', as far as conceptual structure is concerned, process theory comes much closer to being a proper theory.

6.2.5 Questions about quantum processes

Although the theory of quantum processes seems perfectly fit for expressing all established results of standard QM, it is not clear that one can be fully satisfied with it. Processes are usually conceived as histories of systems, but there is no explicit sign of the existence of such systems in the theory outlined above, and it seems compelling to give some further interpretation of what kind of thing a process is supposed to be. At this point one has two options: If one thinks of a physical process in terms of the evolution in time of a well-defined physical system, such as a particle or a field, or some compound made up of particles or fields, then one can hardly be satisfied with an assignment of states only to physical processes and not to these systems themselves. Accordingly, process states should be interpretable in some way in terms of system states. The alternative option is to altogether discard the idea of quantum systems evolving in time, and develop a different notion of a quantum process, in which a process is no longer conceived as system history. More specifically we shall work this out within the framework of a quantum event theory in which physical processes are conceived as clusters of events.

There seems to be a link between, on the one hand, the choice between an 'evolving systems theory' and an event theory, and, on the other hand, the choice between an instantly–causal and a retrocausal approach in QM[11]. We shall discuss an instantly–causal particle theory and a retrocausal event theory, but I have thus far not found a way to construct, in a remotely plausible way, either a retrocausal system theory or an instantly–causal event theory. An exception may be provided by John Dorling, who is working on a revival of Schrödinger's 'matter wave interpretation', in a time symmetric version including retrocausality[12]. The American physicist L. Schulman is working on an approach which should be basically similar[13]. This

[11]See § 4.2.5.

[12]Dorling [67].

[13]E.g., Schulman [189, 188].

would yield an interpretation of QM in terms of classical fields, but I cannot presently survey the complications associated with this approach.

6.3 A particle interpretation

From the point of view of quantum process theory, the initial formalism owes its appearance as a theory of quantum particles, or otherwise systems evolving in time, to an approximation method in which a process is considered as a sequence of processes of infinitesimal duration. Even for isolated processes, however, this method in the form as used above does not really provide an adequate procedure for constructing a particle state space. The problem is that information is lost by crudely taking $\Delta t \to 0$. More specifically, we have $\lim_{\Delta t \to 0} U(\Delta t) = 1\!\!1$, but $U(t)$ cannot be reconstructed from the resulting sequence of $1\!\!1$s. Therefore, particle states must be defined in some smarter way. A procedure which does work is based on using derivatives of operations, rather than the operations themselves, for defining the particle states. For an isolated process, taking place in the time interval $[0, t]$, we have the operation

$$\phi_{(t,0)}(W) = U(t)WU^{+}(t), \tag{6.59}$$

with $U(t) = e^{-iHt}$. Its derivative is given by

$$\dot{\phi}_{(t,0)}(W) = -i[H, \phi_{(t,0)}(W)] = \eta \circ \phi_{(t,0)}(W), \tag{6.60}$$

where η, the *generator* of ϕ, is given by

$$\eta(A) = -i[H, A]. \tag{6.61}$$

A quantum state for the system $\mathcal{S}$ at time t can now be based on $\dot{\phi}$ rather than on ϕ itself. In a time–symmetric pictures, with process states defined by the triples (6.8), this leads to particle states defined by

$$\Theta(t) =< W_{\text{out}}(t), \dot{\phi}, W_{\text{in}}(t) >, \tag{6.62}$$

with $W_{\text{in}}(t) = \phi_{(t,t_{\text{in}})}(W_{\text{in}})$ and $W_{\text{out}}(t) = \phi^{+}_{(t_{\text{out}},t)}(W_{\text{out}})$. On the other hand, in a time–asymmetric view, with process states defined by the pairs (6.35), the system states are defined by

$$\Theta(t) =< \dot{\phi}_t, W(t) > . \tag{6.63}$$

In both cases, the process states can be recovered by integration of $\dot{\phi}$. In either view, all possible states may be collected into a state space $\Gamma_{\mathbf{sys}}$. A given experiment defines, as before, a 'quantum phenomenon', represented by context $\mathbf{u}$. Within the present approaches, there is, for each moment t of the experiment, a contextual state space $\Gamma_{\mathbf{u}}(t) \subseteq \Gamma_{\mathbf{sys}}$, being the set of all possible states which the system might have at time t. A total quantum state space $\Gamma_{\mathbf{qm}}$ may now be defined as the set of all pairs $< \Gamma_{\mathbf{u}}, \Theta >$ considered to be physically possible.

Note: It might seem more elegant to use the operation generators η, rather than the derivatives $\dot{\phi}$, for defining the system states, i.e., one could take

$$\Theta(t) =< W_{\rm out}(t), \eta, W_{\rm in}(t) >, \tag{6.64}$$

instead of (6.62). This, however, would require the general existence of these generators also for nonisolated processes. This requirement is not satisfied because $\dot{\phi}$ cannot generally be written in the form $\eta \circ \phi$.

6.3.1 Dynamics

Clearly, particle states based on operation derivatives can exist only if such derivatives exist, which is not evident *a priori*, especially because it is not even clear that each operation can be broken into phases. We must, therefore, take a closer look at the structure of the operations. According to the dynamical rules of process theory the evolution of a nonisolated process can be represented by a tuple of the form $< W^A_{\rm out}(t_2), \phi^C_{(t_2,t_1)}, W^A_{\rm in}(t_1) >$, with the associated operation (6.32):

$$\phi_{(t_2,t_1)}(X) = {\rm Tr}^A[(W^A_{\rm out}(t_2) \otimes 1\!\!1\,)\phi^C_{(t_2,t_1)}(W^A_{\rm in}(t_1) \otimes X)].$$

A problem with this expression is that, although it is alright for assigning an operation to the evolution from $t = t_1$ to $t = t_2$, it does not describe the evolution of S at intermediate moments $t_1 \leq t \leq t_2$. The problem with breaking the evolution into phases is that *during* the interval $t_1 \leq t \leq t_2$ (6.32) is useless, because a separate operator $W^A_{\rm out}(t)$ does not exist. During this time both $W^C_{\rm in}(t)$ and $W^C_{\rm out}(t)$ are in an entangled state[14] and cannot be factorized into separate W^As and Ws.

Within the time–symmetric approach, I found no immediately apparent solution to this problem. Therefore, we shall here assume the time–asymmetric view, based on the states $< W_{\rm out}(t), \eta, W_{\rm in}(t) >$. As in the initial formalism, this leads to a bifurcating dynamics, in which, at each point in history, we have only one possible past, but many possible futures[15]. In order to work out this option we shall slightly elaborate on the evolution of a nonisolated system, and in order to get a better view on the bifurcational aspect we consider an OVM, rather than a single operation, i.e., instead of (6.32) we start from expression (5.15):

$$\phi_k(W) = {\rm Tr}^A[(R^A(k) \otimes 1\!\!1\,)U^C(W^A \otimes W)U^{C+}].$$

In order to examine the operations in more detail, let us first be more precise about our timekeeping and define the moments $t = 0, t_1, t_2, t_3$ by:

$$\begin{array}{lcl} t = 0 & : & \text{The interaction between } \mathcal{S} \text{ and } \mathcal{S}^A \text{ begins.} \\ t = t_1 & : & \text{The interaction between } \mathcal{S} \text{ and } \mathcal{S}^A \text{ ends.} \\ t = t_2 & : & \text{The measurement on } \mathcal{S}^A \text{ begins.} \\ t = t_3 & : & \text{The measurement on } \mathcal{S}^A \text{ ends.} \end{array} \tag{6.65}$$

[14] See § 4.2.4.

[15] Strangely, this view appears to be more or less self–evident to many people, although, to me, it seems extremely counterintuitive. At this point I am simply trying to oblige.

For times $0 \leq t \leq t_1$, during the interaction between the two systems, we can now assign to $\mathcal{S}$ a state $\Theta(t) = <\dot{\phi}_t, W(t)>$ with:

$$\begin{aligned} W(t) &= \phi_{(t,0)}(W) \\ &= \mathrm{Tr}^A[W^C(t)] \\ &= \mathrm{Tr}^A[(U^C(t)(W^A \otimes W)U^{C+}(t)] \\ \dot{\phi}_t(W) &= \mathrm{Tr}^A[-i[H^C, W^C(t)]]. \end{aligned} \tag{6.66}$$

This interaction with $\mathcal{S}^A$ turns the density operator of $\mathcal{S}$ into an improper mixture[16], because the composite system $\mathcal{S}^C = \mathcal{S} + \mathcal{S}^A$ is in an entangled state.

For $t_1 \leq t \leq t_2$ we get again $\Theta(t) = <\dot{\phi}_t, W(t)>$, but now with:

$$\begin{aligned} W(t) &= \phi_{(t,0)}(W) \\ &= U(t-t_1)\mathrm{Tr}^A[W^C(t_1)]U^+(t-t_1) \\ \dot{\phi}_t(W) &= -i[H, W(t)] = \eta \circ \phi(W), \end{aligned} \tag{6.67}$$

i.e., as the two systems are decoupled the evolution of $W(t)$ is again given by the Schrödinger equation. This only holds, however, up to the time t_2, when the measurement of $\mathcal{S}^A$ begins. The systems are still in an entangled state, and the measurement of $\mathcal{S}^A$ has a profound influence on the state of $\mathcal{S}$. At this point we introduce the assumption that this influence is immediate, and therefore nonlocal, because $\mathcal{S}$ and $\mathcal{S}^A$ may be arbitrarily far separated at this time. More specifically we assume that

> during the measurement on $\mathcal{S}^A$ one of the values $k \in \mathbf{K}$ is realized, leading to a bifurcation in the evolution of $\mathcal{S}^A$ and a reduction of its density operator W^A.
>
> This also leads to an instantaneous bifurcation in the evolution of $\mathcal{S}$;

for times $t_2 \leq t \leq t_3$ this scheme can be elaborated on in the following way.

> We assume that there is an effect $R_t^A(k)$ corresponding to each $k \in \mathbf{K}$.
>
> We assume that for $t = t_2$ we have $R_{t_2}^A(k) \sim R_{t_2}^A(l)$ for all $k, l \in \mathbf{K}$, which means that at this time there is not yet a real bifurcation, because the resulting operations are still proportional. We can let the bifurcation build up gradually to a final result $R^A(k) := R_{t_3}^A(k)$ at $t = t_3$, and we can freely assume that all $R_t^A(k)$ are differentiable during this period.
>
> The operation $\phi_{(t,0),k}$ corresponding to $R_{t-t_2}^A(k)$ is given by

$$\phi_{(t,0),k}(W) = U(t-t_2)\,\mathrm{Tr}^A[(R_t^A(k) \otimes 1\!\!1)W^C(t_2)]\,U^+(t-t_2). \tag{6.68}$$

[16]See § 4.2.

Considering the total operation $\phi_{(t,0),\mathbf{K}}$ we see that at $t = t_2$ it is made up of a number of proportional terms that begin to diverge during the period between t_2 and t_3, resulting in a proper bifurcation of the quantum state. Further, it is clear from the above expression that the time derivative of $\phi_{(t,0),k}(W)$ contains, besides the usual $-i[H, \phi(W)]$ term originating from the Schrödinger equation, an additional dissipative term equal to

$$\dot{\phi}^{dis}_{t,k}(W) = U(t-t_2)\mathrm{Tr}^A[(\dot{R}^A_t(k) \otimes 1\!\!1\,)W^C(t_2)]U^+(t-t_2). \tag{6.69}$$

This term is responsible for the reduction of the density operator. Starting with the density operator $W(t_2)$ at t_2, corresponding to an improper mixture, we end up at t_3 with

$$W_k(t_3) = \phi_{(t_3,0),k}(W),$$

if the value k has actualized. If we do not know which value k has actually occurred then at t_3 we can assign to $\mathcal{S}$ a density operator $\bar{W}(t_3) = \phi_{(t_3,0),\mathbf{K}}$, equal to what the original improper mixture would have been if no measurement on $\mathcal{S}^A$ had taken place. This operator now describes a proper mixture, however.

> **Note:** Comparing the $\dot{\phi}$s during the various phases of the process, we may note that only for $t_1 \leq t \leq t_2$ it explicitly has the form $\eta \circ \phi$. During the other periods, it is given by more complicated expressions, and it is far from clear that generators for ϕ exist at all.

6.3.2 Evaluation

In the above theory of quantum systems the operations exhibit a definite arrow of time: for increasing t there is *first* the generation of an improper mixture between $t = 0$ and $t = t_1$, possibly *followed* by a state reduction between $t = t_2$ and $t = t_3$. Accordingly, we also find that the evolution of W_{out} is by no means comparable to the time reversed evolution of W_{in}, except for isolated systems. By introducing such an asymmetric dynamics we have been able to hold on to the idea of QM as a theory about particles evolving in time, and to avoid the idea of a time–symmetric causality. Moreover, the resulting theory may be regarded as being conceptually close to the initial formalism, with major improvements on some of the most problematic points.

Nevertheless, the benefits of these manoeuvres can be regarded with mixed feelings. In the first place, at least some people, to whom I certainly belong, regard a dynamical time–*asymmetry* as a greater problem than a causal time–*symmetry*. This appears to be a clear example of a case where physicists disagree about the criteria by which they judge their theories. Something like 'taste' seems to be quite important at this point. Secondly, the dynamics we have found looks rather 'ugly' also in other respects than its time asymmetry. It involves regarding the nonlocal 'collapse of the state' as a physically real dynamical process, and it compels us to using such strange constructs as improper mixtures. In the third, and certainly not the least place, the approach introduces the Von Neumann chain problem, which does not exist in this form in process theory, where W_{out} is regarded as an independent state variable. We may recall, however, that in the above system theory the collapse of the state of $\mathcal{S}$ depended on the collapse of the state of $\mathcal{S}^A$, which presumably depended in its turn on the collapse of ... etc. In the fourth place, it might actually be considered a relief to find a theory which is *not* based on the notion of particles, or other systems, evolving in time, because the concept of a quantum particle is quite problematic anyway.

The particle interpretation discussed here gives *one* way of retaining the particle concept in QM. There may be many other ways, which may suffer from fewer of the above 'buts'. However, I have tried to check some of the alternative approaches, but I have thus far not encountered anything which occurred *to me* as substantially more appealing. On the other hand, it has been impossible for me to consider every approach in detail, and moreover, my judgement may be slightly biased because my personal preferences strongly tend to favour event theory.

6.4 Event theory

The standard way to construct a process state space is to start with a system state space $\Gamma_{\mathbf{sys}}$. Let $[t_1, t_2] \subseteq \mathbb{R}$ be the period during which the process takes place, then a process space $\Gamma_{\mathbf{proc}}$ can be defined by taking the set of possible trajectories through $\Gamma_{\mathbf{sys}}$, i.e., the set of functions $[t_1, t_2] \rightarrow \Gamma_{\mathbf{sys}}$, of which those processes are deleted that are considered impossible on theoretical grounds.

We shall now discuss an alternative idea and consider a quantum event theory in which processes are made up of agglomerations of events, rather than of the histories of particle–like entities. Basically, the idea is that spacetime is filled with events, but events which cannot be interpreted as occurring to particles. These events have a tendency, however, to cluster into compounds that have the appearance of world lines. The idea can be illustrated by thinking of a track in a bubble chamber. Such a track consists of many separate bubbles, each of them being the trace of some definite event. We usually think of these events as being caused by a passing particle, but the thought of a particle can in fact be dropped; there is no hard evidence for its existence. Alternatively, one might think of an event in 4-dimensional spacetime as being analogous to a charge in 3-dimensional space. It generates some sort of tension in a field, or in a Kaluza–Klein variant of the theory a distortion of a many–dimensional spacetime manifold, and its presence apparently enhances the probability for the occurrence of other events, i.e., some configurations of events are more likely to occur than other ones. Field lines may now be considered as representing virtual particle trajectories, and density operators, effects, and operations can be regarded as representing fluxes of the fields through 3–surfaces in spacetime.

Example: Consider a process taking place between $t = t_1$ and $t = t_2$ and described by a process state $\Phi(t_2, t_1) =< W_{\text{out}}(t_2), \phi_{(t_2,t_1)}, W_{\text{in}}(t_1) >$. The process takes place within the 4-volume τ, defined by a 3–volume V between the times

t_0 and t_1. From the point of view of an event theory, $W_{\rm in}(t_1)$ can be regarded as representing a flux through the 'bottom' of the spacetime box τ, i.e., the hypersurface given by V at t_1, and $W_{\rm out}(t_2)$ represents a flux through the 'top' of the box, i.e., through V at t_2.

Furthermore, the OVM ϕ represents the conditions at the sides of τ. The intermediate measurement may, for instance, be performed by letting the 'system' interact with a probe which is sent into the volume V, and the measurement result is collected by looking at the probe after it has left V. Accordingly the measurement depends on what goes through the sides of V during the period $[t_1, t_0]$, i.e., through the sides of τ. Another example of how an intermediate measurement may be performed is by placing a screen in the 'path' of an 'electron', in such a way that the electron can either be absorbed by, or pass through a slit in the screen. In this case the absorbing part of the screen is part of the sides of V during the period $[t_1, t_0]$, so again the measurement is determined by conditions on the sides of τ.

I will not claim that this approach immediately solves all conceptual problems of QM. I wish to posit, however, that the concept of a quantum event is by no means more unclear than that of a quantum particle, whereas it does have some evident theoretical advantages. It offers an interpretation of process theory which does not suffer from any of the problems associated with a 'collapse of the state'. Further, it may be expected beforehand that a quantum theory based on events might provide a much smoother link with relativity theory. In ch. 9 I will try to show that indeed the approach does have a major pay-off in relativistic QM.

The kind of picture that will emerge there is roughly the following: the physical contents of a spacetime volume τ can be described in terms of expressions representing something like field fluxes through its boundary $\partial\tau$. Let the expression $\Phi_{\partial\tau}$ include the various contributions. For the present we may take for $\Phi_{\partial\tau}$ a triple $< W_{\rm out}, \phi, W_{\rm in} >$, but within a more fully articulated relativistic event theory there will be the possibility of combining them within a more general type of expression, not restricted to a particular Lorentz frame or a particular shape of τ. Further elaboration on this approach leads to a theory which is manifestly covariant in an elegant way, and which is equivalent to standard quantum field theory in the sense that it can produce the same set of Green functions, or Feynman diagrams. Probabilities are computed by means of (6.15), via a suitably modified version of (6.9), and standard expressions in terms of S–matrix elements are contained as special cases.

The idea of a process which is not the history of a system evolving in time may seem a bit unconventional, but I think it can indeed be argued with sufficient power of conviction that it is by no means inconceivable, and that, on the contrary, it may be extremely fruitful in offering new views on a range of old problems.

6.4.1 The pure state assumption

Because there is no postponed state reduction in the retrocausal approach, no improper mixtures arise during a measurement process. One may, therefore, think about whether they can be avoided entirely, such that Hilbert space vectors rather than density operators would again be the fundamental units in a state description. With the operations given by (6.32):

$$\phi_{(t_{\rm out},t_{\rm in})}(W_{\rm in}) = \\ {\rm Tr}^A[(W^A_{\rm out}\otimes 1\!\!1\,)U^C(t_{\rm out}-t_{\rm in})(W^A_{\rm in}\otimes W)U^{C+}(t_{\rm out}-t_{\rm in})],$$

one would then have to assume that $W_{\rm in}$, $W^A_{\rm in}$, and $W^A_{\rm out}$ are, in fact, proper mixtures, and that, accordingly, there are unique preferred resolutions of ϕ into one–dimensional terms. By considering one such pure state, with

$$W_{\rm in}=|{\rm in}\rangle\langle{\rm in}|,\quad W^A_{\rm in}=|{\rm in}^A\rangle\langle{\rm in}^A|,\quad W^A_{\rm out}(k)=|{\rm out}^A\rangle\langle{\rm out}^A|, \tag{6.70}$$

we then find that the resulting density operator $W_{\rm in}(t_{\rm out})=\phi_{(t_{\rm out},t_{\rm in})}(W_{\rm in})$ will again be pure. We get

$$\begin{aligned} W_{\rm in}(t_{\rm out}) &= \phi_{(t_{\rm out},t_{\rm in})}(W_{\rm in})=V|{\rm in}\rangle\langle{\rm in}|V^+ \\ \text{with}& \\ V &= \langle{\rm out}^A|U^C(t_{\rm out}-t_{\rm in})|{\rm in}^A\rangle \end{aligned} \tag{6.71}$$

We are now free to assume that wherever we encounter mixed density operators we are indeed dealing with statistical mixtures. The truly elementary process states, on the other hand, are tuples of the type

$$\Phi=<\langle{\rm out}|,V,|{\rm in}\rangle>. \tag{6.72}$$

For a state of this type we can define an associated *amplitude* $A(\Phi)$ equal to

$$A(\Phi)=\langle{\rm out}|V|{\rm in}\rangle, \tag{6.73}$$

and the L-value of Φ is now given by

$$L(\Phi)=|A(\Phi)|^2=|\langle{\rm out}|V|{\rm in}\rangle|^2. \tag{6.74}$$

The probability for Φ to occur within a specific context is given, as before, by $p_{\mathbf{u},w}(\Phi)=C_{\mathbf{u},w}L(\Phi)w(\Phi)$.

Note: It may be noted that each Φ can be represented as a vector on $\mathcal{H}^+\otimes\mathcal{H}\otimes\mathcal{H}^+\otimes\mathcal{H}$, but equivalently still as a linear operator $\bar\Phi$ as in (6.48). Using a PVM $\bar\Phi_{\mathbf{u},w}$, as defined in (6.49), probabilities may alternatively be given in the form $p_{\mathbf{u},w}(K)=L(\bar\Phi_{\mathbf{u},w}(K))$, for $K\subseteq\Gamma_{\mathbf{u}}$.

Literature: Čapek [38], Dorling [67], Van Fraassen [95, 90], Hoekzema [121, 122, 120, 123], Von Neumann [165], Levy-Leblond [149], Schulman [189, 188],

Chapter 7

Completeness and locality

Abstract: Several aspects are considered of the completeness of standard QM as well as contextual QM. The nonlocality of QM is investigated, and it is argued that, within QM, nothing stands in the way of the potential observability of nonlocal effects. The causality conditions that are, apparently, effective must have a different origin.

Although the debate about the interpretation of QM is about as old as the theory itself, there seems to be a marked difference in the degree of success in the development of these two subjects. Whereas the history of quantum theory itself is a miraculous success story, the debate about its interpretation appears to be dragging from one stalemate position to another, and a sceptic might even argue that, by and large, one is still chewing on the Bohr–Einstein discussion of the nineteenthirties. Even though nowadays there is a revival of interest in the problems concerning the interpretation of QM, under the circumstances one may regard it as fortunate that these problems appear to be irrelevant to the development of the theory, or, as it is put in a recent textbook[1]:

> The choice between the Copenhagen interpretation and statistical interpretations depends on individual taste only. However, the mathematical formalism of quantum mechanics is independent of these interpretations.

Presumably, this quotation expresses a feeling of relief that is widespread throughout the physical community.

The present work is largely based on the view that progress in the debate on interpretation has been much more significant than is widely assumed. Important results have failed to reach proper acceptation or have been misunderstood, however, partly due to the circumstance that physicists prefer to hold on to the separation of formalism and interpretation, and to the relative irrelevance of the latter. Major fractures between formalism and interpretation, of the kind seen today in QM, are a symptom of disease, however, because they are contrary to what formalisms are supposed to be made for. A formalism is developed especially for the purpose of

[1] Rajput [176] p. 30.

saying as precise as possible what one means. The idea of having a single formalism for expressing the common opinion of a large group of people having contradictory views sounds like a joke to begin with, but it also makes clear what was presumably the point of this montruous construction: major schisms were threatening the physical community, and the development of a common formalism shared by major factions in the population provided a way of suppressing the mutual differences and get along with a common enterprise. Perhaps it is fortunate that this strategy has worked for quite a while, but in due time it *must* become obstructive, and a proper approach to discussing interpretational problems should be based on the recognition that there is not *one* quantum theory, there is a whole bunch of them, and each one, in as far as it is worthwhile at all, deserves its own formalism.

7.1 Quantum 'theory'?

A peculiar feature of the quantum formalism is that it is not an adequate formalization of an object of the type studied in classical logic and called "theory", where "classical logic" relates to two–valued logics (true, false) with Boolean algebras of properties. Accordingly, we must choose between several options:

> One possibility is to hold on to a classical logical analysis of theory structure. In this case the quantum formalism is inadequate. In particular it can then be shown that the formalism is 'incomplete', in a welldefined sense of the word, and that there is some abuse of terminology going on. This leads to the present contextual approaches to QM.
>
> A second possibility is that the formalism is considered basically adequate, but the framework of classical logic is too restrictive. This leads to various possible forms of quantum logic.
>
> A third possibility is to hold on to the quantum formalism and deny the need for a new logical framework. In order to be able to do this one may for instance ignore the logical component of the conceptual problems of QM. This should presumably lead to a discussion about the role of logic with respect to the structure of scientific theories.

The best thing to say about the latter category is that the fact of its origination, in the early days of QM, is understandable, and much of the more interesting work in this class can be considered as protocontextual. For the rest, however, it is an approach that halts between two views, and it would be gratifying if the present work could help to accelerate its disappearance. The origin of this category can be traced to what may be regarded as a combination of 'bugs' in the debate about QM. The first one of these bugs is the above mentioned fracture between the conceptual framework of QM and its formalism. Niels Bohr may be held at least partly responsible for the emergence of this fracture. Although the philosophy of Bohr may be regarded as contextual or protocontextual, Bohr apparently did not see either the possibility or the benefit of developing his ideas into a contextual formalism, hence the fracture between concepts and formal notions.

A second bug can presumably be written on Einstein's account, and concerns a confusion of the concepts of 'completeness' and 'determinism'. Einstein has argued against the completeness of QM. He has also expressed his aversion of the idea of 'God playing dice', i.e., of indeterminism. This has been one of the main factors in initiating a debate between proponents and opponents of 'hidden variables' in QM. Proponents of hidden variables seek to complement QM with hidden variables, in order to make the theory deterministic. Completeness and determinism, however, are two different concepts. One may distinguish various forms of completeness, but, assuming a classical logical framework, a crucial form of *incompleteness* of QM can simply be proven, as a corollary to contemporary 'no hidden variable' proofs, such as Gleason's theorem. The idea is that from these theorems it follows that any hidden variable theory reproducing the empirical content of QM must be contextual, in the sense that quantum mechanical magnitudes must be split. The algebra of physical magnitudes of QM cannot be conserved in a hidden variable theory, and what is, e.g., called "energy" in one context is not necessarily the same as what bears, according to QM, the same name in another context. These theorems are extremely general, however, the only requirements being that the hidden variable theory has a classical logical structure and that it is complete relative to the empirical phenomena it aims to describe. Applying this to QM *itself*, one must either give up the classical logical structure, or conclude that QM is incomplete and its magnitudes must be split. In a contextual approach to QM this incompleteness must be removed, but this does not require the introduction of hidden variables, nor does it automatically lead to a deterministic theory.

The splitting of magnitudes points to a major difference between the contextual and the quantum logical approaches to QM: In a contextual approach the state description of the standard quantum formalism must be regarded as incomplete, and standard usage of the terms "magnitude" and "observable" must be regarded as abusive. The 'physical magnitudes' represented in QM by Hermitean operators are not magnitudes. We may call them *supermagnitudes*, and each such supermagnitude in fact constitutes a *family* of magnitudes. An adequate logical term would be *magnitude type*. The relation between these supermagnitudes and proper magnitudes can be stated as

> supermagnitude + context $\Rightarrow$ magnitude,
>
> if the context has been properly chosen such that a magnitude of this type indeed exists within this context.

Within a quantum logical approach, the standard, classical, meaning of the terms "state" and "magnitude" does not apply, making it possible to redefine them in a manner that agrees with usage of these terms in standard QM. Clearly, however, the desirability of such a move remains questionable.

7.2 Contextual QM and completeness

The history of contextual, or proto–contextual, QM may be considered to begin with Bohr's introduction of the notions of "complementarity" and "quantum phenomenon". The final meaning of these concepts, in as far as Bohr's contribution to the subject is concerned, was presumably reached around 1935, in reaction to the Einstein–Podolsky–Rosen *Gedanken* experiment. The main ingredient that was lacking, apart from a proper formalism, was the realization that contextuality would imply the splitting of magnitudes, but this required insights which clearly were too far ahead at that time.

Three important counter moves against the idea of contextuality were produced by Von Neumann. In the first place, he worked out a mathematical formalism of which the basic structure seemed good enough to be final, at least from a noncontextual point of view[2]. Secondly, he initiated, together with Birkhoff, a quantum logic based on the theory of nondistributive lattices[3]. In the third place he proved a theorem which was used as an argument directed against the possibility of hidden variables[4], thus increasing the air of finality of the existing formalism. This theorem, however, was to play a most curious role in the further debate, although not until decades later.

In 1952 David Bohm published a hidden variable interpretation of QM[5], thus providing a counterexample to Von Neumann's theorem. This has led to new lines of research, not only into new types of hidden variable theories, but also into the hidden assumptions in Von Neumann's theorem. This resulted in various new 'no hidden variable theorems', where especially Gleason's theorem[6] and Kochen and Specker's theorem[7] may be highlighted. The upshot of these theorems is that no classical theory can reproduce both the empirical predictions of QM and its algebra of physical magnitudes. This leads to the realization that either the logical structure of QM is nonclassical, and QM can only be understood by using some quantum-logical framework, or the algebra of physical magnitudes in QM is inadequate, the approach chosen in contextual QM. It is the splitting of magnitudes that gets the latter approach off the ground, and one may, therefore, say that a proper understanding of Von Neumann's theorem contained the germs for the further development of the contextual approach.

For the next step in this history we move on to 1966. In this year an article of John Bell was published[8] in which he investigated the structure of several hidden variable theorems then available. This paper contains the first "adequate analysis of what went wrong" in Von Neumann's proof. Especially clarifying, however, is the

[2] [165].
[3] [28].
[4] [165].
[5] [29].
[6] [100].
[7] [134].
[8] [14], actually written some years earlier, but lost in the drawer of an editor.

end of his section on Gleason's theorem, where he says that[9]:

> "It was tacitly assumed that measurement of an observable must yield the same value independently of what other measurements may be made simultaneously. [...] These different possibilities require different experimental arrangements; there is no *a priori* reason to believe that the results [...] should be the same. The result of an observation may reasonably depend not only on the state of the system [...] but also on the complete disposition of the apparatus."

This boils down to splitting magnitudes. Interesting is also that he then refers to a quotation from Bohr, in which Bohr speaks about "the impossibility of any sharp distinction between the behavior of atomic objects and the interaction with the measuring instruments which serve to define the conditions under which the phenomena appear." I remember a conversation with Roger Cooke in, presumably, 1988 in which we were joking about the smart way in which Bell here abuses Bohr's words, turning them against himself, to argue in favour of the possibility of hidden variables. This conversation kept buzzing through my head for several months until it dawned upon me that perhaps there was less of an abuse of Bohr's words than we had been thinking. What is at stake here is in fact the completeness of the formalism of QM. Although Bohr was opposed to hidden variables and was strongly convinced of the completeness of QM, he cannot be accused of advocating the completeness of the standard *formalism* of QM. On the contrary, he always stressed the need for further interpretation of the formalism in terms of physical concepts. The idea of constructing a complete formalism via the trick of splitting magnitudes, however, required too much logical sophistication at that time, whereas in recent times it has become too closely associated with the search for hidden variables.

The initial formalism

This realization, after hitting target, readily combined with other work I was doing on contextual theories, resulting in a new formal setup, in which, on the one hand, I used the idea of splitting magnitudes for obtaining logical completeness, but where, on the other hand, I tried to stay as close as possible to the standard formalism. The initial contextual formalism was the first result of this approach, and, apart from the above realization about Bell's quotation from Bohr, this work has several other roots, visible in various parts of the present work. In the first place, I must mention my aquaintance with work of Cooke and Hilgevoord[10], who made a detailed study of splitting magnitudes in connection with the Kochen and Specker theorem. Secondly, mostly for reasons relating to problems in philosophy of science, I was at that time studying the 'semantic approach to scientific theories'. In this approach the logical analysis of theories is seen from a slightly different perspective than in

[9] [14] p. 451.
[10] See, e.g., [47].

the more syntactically oriented views that have long been standard[11]. In particular I was studying work of Van Fraassen about a 'modal interpretation of QM', based on a logical theory of 'semiinterpreted languages'[12]. Thirdly, I was interested in ideas about semantical networks, and in particular I was experimenting with a 'contextual logic', constructed from networks of semiinterpreted languages.

A semiinterpreted language is presumably the simplest logical framework suitable for investigating the structure of the state-concept. It was defined in ch. 3 as a triple $< \mathcal{L}, \Gamma, \gamma >$, where $\mathcal{L}$ is a syntax, Γ is a state space, and γ is a 'satisfaction function', giving each sentence $\psi \in \mathcal{L}$ a 'meaning', by means of the set $\gamma(\psi) \subseteq \Gamma$ of states in which ψ is satisfied. A simple 'contextual logic' may be obtained by considering networks of such languages, now called *contexts*, glued together by adding some *connective* structure to the above triples, such that contexts may be 'seen' from within other contexts. In ch. 3 such a network of connected languages is called a "labyrinth of contexts".

'Completeness' of a context relative to another context, in the sense relevant to our purposes, can be defined as the requirement that the context has 'enough states'. For a physical theory to be complete relative to its purported domain of application this means two things. In the first place it implies that the scope of the theory is sufficiently wide, such that all relevant empirical phenomena are represented. We shall call this *weak completeness.* In the second place, it entails *accuracy* of the state description, in the sense that distinct empirical phenomena must have distinct theoretical representations. Comparing logic to a telescope, weak completeness relates to the field of the telescope, and accuracy to its resolving power. The combination of these two demands will be called *strong completeness*, and, evidently, we shall want QM to satisfy this demand. Splitting magnitudes is related primarily to the demand of accuracy.

In QM the contextual approach is used for constructing an interpretation of the concept of "complementarity", by associating a different context with each different 'quantum phenomenon'. Splitting magnitudes is achieved here by defining physical magnitudes context-dependently. For example, there is not assumed to be a single magnitude called "energy", but there is a family of energy-like magnitudes, and in some contexts a magnitude of this type is definable. Each context has its own, classical, state space, and classical logic can be straightforwardly applied. Further, apart from the contextual state assigned to a system, one may also define its *total state.* In the initial formalism, a complete description of the total state includes in the first place a density matrix, as in the standard formalism, in the second place a specification of which context applies, and in the third place the description of the contextual state. The state concept used here is again entirely classical, with properties simply generating Boolean algebras, as usual. As pointed out in ch. 6, the initial formalism has only a transitory significance, because it contains some unacceptable idealizations in the area of measurement theory. It may be noted, however, that logical completeness, in the sense of accuracy of the state description,

[11]See, e.g., [96].
[12]See, e.g., [90].

is achieved without introducing anything having the appearance of a hidden variable.

Process theory

The design for a more final formal setup for QM required including the use of *effects* and *operations*, and the use of conditional probability measures, or ABL measures. Interesting is that both topics relate to the completeness of QM, but here in the sense of weak completeness, i.e., the scope of the theory. At first I assumed that fitting these developments into the initial formalism should be quite straightforward, but after a few months tinkering I found out I had arrived at a theory with a significantly different conceptual structure; the theory introduced here as quantum process theory. The logical structure of process theory is again entirely classical, and in fact quite straightforward, even more so than that of the initial formalism. Further, the theory is complete relative to its domain of application, it is indeterministic, and it does not introduce hidden variables. What remains to be desired? Well, in the first place, there is still some ambiguity with respect to the description of composite systems and locality. The theory is nonlocal, but there remains a choice between nonlocality proper, in the sense of causal influence outside the light cone, and retrocausality, i.e., causality backwards in time. This has consequences for the way in which process quantities are assigned to experiments. Secondly, the theory contains the notion of a "process" as a primitive term. This requires further interpretation of what kind of object a process is supposed to be. Here too one has the choice between various options. This leads in the first place to a division between on the one hand particle theories and quantum field theories, in which a process is interpreted as the history of a quantum system, such as a particle or a quantum field, evolving in time, and, on the other hand, event theories, containing no elementary systems with enduring existence. Both views may be elaborated in different ways, and, accordingly, the right way of looking at quantum process theory is to see it as a general framework, covering a wide spectrum of further possibilities, ranging from more or less orthodox, Bohr–like, views to hidden variable theories, which, after all, must also be contextual, and contextual logic should provide an appropriate approach toward their formalization too.

7.3 Splitting magnitudes

Like in any theory of a classical type, magnitudes in contextual QM are functions on state spaces and serve to label states. Accordingly, a magnitude may be any function on any of the state spaces we have discussed. Of special interest to us, however, are the magnitudes usually referred to as physical observables, and which, in standard QM, are associated with Hermitean operators. This way of representing observables can in contextual QM be generally maintained only within the initial formalism. Therefore, in spite of its restricted applicability, we shall take the initial formalism as the starting point for some considerations about observable magnitudes. We shall introduce them as functions on the contextual state spaces $\Gamma_{\mathbf{u}}$, and then proceed by explaining the relation with their usual representation as operators on $\mathcal{H}$.

Corresponding magnitudes

An important consequence of defining observables primarily as functions on the contextual state spaces $\Gamma_{\mathbf{u}}$ is that they will have a different meaning in different contexts, i.e., the magnitudes of standard QM are split[13] within contextual QM. They appear here not as magnitudes but as equivalence classes of magnitudes, or 'supermagnitudes'.

> **Example:** Let $\mathcal{S}$ be a free particle in one dimension, with a Hamilton operator $\mathsf{H} = \mathsf{p}^2/2m$, where p is the momentum operator, and let $\mathbf{u}$ be a context in which the momentum is measured, i.e., with a state space $\Gamma_{\mathbf{u}} = \bigcup_p \{|p\rangle\langle p|\}$. Within this context momentum itself is a function $p : \Gamma_{\mathbf{u}} \to I\!R$, with $p(|p'\rangle\langle p'|) = p'$. Further, the energy of the system is defined as the magnitude $H = p^2/2m$, i.e., $H(|p'\rangle\langle p'|) = p'^2/2m$. According to our earlier analysis, the meaning of a sentence "$H = E$" within the context $\mathbf{u}$ is then given by
>
> $$\gamma_{\mathbf{u}}(H = E) = \{|p\rangle\langle p| \mid p^2/2m = E\}, \tag{7.1}$$
>
> i.e., it is a set of two one–dimensional projections. Let us now compare this to what we get in a different context $\mathbf{v}$, in which the energy is measured instead of the momentum. The eigenspaces of the Hamilton operator are the two-dimensional subspaces spanned by a vector $|p\rangle$ and the corresponding $|-p\rangle$. The projection operators on these eigenspaces are equal to $|p\rangle\langle p| + |-p\rangle\langle -p|$, and, accordingly, we get $\Gamma_{\mathbf{v}} = \bigcup_p \{|p\rangle\langle p| + |-p\rangle\langle -p|\}$. The meaning of a sentence "$H = E$" within this new context is given by
>
> $$\gamma_{\mathbf{v}}(H = E) = \{|p\rangle\langle p| + |-p\rangle\langle -p| \mid p^2/2m = E\}, \tag{7.2}$$
>
> i.e., it is a set consisting of one two-dimensional projection. Clearly $\gamma_{\mathbf{u}}(H = E)$ and $\gamma_{\mathbf{v}}(H = E)$ are not identical, so indeed the standard quantum mechanical magnitudes are split, i.e., their meaning is context dependent.

The splitting of magnitudes raises the point of what is the precise relation between a magnitude in standard QM and the set of contextual magnitudes that corresponds to it in contextual QM. Therefore, let us consider two contexts $\mathbf{u}, \mathbf{v} \in \Psi$ and two magnitudes $X : \Gamma_{\mathbf{u}} \to \mathbf{R}(X)$ and $Y : \Gamma_{\mathbf{u}} \to \mathbf{R}(Y)$ in $\mathbf{u}$ and $\mathbf{v}$. We shall define them to be *corresponding* magnitudes if the following two conditions are satisfied:
Corresponding magnitudes:

$$\begin{aligned} \mathbf{R}(Y) &= \mathbf{R}(X) \\ \sum_{P \in \gamma_{\mathbf{u}}(X=x)} P &= \sum_{P' \in \gamma_{\mathbf{v}}(Y=x)} P' \quad \text{for all } x \in \mathbf{R}(X). \end{aligned} \tag{7.3}$$

[13] See Van Fraassen [95], Cooke and Hilgevoord [47], Redhead [177] pp.134–134.

These conditions specify an equivalence class of magnitudes to which an operator X can be assigned by means of

$$\mathsf{X} = \sum_{x\in\mathbf{R}(X)} \sum_{P\in\gamma_{\mathbf{u}}(X=x)} xP. \tag{7.4}$$

The correspondence conditions ensure that the resulting operator is context independent; it is the same in every context in which a magnitude of this class can be defined.

Let us further note that if we use the probability measure (6.7) of standard QM, leading to

$$p_{\mathbf{u},W}(X = x) = \sum_{P\in\gamma_{\mathbf{u}}(X=x)} \mathrm{Tr}[PW], \tag{7.5}$$

then for the two corresponding magnitudes X and Y we have, by virtue of (7.3):

$$p_{\mathbf{u},W}(X = x) = p_{\mathbf{v},W}(Y = x), \tag{7.6}$$

i.e., probabilities are to some extent context independent and corresponding magnitudes are also statistically equivalent, *at least under this measure.*

Note: This statistical equivalence is evidently a prerequisite for identifying the magnitudes in a given correspondence class and for considering them as a single magnitude, as in standard QM. It is questionable, however, whether statistical equivalence is also sufficient for such an identification to be logically coherent. In standard QM, what makes it possible *not* to split the magnitudes is that they are not magnitudes in the sense of classical logic.

Note: The condition that statistical equivalence is required to identify the contextual magnitudes in a given correspondence class also implies that in standard QM one can admit only probability measures for which this statistical equivalence is satisfied. A strange consequence of this is that the ABL measure, which is obtained from the standard measure simply by closing it under conditionalization, must be excluded. This may explain to some extent why it lasted until 1964 (!!) before this measure appeared in the literature, and why it has had so little impact even after its publication. This fact also marks a clear instance of how the development of QM has been obstructed by holding on to the notion of the Hilbert space as representing the complete state space of QM.

Example: In order to see that corresponding magnitudes are not generally statistically equivalent under the ABL measure, let us consider a Hilbert space spanned by three orthogonal vectors $|1\rangle$, $|2\rangle$, and $|3\rangle$. Let $\mathbf{u}$ and $\mathbf{v}$ be two contexts, with $\Gamma_{\mathbf{u}}$ and $\Gamma_{\mathbf{v}}$ given by

$$\Gamma_{\mathbf{u}} = \{|1\rangle\langle 1|\, , |2\rangle\langle 2|\, , |3\rangle\langle 3|\},\ \Gamma_{\mathbf{v}} = \{|1\rangle\langle 1| + |2\rangle\langle 2|\, , |3\rangle\langle 3|\}. \tag{7.7}$$

Let $X_{\mathbf{u}}$ and $X_{\mathbf{v}}$ be two corresponding magnitudes in $\mathbf{u}$ and $\mathbf{v}$, with

$$\begin{aligned} X_{\mathbf{u}}(|1\rangle\langle 1|) &= X_{\mathbf{u}}(|2\rangle\langle 2|) = X_{\mathbf{v}}(|1\rangle\langle 1| + |2\rangle\langle 2|) = a \\ X_{\mathbf{u}}(|3\rangle\langle 3|) &= X_{\mathbf{v}}(|3\rangle\langle 3|) = b. \end{aligned} \tag{7.8}$$

If we now take a W_{in} and a W_{out} given by

$$\begin{aligned} W_{\text{in}} &= \frac{1}{3}(|1\rangle + |2\rangle)(\langle 1| + \langle 2|) + \frac{1}{3}|3\rangle\langle 3| \\ W_{\text{out}} &= \frac{1}{3}(|1\rangle - |2\rangle)(\langle 1| - \langle 2|) + \frac{1}{3}|3\rangle\langle 3|, \end{aligned} \tag{7.9}$$

then substitution in the ABL measure straightforwardly gives

$$\begin{aligned} & p_{\mathbf{u},W_{\text{out}},W_{\text{in}}}(X = a) \\ &= C_{\mathbf{u},w}(\text{Tr}[W_{\text{out}}|1\rangle\langle 1|W_{\text{in}}] + \text{Tr}[W_{\text{out}}|2\rangle\langle 2|W_{\text{in}}]) \\ &= 2/3 \\ & p_{\mathbf{u},W_{\text{out}},W_{\text{in}}}(X = a) \\ &= C_{\mathbf{v}}\text{Tr}[W_{\text{out}}(|1\rangle\langle 1| + |2\rangle\langle 2|)W_{\text{in}}] \\ &= 0. \end{aligned} \tag{7.10}$$

Consequently, $X_{\mathbf{u}}$ and $X_{\mathbf{v}}$ are not statistically equivalent.

Note: The present way of ordering the projection operators within contexts is based on taking the disjoint union of the contextual state spaces. One may consider the idea of simply taking their union, and collect all projection operators on $\mathcal{H}$ in a set $\mathcal{P}$. The quantum observables may now be defined as *partial* functions on $\mathcal{P}$, with

$$\text{Dom}(X) = \{P \in \mathcal{P} \mid \mathsf{X}P = xP, \text{ for some } x \in \mathbf{R}(X)\}.$$

One may, thus, define them as ordinary functions even without splitting them. However, the magnitudes defined this way do not directly relate to observable quantities, because performing a measurement and assigning probabilities in QM always implies imposing a particular context. Therefore, instead of measuring such a magnitude one always measures some particular 'submagnitude' derived from it, and in this sense they remain split.

7.3.1 Semiseparability

The statistical equivalence of corresponding operators in standard QM rests upon a condition which we shall call semiseparability and which has played an important role in various discussions about QM, concerning subjects like locality and hidden variables. The semiseparability condition, which expresses a limited form of context independence, can be formulated as follows:

Definition: Consider a set Ψ of contexts $\mathbf{u}$ with state spaces $\Gamma_{\mathbf{u}}$, and let there be equivalence classes of *corresponding states* in different contexts, i.e., if $P \in \Gamma_{\mathbf{u}}$ then for some $\mathbf{v} \in \Psi$ there may be a corresponding state $P \in \Gamma_{\mathbf{v}}$. Now let p be a family of probability distributions $p_{\mathbf{u}}$ on the algebras $\mathcal{F}_{\mathbf{u}}$ of measurable subsets of the contextual state spaces $\Gamma_{\mathbf{u}}$. This family will be called *semiseparable* iff (if and only if) for all sets of corresponding states we have

$$p_{\mathbf{u}}(P) = p_{\mathbf{v}}(P). \tag{7.11}$$

It is readily verified that for the probability measures of standard QM, $p_{\mathbf{u},W}(P) = \mathrm{Tr}[PW]$ if $P \in \Gamma_{\mathbf{u}}$, this condition is satisfied for arbitrary W, i.e., for every $W \in \mathcal{D}$:

$$p_{\mathbf{u},W}(P) = p_{\mathbf{v},W}(P) \text{ if } P \in \Gamma_{\mathbf{u}} \text{ and } P \in \Gamma_{\mathbf{v}}. \tag{7.12}$$

The relevance of this condition for the hidden variable issue can be explored by using Gleason's theorem[14], which says that if $\mathcal{H}$ is a complex Hilbert space of dimension greater than or equal to 3 and μ is a measure on the subspaces of $\mathcal{H}$ which is additive in the sense that, for mutually orthogonal P_i:

$$\sum_i \mu(P_i) = \mu(\sum_i P_i), \tag{7.13}$$

then there is a self–adjoint operator W such that

$$\mu(P) = \mathrm{Tr}[PW]. \tag{7.14}$$

Let us now imagine a hidden variable theory $\mathcal{HV}$, underlying QM in the sense that all predictions of QM can be reproduced by $\mathcal{HV}$. We assume it to be a theory with a classical logical structure, with a state space $\Gamma_{\mathbf{hv}}$ and with physical magnitudes defined as functions on $\Gamma_{\mathbf{hv}}$. The empirical predictions of standard QM are given in the form of probabilities $p_{\mathbf{u},W}(P)$, and, therefore, we assume that $\mathcal{HV}$ can produce matching probabilities $p_{\rho,\mathbf{u}}(P)$, where ρ is a probability distribution over the hidden variable states $\xi \in \Gamma_{\mathbf{hv}}$. The contexts $\mathbf{u} \in \Psi$ are generated, as before, by the resolutions of the unity operator $1\!\!1$ on a Hilbert space $\mathcal{H}$, and in order to be able to apply Gleason's theorem we assume that $dim\ \mathcal{H} \geq 3$. We now prove the following lemma:

Lemma: Assume that Ψ is a complete set of contexts, in the sense that to every resolution of $1\!\!1$ there is a matching $\mathbf{u} \in \Psi$, i.e., with $\Gamma_{\mathbf{u}}$ generated by this resolution. Assume, furthermore, that the probabilities produced by $\mathcal{HV}$ satisfy the semiseparability condition

$$p_{\rho,\mathbf{u}}(P) = p_{\rho,\mathbf{v}}(P). \tag{7.15}$$

On these premises it follows that there is a density operator W_ρ such that $p_{\rho,\mathbf{u}}(P) = \mathrm{Tr}[PW_\rho]$.

Proof: Let $\Lambda \subseteq \Gamma_{\mathbf{u}}$ be some state in a context $\mathbf{u} \in \Psi$. We now define

$$\begin{aligned} P_\Lambda &:= \sum_{P\in\Lambda} P, \\ P_\Lambda^\perp &:= 1\!\!1 - P_\Lambda, \\ \mathbf{v} &:= \text{the context with } \Gamma_{\mathbf{v}} = \Lambda \cup \{P_\Lambda^\perp\}, \\ \mathbf{w} &:= \text{the context with } \Gamma_{\mathbf{w}} = \{P_\Lambda, P_\Lambda^\perp\}. \end{aligned} \tag{7.16}$$

[14]Gleason [100], Cooke, Keane and Moran [45].

From the semiseparability condition (7.15) it follows that

$$\begin{aligned} p_{\rho,\mathbf{u}}(\Lambda) &= \sum_{P\in\Lambda} p_{\rho,\mathbf{u}}(P) \\ &= p_{\rho,\mathbf{v}}(\Lambda) \\ &= 1 - p_{\rho,\mathbf{v}}(P_\Lambda^\perp) \\ &= 1 - p_{\rho,\mathbf{w}}(P_\Lambda^\perp) \\ &= p_{\rho,\mathbf{w}}(P^\Lambda). \end{aligned} \tag{7.17}$$

This construction can be copied for any arbitrary state Λ, existing in any arbitrary context $\mathbf{u}$. We can, therefore, construct a context independent measure p_ρ such that, for every $\mathbf{u} \in \Psi$ and every $\Lambda \in \Gamma_{\mathbf{u}}$:

$$\begin{aligned} \sum_{P\in\Lambda} p_{\rho,\mathbf{u}}(P) &= \sum_{P\in\Lambda} p_\rho(P) \\ &= p_\rho(\sum_{P\in\Lambda} P). \end{aligned} \tag{7.18}$$

This p_ρ is an additive measure on the subspaces of $\mathcal{H}$ and therefore we can apply Gleason's theorem and conclude that there is a W_ρ such that $p_{\rho,\mathbf{u}}(P) = \mathrm{Tr}[PW_\rho]$.
□

Corollary: From this lemma we can conclude that magnitudes in $\mathcal{HV}$ must be split.

Proof: Let $\eta(\xi)$ be the distribution on $\Gamma_{\mathbf{hv}}$ which is concentrated entirely on ξ. Because magnitudes in $\mathcal{HV}$ are simply functions on $\Gamma_{\mathbf{hv}}$, the probabilities $p_{\eta(\xi),\mathbf{u}}(P)$ can take only the values 0 or 1, for all $\mathbf{u}$ and all P. For a probability given by $\mathrm{Tr}[PW]$ this is not possible. Therefore semiseparability must be violated and, consequently, corresponding magnitudes cannot be generally statistically equivalent and must be split. □

Note: The statement that contextual magnitudes must at least be statistically equivalent if they are not split, contains a slight complication. Let $\Gamma_{\mathbf{hv}}$ be a state space on which magnitudes are defined as functions $X : \Gamma_{\mathbf{hv}} \to \mathbf{R}(X)$ and let ρ be a probability distribution on $\Gamma_{\mathbf{hv}}$. Now consider the case that a change of context leaves the magnitudes unaltered, but affects ρ instead, i.e., instead of a single distribution ρ we have a set of contextual hidden variable distributions $\rho_{\mathbf{u}}$. In the terminology of an earlier publication[15] such a theory would be called a theory with contextual states, and it seems to provide a counterexample of the idea that *magnitudes* must be split if they are not statistically equivalent. It seems as though the probability distributions can be split instead. This view, however, was based on less strict ideas about the relation between states and magnitudes than those propounded in the present work. Within the present framework the situation may be analyzed by noting that, within a theory with such a family of

[15] Hoekzema [121].

distributions $\rho_{\mathbf{u}}$, $\Gamma_{\mathbf{hv}}$ is not the complete state space. The full specification of the state of a system would also require a specification of the applying context, and, accordingly, the contexts must be represented in the state space. Therefore, the complete state space must minimally be something like $\Psi \times \Gamma_{\mathbf{hv}}$, or in a worse case perhaps $\Psi \rightarrow \Gamma_{\mathbf{hv}}$. On such a state space one can have a probability distribution $\sigma(\mathbf{u}, \Lambda)$ such that the contextual distribution $\rho_{\mathbf{u}}$ is expressed by

$$\rho_{\mathbf{u}}(\Lambda) = \sigma(\mathbf{u}, \Lambda)/\sigma(\mathbf{u}, \Gamma_{\mathbf{hv}}). \tag{7.19}$$

Evidently, however, magnitudes on this new state space are split after all, because X is, in fact, composed of many contextual magnitudes $X_{\mathbf{u}} : \{\mathbf{u}\} \times \Gamma_{\mathbf{hv}} \rightarrow \mathbf{R}(X)$.

7.3.2 Generalizations

The above considerations about magnitudes were based on the initial quantum state space of triples $< \Gamma_{\mathbf{u}}, P, W >$. Many details must be revised, however, in a quantum process theory where EVMs, OVMs and the ABL measure enter the scene. The basic idea of defining physical observables as functions on the contextual state spaces can be retained, but the representation of these magnitudes by self-adjoint operators, one of the central features of standard QM, no longer holds, at least not in quite the same way. To a given contextual magnitude $X_{\mathbf{u}}$, defined as a function on the set of operations in a given state space $\Gamma_{\mathbf{u}}$, one might still define, in analogy to (7.4), a corresponding operator X by using the effects $R(k)$ associated with the operations ϕ_k and take

$$\mathsf{X} = \sum_{x \in \mathbf{R}(X)} \sum_{\phi_k \in \gamma_{\mathbf{u}}(X=x)} x R(k). \tag{7.20}$$

Starting from a probability measure $p(X = x) = \mathrm{Tr}[\sum_{\phi_k \in \gamma_{\mathbf{u}}(X=x)} R(k)]$, the expression $\mathrm{Tr}[\mathsf{X}W]$ gives the correct expectation value for the magnitude X, but this is about as far as the analogy with standard QM goes. An important difference is, for instance, that the set of eigenvalues of X is not equal to the set of values of the magnitude X. Moreover, there can be quite different magnitudes within a given context which, nevertheless, correspond to the same operator. Apparently, the relation between magnitudes and operators is not as fundamental as it seemed to be in standard QM. Such things might at first seem a bit strange and certainly require getting used to. Moreover, there may be a range of further consequences which will have to be examined and sorted out. We shall not presently go into more detail, however. All this should be regarded as part of a major change of attitude with respect to QM, and the main lines of a new approach should be clear before we settle down on the details.

7.4 The incompleteness of the standard formalism of QM

Gleason's theorem is generally used as an argument against hidden variable theories, but, as noted earlier, it is worthwhile to slightly extend the argumentation to draw conclusions about QM itself. Hidden variable theories are often thought of as theories without strange features like indeterminism or complementarity, but it may be noted that the demands imposed above on hidden variable theories are extremely minimal. No effort was made at all to exclude either a nondeterministic dynamics or complementarity. In fact, the only thing required is that the theory has a state space in which observed facts can be accurately represented by subsets. Within the classical logical framework this expresses merely the condition that the theory is complete[16] with respect to its intended domain of empirical phenomena. Therefore, Gleason's theorem can be understood to imply that standard QM is incomplete, and the only way out would be to say that QM has a nonclassical logical structure and the classical definition of completeness does not apply to it. Presumably, the alternative would then be quantum logic[17], but I would not know an adequate answer to a sceptic arguing that quantum logic is merely a way to incorporate incompleteness into the logical framework itself. Contextual QM, on the other hand, provides a direct and simple way to *make* QM complete, by adding sections to the state space in a manner which amounts to little more than pasting Boolean algebras in a rather straightforward manner. Gleason's theorem then makes clear why, inevitably, magnitudes must be split in contextual QM. According to this analysis, one may say that the main conclusion to be drawn from Gleason's theorem is that standard QM is incomplete and its magnitudes must be split.

Note: The addition of the contextual variables creates a resemblance between contextual QM and hidden variable theory, besides the fact that the additional contextual variables are not hidden but, on the contrary, serve precisely to represent the observable facts themselves. In agreement with this, probability distributions violating the semiseparability condition are constructed as easily as in a hidden variable theory. One may, for instance, define pure probability distributions

$$\begin{aligned} p_{\mathbf{u},P}(P') &= 1 \quad \text{iff } P' = P \\ &= 0 \quad \text{otherwise,} \end{aligned} \tag{7.21}$$

existing only within a given context $\mathbf{u}$. The circumstance that the probabilities produced by the dynamical equations of standard QM are not of this type does not affect the existence of such distributions as well–defined mathematical entities. Further, we may note that the ABL measure also produces probabilities violating semiseparability.

[16]'Strongly complete', in the terminology of § 10.1.4.
[17]See § 8.2.

Completeness and determinism

At earlier occasions in this book I used the term "malignantly incomplete" rather than merely "incomplete". The reason for regarding the particular form of incompleteness of QM as particularly malignant is, in the first place, that it affects central parts of the theory which are crucial for its empirical interpretation, namely the interpretation of all physical magnitudes. Furthermore, however, it has contributed to the survival of some highly misleading illusions; first of all, the idea of a presumed close analogy between classical mechanics and QM based on the conception of Hilbert space as the conceptual equivalent of a classical phase space. In my view, the low degree of acceptation of OVMs and of the ABL measure among physicists may be directly related to the desire to hold on to this analogy. The theory of OVMs indicates that the conception of a state as a subspace of the Hilbert space is inadequate, whereas the ABL measure points at the necessity of splitting magnitudes in QM. This makes both of them inconsistent with the existing quantum mythology.

It is worthwhile observing how, in fact, the concept of completeness goes through the debate on QM as Ariadne's thread, and also through our own analysis. In order to see this, let us first draw a proper distinction between 'completeness' and 'determinism'. A suitable concept of completeness in the present context would be to say that a theory is *complete*, relative to the phenomena it is aimed to represent, if 'it has enough states' in the sense that there should be a matching theoretical possible world corresponding to every phenomenological possible world; and let us further call the theory *strongly complete* if different phenomenological worlds correspond to different theoretical worlds[18]. Determinism, on the other hand, is generally understood to mean that the state of a system at one moment in time determines the state at any other moment.

The most ardent defender of the completeness of QM was presumably Niels Bohr. Bohr did not defend the completeness of the quantum *formalism*, however. On the contrary, he always stressed that the description of a physical experiment should take place in *classical terms*, and he had his reservations about the formalism. In practice, however, QM is often assumed to be complete in the sense that the state vector gives a maximally complete description of the state of a system, which boils down to assuming the completeness of the quantum formalism after all. A problem one then encounters directly is that a state vector generates a family of probability measures, assigning probabilities to the values of physical magnitudes pertaining to the system, but it does not say which magnitudes pertain to the system, nor which values these magnitudes have. This means, in fact, that the state description by means of a state vector is *incomplete*, but in practice this state of affairs is usually discussed as a symptom of the *indeterminism* of QM, i.e., QM is considered indeterministic because the 'state' does not determine the assignment of values to magnitudes.

With the standard meaning of the notion of completeness taken over by the concept of determinism, what is "complete" supposed to mean? Presumably, it

[18]See also § 10.1.4.

still means that QM has enough states, but now in the sense that it has enough probability measures, i.e., if the state vector provides a complete description of a system, this should mean that every theoretically possible probability measure in QM can be described by a vector in $\mathcal{H}$. This is not true, however, because the occurrence of improper mixtures, which cannot be interpreted as classical mixtures, makes it necessary to use the wider class of density operators. Further, using density operators is not enough, because the resulting set of probability measures is not closed under conditionalization and must be extended to include the ABL measures.

A short-cut argument

Surveying our own discussion of the previous chapters, it can be noted that various major topics are, in fact, intimately connected with the issue of completeness, either of the set of probability measures or of the set of state descriptions. Consider for instance the following argument, which provides an interesting short cut through various parts of our analysis.

Let us start with the set of measures coming from Gleason's theorem, i.e., the density operator measures of type (4.4):

$$p_W(P) = \mathrm{Tr}[PW].$$

This set can be considered incomplete in the sense that it is not closed under various operations we may want to perform on it. In the first place, it is not closed under composition. If a sequence of successive measurements is performed, then the probability for finding a result $< P_N, t_N; \cdots; P_1, t_1 >$ is given by

$$p_W(P_N, t_N; \cdots; P_1, t_1) = \mathrm{Tr}[P_N U_N \cdots P_1 U_1 W P_1 U_1^+ \cdots P_N U_N^+],$$

which cannot be written in the above form, because generally there is no projection equal to $P_N U_N \cdots P_1 U_1$. One may overcome this problem by admitting, besides projections, more general operators V; and this boils down to introducing OVMs, which leads to admitting measures of type (5.17):

$$p_W(K) = \mathrm{Tr}[\phi_K(W)].$$

It then turns out that introducing OVMs has greater benefits than merely closing the measure under sequencing: OVMs are applicable for describing any interactions between a quantum system and a black box producing measurement outcomes.

As we have discussed earlier, however, this larger set of measures is itself incomplete because it is not closed under conditionalization. Closure of the set of measures in this case leads to the ABL measure, i.e., to admitting measures of type

$$p(K) = C\,\mathrm{Tr}[W_{\mathrm{out}}\phi_K(W_{\mathrm{in}})].$$

Once more it turns out that this has greater benefits than merely closing the set of measures under, in this case, conditionalization. As a side effect, considerations about the time-symmetric character of the theory are no longer complicated by undue restrictions on the use of probability theory. There is more, however. By merely looking at the above expression we obtain a list of what is apparently the minimal set of parameters needed to describe a physical process, such as an experiment. Tuples of the form $< W_{\text{out}}, \phi, W_{\text{in}} >$ seem appropriate to do the job, i.e., with practically no trouble at all we have arrived at the process states (6.8).

Evidently, the above argument ignores many problems of interpretation and, on its own, does not provide a satisfactory basis for setting up a new QM. Even so, it has at least one major benefit: in a few simple steps it leaves the standard formalism way behind, and this is by itself a big step forward. In the standard formalism QM is made to work by means of a magical recipe, establishing a sort of analogy with classical mechanics. This analogy has certainly brought us a memorable episode in the history of physics, and although it breaks down as soon one tries to touch it, it has proved to be immensely enchanting and delusive. The sooner we leave it behind, the better.

7.5 Locality

Clearly, the term "hidden variable theorem", used for results as obtained by Von Neumann, by Gleason, and by Kochen and Specker, is misleading. These theorems tell us something about the algebraic structure of magnitudes in any classical theory which is empirically equivalent to QM. However, this includes QM itself if it is interpreted classically, as in contextual QM. A similar conclusion holds with respect to Bell's theorem. Consequently, any classical theory which is empirically equivalent to QM must be nonlocal, included QM itself, if interpreted classically, leaving the choice between a theory with instantaneous causality, i.e., influence outside the lightcone, or a theory with retrocausality, i.e., influence back in time.

It may be noted that, if one holds on to a classical state concept, it is not necessary to use Bell's theorem for showing the nonlocality of a theory in which a wave function or a density operator is part of the state description of a system. State reduction is a nonlocal affair, as was noted in § 4.2.5. Therefore, in order to hold on to the idea of QM itself as a local theory, one must either resort to some weak form of locality, or turn down the interpretation of the wave function as describing, either completely or incompletely, the physical state of a quantum system, at least in a classical sense. The latter possibility leads either to a nonclassical state concept, and, thus, to a nonclassical logical framework, or to a view in which the wave function is regarded as describing our *knowledge* about a system, rather than the *state* of the system. The latter raises the question of how the state of the system must be described if, apparently, the wave function does not do the job. However, nonclassical logic is not our present business, whereas, for any clasical state concept whatsoever, Bell's theorem provides proof that nonlocality cannot be avoided anyway, unless one adopts a 'conspiracy theory', in which there are systematic correlations between the state of the systems we measure and our decisions about what quantities we decide to measure. We shall reject the latter alternative and, accordingly, we have no way of keeping up the illusion of QM as a local theory.

In view of the fact that so little is noticable of the nonlocality we appear to be facing, introducing some weaker form of locality seems an obvious strategy, regardless of whether such a concept might be fully satisfactory in a philosophical sense, which it is not. If, apparently, strict locality is violated by QM, what coincidence causes the weak, observational form of locality which seems to prohibit either out of lightcone signalling between distant observers, or observation of future events? It will be argued that the answer to this question is not contained in QM itself. The theory actually places no restrictions whatsoever on the occurrence of observably nonlocal effects. Instead, the natural place to look for a solution of the problem of why weak locality apparently holds is presumably in the cosmological boundary conditions. Some speculation about this subject will be given below, including a tentative reason for a form of weak locality, based on cosmological considerations within a retrocausal framework. First, however, we take a look at Bell's theorem and make sure that, indeed, nonlocality is unavoidable also within a contextual approach.

7.5.1 Bell's theorem

Bell's original theorem was applicable only to deterministic hidden variable theories, but various generalizations have been given later. Still one might conceive of the possibility that Bell's theorem might not be valid for contextual theories. In fact, I have held such a view myself[19], at a time when I had less strict ideas about logical structure. In a contextual approach based on classical logic, however, nothing interferes with deriving a version the Bell inequalities, and the same conclusion holds with respect to theories based on process states, rather than states of systems evolving in time. All that is needed is some suitably formulated version of the required locality criteria. These point may be illustrated by giving a derivation of the Bell inequalities in a general type of contextual process theory.

Without assuming a quantum mechanical approach, consider two processes $\mathcal{P}^A$ and $\mathcal{P}^B$ in two separate regions of spacetime. In the process $\mathcal{P}^A$ a measurement is performed. It will be assumed that the corresponding observer $\mathcal{O}^A$ can choose between two modes of the measurement apparatus, denoted by $a \in \{1,2\}$, and the measurement outcome is either $A = 1$ or $A = -1$, in either mode of the apparatus. Process $\mathcal{P}^B$ is entirely similar; the observer $\mathcal{O}^B$ may choose between two modes $b \in \{1,2\}$ and obtains outcomes $B = 1$ or $B = -1$. In contextual terms this may be interpreted in the following way. The different apparatus modes of $\mathcal{P}^A$ and $\mathcal{P}^B$ determine different contexts $\mathbf{u}_a^A$ and $\mathbf{u}_b^B$. Each context $\mathbf{u}_x^X$ has a corresponding state space $\Gamma_x^X = \{1,-1\}$, each one with two states, corresponding to the two possible outcomes.

Let us assume that some theory $\mathcal{T}$ assigns probabilities to these outcomes. One may distinguish different types of probabilities, and not all of them are necessarily

[19]See [121].

well–defined in one theory. For instance, one may have a theory about probabilities $p_a(A)$ and $p_b(B)$ in which a joint probability $p_{ab}(A,B)$ does not exist, or *vice versa.* We shall assume that at least probabilities of the type $p_{ab}(A,B)$ are well–defined in $\mathcal{T}$. In this case one may define a correlation function C_{ab} by

$$C_{ab} = \sum_{A,B=\pm 1} AB p_{ab}(A,B) \tag{7.22}$$

and a *Bell expression* S by

$$S = |C_{11} + C_{12} + C_{21} - C_{22}|. \tag{7.23}$$

Locality assumptions

Let us assume a spacelike separation of the two regions of spacetime in which the processes occur, i.e., $\mathcal{P}^A$ and $\mathcal{P}^B$ are outside each others lightcone. In this case, one may impose conditions on the probability distributions of $\mathcal{T}$, on the basis of some locality or causality principle. As a minimal demand one may require that superluminal communication between $\mathcal{O}^A$ and $\mathcal{O}^B$ should be impossible. This leads to the requirement that $p_a(A)$ must exist and be equal to

$$p_a(A) = \sum_B p_{ab}(A,B), \tag{7.24}$$

and similarly for $p_b(B)$. If this condition is not satisfied, then $\mathcal{O}^A$ may determine which mode b is selected by $\mathcal{O}^B$, by measuring the frequencies obtained in an ensemble of realizations of this experiment. This could be used for signalling at superluminal velocities.

There are, besides the impossibility of superluminal communication, stronger locality demands which should be imposed. One would, in particular, like to express the idea that processes occurring in spacelike separated regions are in some way physically independent. A very strong condition which would, indeed, express this idea is the factorization of the joint probability, i.e.

$$p_{ab}(A,B) = p_a(A)p_b(B). \tag{7.25}$$

This requirement is evidently too strong, however, as it would imply that there can be no correlations at all between the measurement results in the processes $\mathcal{P}^A$ and $\mathcal{P}^B$. A more adequate representation of customary ideas about causality would be to require that, if there are such correlations, these should have their origin in physical processes occurring in the shared past of $\mathcal{P}^A$ and $\mathcal{P}^B$, i.e., within the intersection of their past lightcones. In this case, one may introduce a context **sp** (shared past), in which all relevant aspects of all relevant processes in this region are represented[20].

The context **sp** has a state space $\Gamma_{\mathbf{sp}}$, and causality may now be expressed by the demand that

$$p_{ab}(A,B) = \int_{\lambda \in \Gamma_{\mathbf{sp}}} p_a(A,\lambda) p_b(B,\lambda) d\rho(\lambda), \tag{7.26}$$

[20]See also § 10.2.4.

where ρ is a probability measure on the subsets of $\Gamma_{\mathbf{sp}}$. Using this condition, while defining $A_a(\lambda) = \sum_A Ap_a(A,\lambda)$ and $B_b(\lambda) = \sum_B Bp_b(B,\lambda)$, the correlation function (7.22) becomes equal to

$$C_{ab} = \int_{\lambda\in\Gamma_{\mathbf{sp}}} A_a(\lambda)B_b(\lambda)d\rho(\lambda). \tag{7.27}$$

For the Bell expression this gives

$$S = \left| \int_{\lambda\in\Gamma_{\mathbf{sp}}} \left[A_1(\lambda)\big(B_1(\lambda) + B_2(\lambda)\big) + A_2(\lambda)\big(B_1(\lambda) - B_2(\lambda)\big)\right] d\rho(\lambda) \right|. \tag{7.28}$$

This immediately leads to a Bell inequality $S \leq 2$, that is well known to be violated by spin experiments in QM.

Clearly, the above derivation closely follows standard routines, involving only a minor reformulation into a slightly different terminology.

7.5.2 The observability of nonlocal effects

Having established the nonlocality of QM, the next question is why so little of this is observed in the laboratory, or even in daily life. For instance, why is superluminal signalling apparently forbidden? Is there, anywhere in physics, some 'weak locality' principle at work, which constrains signalling, although without inhibiting the existence of Bell correlations? If so, how can we formulate such a principle more precisely, and in which part of physics should we be looking for it?

In standard QM one has formulated such principles; the best known, presumably, concerning the commutation of spacelike separated observables in quantum field theory. However, there are direct links between the set of probability measures admitted by the theory and the validity of locality principles. Contextual quantum process theory admits a more general class of measures than the standard formalism. Do these measures satisfy the same locality principle; is weak locality valid in contextual QM? The answer to this question is a definite *no*. Within contextual process theory itself nothing prohibits the occurrence of observable nonlocality effects, as may be illustrated by an example.

Example: Three physicists $\mathcal{O}^A$, $\mathcal{O}^B$, and $\mathcal{O}^C$ collaborate in an experiment. At the time of their relevant activities, $\mathcal{O}^A$ and $\mathcal{O}^B$ are spacelike separated, whereas $\mathcal{O}^C$'s contribution to the experiment is in their shared past. For simplicity, we shall here use a particle language for describing the experiment, even though, incidentally, the three physicists themselves use event theory, and have adopted a corresponding terminology.

As agreed beforehand, $\mathcal{O}^C$ carries out a number of state preparation procedures of pairs of correlated spin–1/2 systems, of the kind appearing in Bohm's version of the EPR experiment, i.e., with spin states given by

$$|\psi\rangle = 2^{-1/2}\left(|+,-\rangle - |-,+\rangle\right). \tag{7.29}$$

After the state preparation, one of the particles of each pair is sent to $\mathcal{O}^A$ and the other to $\mathcal{O}^B$. A second arrangement agreed beforehand, is that $\mathcal{O}^B$ will carry out either one of two retroparation procedures, on the entire set of particles he receives. He has agreed to choose between either finding a particle with spin up in **z**–direction, i.e., described by $\langle +|$, or a particle with spin up in **x**–direction, i.e., described by $2^{-1/2}(\langle +| + \langle -|)$. Because the particles received by $\mathcal{O}^A$ and $\mathcal{O}^B$ are correlated pairs, it is possible for $\mathcal{O}^A$ to find out which decision $\mathcal{O}^B$ has made. If $\mathcal{O}^B$ has decided to find spin up in **z**–direction, then all his own particles should have spin down in **z**–direction. If, on the other hand, $\mathcal{O}^B$ has decided to find spin up in **x**–direction, then his own particles should have spin down in **x**–direction. The difference between the two possibilities is easily measured. In this way, $\mathcal{O}^A$ and $\mathcal{O}^B$ may establish superluminal communication, thus contradicting views of earlier generations of physicists

Obviously, there is something rather unusual about this example. The idea of $\mathcal{O}^B$ *choosing* to find particles with spin up in **z**–direction is not entirely standard. However, regardless of whether or not it is physically possible for him to do this, the point is that there is nothing *within quantum process theory* which would inhibit it. Quantum process theory is a basically time–symmetric theory, in which it is not *a priori* stranger for $\mathcal{O}^B$ to decide *retro*paring particles with spin up in **z**–direction, than it is for $\mathcal{O}^C$ to decide *pre*paring pairs of particles in a state $2^{-1/2}(|+,-\rangle - |-,+\rangle)$. What goes wrong here, is that common sense ideas about free will are not appropriate for using then within the framework of a time–symmetric theory. If we want to do justice to the time–symmetric character of a theory, however, this means that we must build intuitions which respect this, and that common sense ideas about time–asymmetry and about constraints on free will may be entered only at a later stage. In order to help building such intuitions, one may, for example, imagine that certain experiments carried out two decades ago had a different result.

Example: If we shine at the sky with a flashlight, or a spotlight, or a laser beam, and we measure the power output by means of an amperemeter than what we expect to find is that the output is independent of the direction in which we shine. It is not *a priori* evident, however, that this is physically necessary. In fact, Wheeler and Feynmans time–symmetric theory of electromagnetism[21] contains indications that this condition might not hold at all, and for this reason experiments have been performed to test it[22]. Results agreed with common sense expectations.

Imagine, however, that one day the test is repeated more accurately, and with a different outcome. Suppose, in particular, that spots are found in the sky which do not accept any light we try to send there, i.e., shining in those directions, power output drops to zero. Such a result may immediately be put to use, both for superluminal signalling and for signalling back in time. Consider the three

[21] [221, 220].
[22] Partridge [166].

physicists of the previous example. They set up an experiment similar to the one with the particles, but, in the present case, $\mathcal{O}^C$ prepares correlated photon pairs, say a billion pairs per second. Of each pair, one photon is sent to $\mathcal{O}^A$, and the other one to $\mathcal{O}^B$. However, $\mathcal{O}^B$ has a 'transmitter' which works in either of two modes. In 0–mode, the beam of photons received from $\mathcal{O}^C$ is redirected towards one of the 'dark spots' in the sky. This spot will not absorb them, however, and, therefore, $\mathcal{O}^C$'s preparation apparatus cannot emit them, and the power output of his apparatus drops to zero, which he may readily notice. Consequently, observer $\mathcal{O}^A$ does not receive any photons either, and this too is easily observable for him. In 1–mode, on the other hand, $\mathcal{O}^B$'s beam transmitter works as a receiver, absorbing photons and transforming them into a signal. By switching between the modes, say some thousands or millions of times per second, $\mathcal{O}^B$ can send signals both to $\mathcal{O}^C$ and to $\mathcal{O}^A$. Moreover, if $\mathcal{O}^A$ has a similar transmitter and $\mathcal{O}^C$ uses the power switch of his apparatus, then the three of them may communicate as much as they like in whatever direction they want.

Clearly, in as far as experiments like these are inhibited at all, QM does not seem the proper place to apply the corresponding constraints, at least not if one takes the retrocausal approach to nonlocality. Indeed, cosmology seems a much better area to search for an explanation of the apparent time–asymmetry we, in fact, observe.

7.6 Speculations on time–symmetry, causality, and quantum gravity

7.6.1 ABL associated cosmology

A difference between the retrocausal and the nonlocal approaches presented in ch. 6 is that in a retrocausal theory the operator W_{out} represents independent degrees of freedom, whereas in the nonlocal approach it is an artefact, defined in terms of the history of other state variables. In the latter case W_{out} can serve to be used in the ABL measure, for describing experiments in which one desires to conditionalize on the values of arbitrary measurement outcomes rather than on the prepared state only. However, it lacks the more fundamental relevance which is present in retrocausal theory. We shall proceed by discussing some tentative ideas about the cosmological significance of the ABL measure in a retrocausal theory, focussing in particular on finding an explanation for the apparent absence of observable retrocausality effects.

Let us briefly reflect on what it means for us to look at the outside world, for instance by means of astronomical observation. Instead of making measurements on a system, fixed into a state W_{in} by means of laboratory equipment, we are then looking directly at the structure of a cosmic W_{in}, prepared by nature itself; and we may describe our observations in terms of a statistically mixed operator $\bar{W}_{\text{in}}$, giving for instance the statistical distribution of photon states if we point a telescope at a certain star. What we get to see, for instance in the visible part of the electromagnetic spectrum, is that there are dots of light in a sky which, for the rest, is practically black, i.e., $\bar{W}_{\text{in}}$ varies widely in response to our moving the equipment. We may, likewise, investigate the structure of the cosmic $\bar{W}_{\text{out}}$. Therefore, let us consider what is more or less the timereversed situation. We take a flashlight, measure its output by means of an amperemeter and shine at the sky. What we get to see is that the output is independent of the direction in which we shine. This is a remarkable experimental fact, in need of explanation. As far as we presently know, the observations are in good agreement with the assumption that

$$\bar{W}_{\text{out}} = \mathbb{1} \, , \tag{7.30}$$

which we might interpret as saying that, in the negative energy part of the spectrum, the sky is equally bright in all directions and at all wavelengths, and in fact for all types of fields.

For a tentative explanation of this extraordinary state of affairs, a natural starting point is, what has become known as, Olbers' paradox: imagine an infinite, homogeneous, star–filled steady state universe. Elementary computations then imply that, for arbitrary finite densities of stars, the sky is homogeneously bright with a mean radiation temperature equalling the mean stellar surface temperature. In our own universe the blackness of the night sky is thought to be due to the finite age of the universe[23]. It is further assumed that we live in approximately an expanding Friedmann–Robertson–Walker universe in which the expansion has redshifted the primordial radiation, left over from the Big Bang, to a temperature of somewhat below $3K$. Owing to these conditions, the present universe is in a state far removed from thermodynamic equilibrium, and the inhomogeneous radiation we observe, bright stars in a dark sky, is one of the symptoms of this fact.

In the reversed direction of time these arguments do not apply. The timereversed age of the universe is likely to be infinite, and timereversed expansion is contraction. Owing to the latter, negative energy radiation reaching us from the far future is highly blue shifted. Suppose for a moment that the state $\bar{W}_{\text{out}}$ describes a universe in thermodynamic equilibrium. We may then tentatively write it in the form

$$\bar{W}_{\text{out}} = e^{-H/kT} . \tag{7.31}$$

For sufficiently high temperatures the difference between (7.31) and (7.30) may become unobservable.

Of course, the idea that $\bar{W}_{\text{out}}$ would describe a universe in thermodynamic equilibrium is highly disputable. A condition for a universe, or for a region in a universe, to eventually reach thermodynamic equilibrium is that the optical depth is finite. Computations[24] suggest that this condition may not be satisfied. Furthermore, the whole idea of a homogeneous universe may be rather far fetched. Instead of an open Friedmann universe, dominated by a single Big Bang, or a closed Friedmann universe, dominated by two equally large singularities, we may imagine a chaotic

[23] Bondi [30], Harrison [111], Wesson [216].

[24] See, e.g., Hoyle and Narlikar [127].

universe, with arbitrary numbers of singularities of any size. It would be interesting to know whether (7.30) could be correct in some approximation, even in very general circumstances.

An explanation for how this might be possible could presumably be based only on our proximity to a 'large' (in some suitable measure) singularity, producing, for some reason, a geometrically approximately flat region in the universe, with a high expansion rate over a prolonged period of time. Accordingly, if we are deep enough in the well, any negative energy radiation reaching us from the far future is highly compressed and blue shifted. If this effect is large enough, for the wavelengths at which we are able to look, then the difference between (7.30) and (7.31) may be negligible in any direction, in spite of the fact that this radiation may be inhomogeneous and not in thermodynamic equilibrium at all. In other words, at relatively low wavelengths $\bar{W}_{\text{out}}$ may become homogeneously saturated, and in this case the exchange rates of energy quanta are determined exclusively by $\bar{W}_{\text{in}}$.

This picture has major implications for the subject of time–asymmetry in physics, which may be illustrated by considering the behaviour of clocks. For simplicity, imagine a closed Bang–Crunch universe and a clock of which the worldline begins deep in the well of the Bang and ends deep in the well of the Crunch. Assume that, like all clocks we know, its functioning depends on an internal energy source where electromagnetic energy, in some form, is stored. As the clock ticks on, this energy slowly leaks into the environment. Accordingly, the functioning of the clock depends on its ability to radiate photons into the environment. Within an ABL associated cosmology this depends not only on the state of the clock and its energy source, but also on the ability of the environment to absorb photons. Deep in the well of the Bang, this condition may be satisfied by virtue of the saturation of $\bar{W}_{\text{out}}$. As the clocks begin to 'surface', however, we may find that the temperature of $\bar{W}_{\text{out}}$ drops and its inhomogeneity becomes apparent. As a result, the clock slows down and, in fact, its speed may even come to depend on its orientation. Finally, as the clock comes closer to the Crunch, the temperature of $\bar{W}_{\text{in}}$ begins to rise. Whereas the clock will have more and more difficulty in emitting photons, the pressure on it to absorb them increases. The clock will begin to run backwards and, as W_{in} becomes homogeneously saturated at the working wavelengths of the clock, its speed becomes regular and independent of orientation. This particular type of temporal symmetry will be hard to achieve by other means than by use of the ABL probability measure, as is well illustrated by Hawking's change of mind on this subject[25].

Evidently, in due time the above argument should be replaced by a more thorough analysis, encompassing more detailed considerations about cosmology and cosmic thermodynamics; but this is far outside the scope of our present considerations.

[25] [112] ch. 9.

7.6.2 Covariant gravitational force fields; the Equivalence Principle revisited

An immediate difficulty encountered upon trying to quantize general relativity is that, on the one hand, the metric field seems to be the proper thing to quantize, but, on the other hand, it is quite unclear how one should try to find solutions to field equations in which the metric field is itself a quantum field, rather than a numerical field. We shall link this problem to another topic in general relativity, concerning the question of how to define an energy–momentum tensor for the gravitational field. It is widely assumed that such a tensor does not exist, and that one can at most define a pseudotensor, lacking the proper transformation properties under nonlinear coordinate transformations.

It may be noted however that, in view of the equivalence principle, the nonexistence of a fully covariant gravitational energy–momentum tensor could be regarded as quite remarkable. The main historical role of Einstein's equivalence principle has been its application in establishing a fully geometric description of gravity. Equivalence works both ways, however, and, accordingly, it should also be possible to obtain a description of gravity as a pure force field. Curiously, it is widely suggested in the literature that a fully covariant treatment of gravitation as a force field cannot be given, owing to the, presumed, circumstance that the resulting gravitational energy–momentum tensor would not be a proper tensor but a pseudotensor, which does not transform covariantly under nonlinear transformations and which is, moreover, not uniquely determined. Wald writes, for instance[26]:

> "[...] As a result, we have no meaningful way of describing gravity as a force field"

> "However, [...], there is no known meaningful notion of the energy density of the gravitational field in general relativity."

It will be my aim to contradict this suggestion by using the backward equivalence principle for obtaining a bimetric reformulation of general relativity, in which gravity can be described either as a geometric phenomenon or as a force field, or as an arbitrary mixture of both. Gravitation as a force is generated by a tensor field, for which a fully covariant energy–momentum tensor is readily calculated. A further advantage of this approach is that it generates a theory with two metrics, one true metric, in the geometrical sense of the word, and a mathematically similar object, which can be given the interpretation of a gravitational potential. This yields the possibility of quantizing the force field, while retaining a numerical metric field. The equivalence principle itself can then be formulated as a gauge freedom for the gravitational field, expressing an invariance of the theory under metric transformations.

In general relativity the geometric treatment of gravity leads to a metric tensor $g_{\mu\nu}$. Clearly, if gravity is treated as a force field we can no longer use this particular metric because, otherwise, we obtain twice as much of it. In a pure force field

[26] [214], p. 67 and p. 286.

description of gravity the metric must be flat, i.e., must have a vanishing Riemann tensor. Let us call this new metric $h_{\mu\nu}$. In a force field description the gravitational field and the gravitational potential appear as functions of $g_{\mu\nu}$ and $h_{\mu\nu}$. By varying $h_{\mu\nu}$ we obtain complete freedom in choosing how much gravity we want to describe as a force field and how much is put on the account of geometry. The equivalence principle, thus, translates into a freedom of mutual exchange between the metric and the gravitational potential. Now let us look at this in more detail.

Gravitational force field equations

From the Einstein field equations

$$-kT^{\mu\nu} = G^{\mu\nu}, \tag{7.32}$$

and from the definition of the Einstein tensor $G^{\mu\nu}$, it follows that the covariant divergence of the energy–momentum tensor vanishes:

$$D_\nu T^{\mu\nu} = 0, \tag{7.33}$$

where D_ν denotes covariant differentiation relative to $g_{\mu\nu}$. In a purely geometric description, $T^{\mu\nu}$ contains no gravitational terms. All contributions of gravity in the conservation of energy and momentum originate from the covariant differentiation. In a force field description, $g_{\mu\nu}$ is replaced by $h_{\mu\nu}$ and, as a consequence, equations (7.32) and (7.33) are no longer valid. They can be made to hold again, however, by a suitable modification of the energy–momentum tensor. The new energy–momentum tensor $S^{\mu\nu}$ will contain gravitational terms balancing the change of metric. The two metrics each generate their own field of connections:

$$\{_\beta{}^\alpha{}_\gamma\} = \frac{1}{2}g^{\alpha\mu}(\partial_\gamma g_{\mu\beta} + \partial_\beta g_{\gamma\mu} - \partial_\mu g_{\beta\gamma}) \tag{7.34}$$

$$[_\beta{}^\alpha{}_\gamma] = \frac{1}{2}h^{\alpha\mu}(\partial_\gamma h_{\mu\beta} + \partial_\beta h_{\gamma\mu} - \partial_\mu h_{\beta\gamma}), \tag{7.35}$$

where $h^{\mu\nu}$ is the inverse of $h_{\mu\nu}$, and not equal to $g^{\mu\alpha}g^{\nu\beta}h_{\alpha\beta}$. Contrary to the connections themselves, their difference $N^\alpha_{\beta\gamma}$:

$$N^\alpha_{\beta\gamma} = \{_\beta{}^\alpha{}_\gamma\} - [_\beta{}^\alpha{}_\gamma], \tag{7.36}$$

is again a tensor field.

In relation to the new metric, test bodies do not follow geodesics. They experience a gravitational force given by

$$F^\mu = N^\mu_{\beta\gamma}\frac{dx^\beta}{dt}\frac{dx^\gamma}{dt}. \tag{7.37}$$

Denoting covariant differentiation according to $h_{\mu\nu}$ by $\mathcal{D}_\mu$, we may write

$$\begin{aligned} D_\beta\xi^\alpha &= \partial_\beta\xi^\alpha + \{_\beta{}^\alpha{}_\gamma\}\xi^\gamma \\ &= \mathcal{D}_\beta\xi^\alpha + N^\alpha_{\beta\gamma}\xi\gamma. \end{aligned} \tag{7.38}$$

Accordingly, (7.33) can be rewritten into

$$\begin{aligned} \mathcal{D}_\nu T^{\mu\nu} + N^\mu_{\nu\alpha} T^{\alpha\nu} + N^\nu_{\nu\alpha} T^{\mu\alpha} &= D_\nu T^{\mu\nu} \\ &= 0. \end{aligned} \tag{7.39}$$

Using[27] $\{_\nu{}^\nu{}_\alpha\} = \partial_\alpha(log\sqrt{-g})$, where g is the determinant of $g_{\mu\nu}$, combined with the similar equation for the other metric $h_{\mu\nu}$, we obtain

$$N^\nu_{\nu\alpha} = \partial_\alpha\Big(log\sqrt{g/h}\,\Big). \tag{7.40}$$

Substitution in (7.39) yields

$$\mathcal{D}_\nu\Big(T^{\mu\nu}\sqrt{g/h}\,\Big) + N^\mu_{\nu\alpha} T^{\alpha\nu}\sqrt{g/h} = 0. \tag{7.41}$$

Let $t^{\mu\nu}$ be one of the tensors satisfying

$$\mathcal{D}_\nu\Big(t^{\mu\nu}\sqrt{g/h}\,\Big) = N^\mu_{\nu\alpha} T^{\alpha\nu}\sqrt{g/h}. \tag{7.42}$$

We may then rewrite (7.41) into

$$\mathcal{D}_\nu\Big((T^{\mu\nu} + t^{\mu\nu})\sqrt{g/h}\,\Big) = 0. \tag{7.43}$$

In as far as we consider only (7.33), the tensor

$$S^{\mu\nu} = (T^{\mu\nu} + t^{\mu\nu})\sqrt{g/h}, \tag{7.44}$$

is acceptable as the new energy–momentum tensor, and $t^{\mu\nu}\sqrt{g/h}$ is the gravitational term. Owing to the fact that (7.42) is a differential equation, which fixes only the divergence of $t^{\mu\nu}$, this term is not uniquely determined. In order to stay completely within the context of general relativity, however, it is natural also to require the validity of the field equations (7.32) themselves. This requirement is sufficient to fix the choice of $t^{\mu\nu}$ uniquely. Under a transformation $g_{\mu\nu} \to h_{\mu\nu}$ of the metric, the field equations must transform into

$$-\,kS^{\mu\nu} = H^{\mu\nu}, \tag{7.45}$$

where $H^{\mu\nu}$ is the Einstein tensor of the new metric. We must now have

$$\begin{aligned} -kS^{\mu\nu} &= -k(T^{\mu\nu} + t^{\mu\nu})\sqrt{g/h} \\ &= (G^{\mu\nu} - kt^{\mu\nu})\sqrt{g/h} \\ &= H^{\mu\nu}, \end{aligned} \tag{7.46}$$

which leaves as our only choice for $t^{\mu\nu}$:

$$kt^{\mu\nu} = \Big(G^{\mu\nu} - H^{\mu\nu}\sqrt{h/g}\,\Big). \tag{7.47}$$

This equation provides an elegant formulation of the equivalence principle as an invariance under metrical transformations[28]. We are free to apply any change of metric, provided that it is counteracted by a suitable gravitational field.

[27]See, e.g., Adler *et al.* [5] p.76.

[28]It may now be appreciated that a bimetric theory is in fact needed to obtain an adequate formulation of the equivalence principle within general relativity. Previously, Natan Rosen has formulated a bimetric theory in which the equivalence principle is violated, see, e.g, Rosen [183].

Covariance

How do these results compare to what is more or less the standard analysis, producing energy–momentum pseudotensors that do not transform properly under nonlinear transformations? The usual approach[29] is aimed at replacing (7.33) by an equation containing only normal, instead of covariant derivatives. One readily arrives at the equation

$$\begin{aligned}\partial_\nu S^{\mu\nu} &= \partial_\nu((T^{\mu\nu}+t^{\mu\nu})\sqrt{-g}\,) \\ &= 0,\end{aligned} \tag{7.48}$$

where $t^{\mu\nu}$ is a matrix satisfying

$$\partial_\nu(t^{\mu\nu}\sqrt{-g}) = \{_\nu{}^\mu{}_\alpha\}T^{\alpha\nu}\sqrt{-g}, \tag{7.49}$$

which is precisely what we obtain in our former analysis if we take for $h_{\mu\nu}$ the global Lorentz metric. This also makes clear why $t^{\mu\nu}$ cannot be a proper tensor. The lorentzian form of $h_{\mu\nu}$ is not conserved under nonlinear transformations. Accordingly, $t^{\mu\nu}$ behaves properly only under the Poincare group of transformations. It may now be recognized that the origin of the difficulties in the standard analysis is that it does not properly account for the fact that it is implicitly based on a bimetric theory.

It may be clarifying at this point to recall that the theory of general relativity consists of two separate parts. The first part is a theory of general coordinate transformations. This is a strictly mathematical addition to special relativity. It should, and probably would, have been added irrespective of any ideas about gravitation. The second part of general relativity is the part that does concern gravitation. This part does have physical content. It is based on the equivalence principle, and it says that a fully geometric treatment of gravitation can be given by admitting metrics of nonvanishing curvature.

Now let us, again, consider (7.48) and (7.49). Within the framework of special relativity without general coordinate transformations, where the set of admissible coordinate transformations is restricted to the Poincare group, these equations provide an acceptable basis for a theory of gravitation, with gravitation interpreted as a force field in a spacetime with a global Lorentz metric. Clearly, however, the form of these equations is not conserved under more general transformations. From the theory of general coordinate transformations it follows that, in this case, the normal derivatives must be replaced by covariant derivatives. This holds for the above equations, describing a gravitational force field, but also for any other equations, describing any other force field. Of course, however, these covariant derivatives must be taken with respect to the metric $h_{\mu\nu}$ on which the force field picture of gravitation is based. Taking covariant derivatives with respect to the curved metric $g_{\mu\nu}$ of general relativity amounts to confusing two different pictures of gravitation.

[29] See, e.g., Landau et al. [143] §96, Misner *et al.* [162] §20.3, Weinberg [215] §7.6, Adler *et al.* [5] §11.2.

Quantum gravity

Applying the above analysis to quantum gravity, it seems natural to try to obtain a manifestly covariant theory of gravitation in which the metric is treated classically, and the force field component of gravitation is quantized. The equivalence principle boils down to the the demand that the theory must be invariant under a change in metric, provided that an appropriate compensation is included in the force field. In the quantized theory this invariance should be expressible as a metric gauge freedom.

7.6.3 A thought about renormalization

It cannot be expected that the above tentative ideas about gravitation spontaneously lead to a solution of, what is often conceived as, the main problem in quantum gravity: nonrenormalizability. Perhaps the most speculative of our present thoughts about the subject will be devoted to this topic.

Renormalization theory is a subject which concerns the description of systems of infinite complexity. In § 2.2 the word "compositał" was introduced for a general class of systems composed of parts in a possibly infinitely complex manner, and renormalization was described as a way of making comprehensible pictures of such a system, in spite of its complexity. The idea was illustrated with the example of a coil, made of coiled thread, made of coiled thread....

In quantum electro dynamics (QED), infinite complexities enter the scene because amplitudes are infinite sums of ever more complex Feynman diagrams. QED owes its renormalizability to a scale invariance, and to the fact that renormalization factors remain finite for finite scale transformations. Quantum gravity seems not to be renormalizable. Can we understand what goes wrong here, and is there any hope that we might be able to understand enough of the theory to be able to use it in some fruitful way if renormalization indeed does not apply? I shall try to elaborate on these questions, although admitting in advance that my analysis will be highly speculative, without even providing clearcut answers.

One problem which may play a role is that in quantum gravitation, contrary to QED, we are dealing with a twofold compositionality. Gravitating systems are composite in two different ways and this breaks a scale invariance that was a main ingredient in the renormalizability of QED. One type of compositionality is the type that we know from electrodynamics. All particles in the theory should be considered as dressed particles and the bare particles are not really there. For the other type of compositionality, we must take a closer look at the theory of general relativity itself.

The field equations admit solutions, such as the Schwarzschild black hole, describing a bridge between two spacetimes. There may be good reasons for excluding the interior Schwarzschild solution, including the bridge part, as describing a physically meaningful situation. The present theory, however, seems to admit these objects, and, moreover, the theory might contain other types of bridge–like solutions. In a quantized theory we should then expect that the field is able to create them spontaneously. A spacetime in a quantum gravitational theory should therefore be a structure full of real and virtual bridges to other spacetimes. Such another spacetime may be empty, in as far as a field theoretical vacuum can be said to be empty, but of course there should also be solutions of the field equations in which the other spacetime is a universe full of matter and energy of its own. Furthermore, there will be bridges to other universes in which there are bridges to other universes etc. In other words, we obtain a composital of universes, and in order to obtain some sort of comprehensible theoretical description we must find a way to get rid of excess structure. It is not so clear precisely how to achieve this, as physics in our own universe is smoothly connected to the physics in other universes.

> **Example:** In order to find a more or less comparable situation for the coil compositals of § 2.2, imagine a space filled with small bridges to other spaces. In the process of coiling a thread we now allow the thread, if it is thin enough, to cross these bridges and make loops of arbitrary length through other spaces. Up to a certain value of the scale parameter n, our theory about these coils can be the same as before. Above this value, however, the thread will be thin enough to enter into other spaces and the situation becomes much more complicated. This may be sufficient to cause renormalization via the standard procedures to fail, in the first place because scale invariance is broken, in the second place because the renormalization factors may become infinite even for finite increases of the usual scale parameter n. A new theory of coils of this type should not only deal with the coil composital aspect of the situation, but also with the composital of spaces. A similar situation may exist in quantum gravity.

Suppose that the above analysis comes anywhere close to being correct, then what we are looking for is a way to cope with the new situation, for instance by trying to describe the physics in our own universe in isolation of whatever may be going on in the universes to which we are connected. A result of renormalization is that some phenomena which can be considered as physically real are nevertheless effectively invisible. If renormalization does not work, there may still be another mechanism which hides these phenomena. In the present case, the event horizons around black holes might provide such a mechanism, by precluding the occurrence of causal relations between the universes involved. Quantum gravitation may give rise to infinite self–energy processes, but if these processes are causally disconnected from our universe we may simply neglect them. This, however, raises the problem that we need a causality criterion, in some sufficiently precise formulation, and this is by no means unproblematic. The meaning of "causality", and of related terms like "locality" and "determinism", is much disputed in QM. In fact, we might have to consider the possibility that causal relations between two universes joined by a black hole are not impossible at all. Especially if, in accordance with our earlier analysis, time–symmetric causality is admitted in the theory, there may well be causal relations of some sort between the two universes, and effects of this may be or become visible at some point in history.

> **Example:** Imagine a space traveling species, the Orionese, who have very fast ships and lots of patience. They do not mind setting themselves goals billions

of years in the future; in the meantime they just go about doing their business, or they take a nap. The Orionese have the following strategy: their astronomers observe the mass distributions in space, searching for large regions which will eventually take part in a gravitational collapse. After locating such a region they send ships over there, in order to be included in the process. What they expect to happen is that during their journey the arrow of time in these ships will be reversed, along with that of the entire collapsing region in which they are then traveling. The ship will, in a sense, have two histories: one in our universe, i.e., the part of spacetime dominated by the singularity which we call the Big Bang, and one in the other universe, dominated by a singularity which we would regard as a Crunch. The upshot is that history in this other universe must provide an evolution producing Orionese, travelling about in space ships, some of which coming our way. In the other universe the Orionese who come our way may have cousins who have chosen a different collapsing region to travel into, enabling the Orionese species to reach a third universe, which may not be directly connected at all to ours. In this way the Orionese, who like a bit of *Lebensraum*, hope to spread from universe to universe.

The Orionese may be wrong. For us it remains to be investigated whether our physical theories will, indeed, admit solutions of this type. Suppose they are right, however, then this would entail the existence of definite causal chains between different universes.

What I have tried to indicate in the present section about renormalization is that renormalization may not be our only hope for obtaining a manageable gravitational theory; there may be other suitable hiding mechanisms. Causality may provide such a mechanism, but it will require a careful approach in working it out, especially within a time–symmetric picture. This may not be easy. Common intuitive pictures about time seem rather inept for dealing with causal symmetry, and the idea requires extensive training to get used to. Eventually, the best approach to pursue might be a picture in which causality can locally provide a well functioning hiding mechanism, whereas, on a global scale, we are dealing with phenomena which are not hidden at all; they might simply be physically real, and visible under the appropriate cosmological conditions.

Literature: Adler *et al.* [5], Belinfante [12], Bell [14], Birkhoff and Von Neumann [28], Bohm [29], Bondi [30], Cooke and Hilgevoord [47], Cooke [45], Davies [60], Demopoulos [62], Van Fraassen [95, 90, 96], Gleason [100], Harrison [111], Hawking [112], Hoekzema [121, 122], Hoyle and Narlikar [127], Jauch and Piron [131], Kochen and Specker [134], Landau *et al.* [143], Maczynski [156], Misner *et al.* [162], Von Neumann [165], Partridge [166], Rajput [176], Redhead [177], Selleri [191], Rosen [183], Shimony [192], Wesson [216], Wheeler and Feynman [221, 220].

Chapter 8

A maze of QMs

Abstract: There is a multitude of different quantum theories, differing subtly or dramatically. We shall try to catch a glimpse of this universe of possibilities.

The standard formalism of QM provides a framework, embracing many different views on the physical meaning of QM. Contextual QM provides a similar framework, in which, however, some deficiencies of standard QM have been repaired. In particular the formalism of quantum process theory[1] should provide an excellent starting point for developing whatever ideas one might have about the interpretation of QM. In the previous chapter, two major alternative approaches were introduced, leading to theories of quite different types: a time–asymmetric approach with instantaneous causality, featuring an ontology including quantum systems evolving in time, and a time–symmetric, retrocausal approach, with an ontology in terms of events. The difference between these two approaches is an important one, but it covers by no means the full spectrum of possible variation. There is ample opportunity for constructing further distinctions, and one may use this for trying to give contextual reconstructions of views which have occurred in the historical debate on QM.

There are a number of different ways in which the latter subject may be approached. One way is to start from a contextual formalism which remains close to the standard theory, such as the initial formalism. As a next step one may introduce modifications, in particular expressing ideas corresponding to positions taken in the debate about the interpretation of standard QM. And finally, one should note the deficiencies in the initial formalism, deal with these deficiencies within the modified formalism, and, presumably, obtain a well–functioning version of contextual QM. This approach is illustrated by a few preliminary case studies in § 8.3.

A second approach is that one takes a problem which, in practice, has turned out to generate many different ideas about interpretation. Von Neumann's problem of infinite regress in measurement theory is such a typical 'crossroads–generator'. One may study such a problem and look at the alternatives to which it leads in standard QM. Subsequently, one takes a contextual formalism, such as process theory, and

[1] See 6.2.

tries to find out whether a similar set of alternatives may be constructed. This approach is illustrated by some general considerations about the Von Neumann chain in § 8.4, whereas § 8.5 contains some further elaborations about a relative state approach in process theory.

Finally, of course, one may take a formalism one likes, and develop it regardless of connections with historical positions. Ch. 9 contains an example of this kind, a time–symmetric event field theory. For the individual scientist with a genuine enthusiasm for a particular view, this is presumably the most pleasant and effective way of working. At the same time, however, it is good to realize that any particular approach lives in a sea of alternatives, and that other scientists may have quite different preferences.

The present chapter is not meant to present anything like a complete inventory of what possibilities there may be in interpreting QM. However, the examples presented here should provide some idea of what types of freedom there may be, and how difficult it will be to properly survey the alternatives and compare their merits. Before moving on to the subject of contextual interpretations of QM, a few pages will be devoted to discouraging the pursuit of nonclassical alternatives, and, before moving on to this, a few general remarks will be made about methodological matters.

8.1 General methodological remarks

If one compares different views, then one of the questions one may encounter is to what extent these views are indeed different, as, perhaps, they may also be regarded as different formulations of the same view. For instance, in standard QM one may apply transformations to different 'pictures', such as the Schrödinger picture and the Heisenberg picture. Generally, one would like to regard these pictures as equivalent. They represent different formulations of the same theory, and one may choose between them merely as a matter of convenience. A decision to regard two views as equivalent may have many aspects, including highly personal and subjective sides. For instance, a Protestant groom marrying a Catholic bride may choose to think of the difference between the two religions as being irrelevant, whereas in other parts of the world, or in other ages, people have been killing each other for it. Often, however, there also are more objective arguments. Especially in science, one may argue for the equivalence of two views by exhibiting the relations between them. However, this presupposes that both views are sufficiently sharply defined, and, moreover, it may turn out that there are many different equivalence relations, depending on whether specific differences between the views are counted as relevant or irrelevant. A generally important, although by no means the only criterion for the equivalence of two theories concerns the structure of their state spaces.

Example: In process theory, a quantum process is represented by a triple $\Phi =< W_{\rm out}, \phi, W_{\rm in} >$. For a specific experiment there is a contextual state space, consisting of the set of processes which may be realized in it; and for a class of experiments we may construct a quantum state space[2] $\Gamma_{\mathbf{qm1}}$, consisting of the set of pairs $< \Gamma_{\mathbf{u}}, \Phi >$, where $\Gamma_{\mathbf{u}}$ is a contextual state space and $\Phi \in \Gamma_{\mathbf{u}}$.

Now consider two invertible operations θ and χ on the algebra of operators on the Hilbert space $\mathcal{H}$. These may be used to define a transformation of the above process states Φ into

$$\Phi' =< \chi^+(W_{\text{out}}), \chi \circ \phi \circ \theta^{-1}, \theta(W_{\text{in}}) > . \tag{8.1}$$

The transformed states may be collected in a transformed quantum state space $\Gamma_{\mathbf{qm2}}$. Because this transformation leaves the probability measure unchanged, one may use $\Gamma_{\mathbf{qm2}}$ instead of $\Gamma_{\mathbf{qm1}}$ for describing the same class of experiments while obtaining a fully equivalent description. The difference between the Schrödinger picture and the Heisenberg picture in standard QM, for instance, can be analyzed in these terms within process theory. Starting from a process state $\Phi =< W_{\text{out}}, \phi, W_{\text{in}} >$, a transformed state corresponding to the Schrödinger picture would be $\Phi' =< W_{\text{out}}, \phi_{\mathcal{I}\!\mathcal{I}}, \phi(W_{\text{in}}) >$, whereas the Heisenberg picture is obtained by taking $\Phi' =< \phi^+(W_{\text{out}}), \phi_{\mathcal{I}\!\mathcal{I}}, W_{\text{in}} >$.

Connective morphisms

The above state spaces $\Gamma_{\mathbf{qm1}}$ and $\Gamma_{\mathbf{qm2}}$ can be considered equivalent, because of the existence of an isomorphism between them. Because different versions of contextual QM all have a classical logical structure, the relations between them may be investigated in standard ways. This includes, first of all, the possibility of studying morphisms between their state spaces. These morphisms may be defined in the following way[3]: assuming that $\Lambda_{\mathbf{qm1}} \subseteq \Gamma_{\mathbf{qm1}}$ is a state in a version QM1 of contextual QM, one may ask which set of states in another version QM2 is considered to be possible, knowing that according to QM1 the system is described by $\Lambda_{\mathbf{qm1}}$. Reading $\Lambda_{\mathbf{qm1}}$ as a variable, this procedure defines a mapping

$$\mathcal{J}_{\mathbf{qm2|qm1}} : \wp(\Gamma_{\mathbf{qm1}}) \to \wp(\Gamma_{\mathbf{qm2}}),$$

assigning to each state $\Lambda_{\mathbf{qm1}}$ in QM1 a corresponding state $\mathcal{J}_{\mathbf{qm2|qm1}}(\Lambda_{\mathbf{qm1}})$ in QM2. *Vice versa* there is a similar mapping

$$\mathcal{J}_{\mathbf{qm1|qm2}}(\Lambda_{\mathbf{qm2}}) : \wp(\Gamma_{\mathbf{qm2}}) \to \wp(\Gamma_{\mathbf{qm1}}).$$

The equivalence of the QM1 and QM2 in the above example can now be expressed by the pair of equations

$$\begin{aligned} \mathcal{J}_{\mathbf{qm1|qm2}} \circ \mathcal{J}_{\mathbf{qm2|qm1}}(\Lambda_{\mathbf{qm1}}) &= \Lambda_{\mathbf{qm1}}, \text{ for all } \Lambda_{\mathbf{qm1}} \subseteq \Gamma_{\mathbf{qm1}} \\ \mathcal{J}_{\mathbf{qm2|qm1}} \circ \mathcal{J}_{\mathbf{qm1|qm2}}(\Lambda_{\mathbf{qm2}}) &= \Lambda_{\mathbf{qm2}}, \text{ for all } \Lambda_{\mathbf{qm2}} \subseteq \Gamma_{\mathbf{qm2}}. \end{aligned} \tag{8.2}$$

[2]See § 6.2.

[3]See also § 10.1.2.

Different types of state spaces

It may further be noted that many theories generate a number of different state spaces, rather than only a single one, and these may all be interesting to be involved in comparing the theory with another one. For instance, let $\mathcal{T}_1$ and $\mathcal{T}_2$ be two probabilistic theories, describing the dynamics of a system of particles. For describing the state of some given system, $\mathcal{T}_1$ has a state space $\Gamma_{\mathbf{1.sys}}$, and $\mathcal{T}_2$ a corresponding state space $\Gamma_{\mathbf{2.sys}}$. These may be compared by means of morphisms $\mathcal{J}_{\mathbf{1.sys|2.sys}}$ and $\mathcal{J}_{\mathbf{2.sys|1.sys}}$.

The dynamical laws of the two theories, however, are contained in the structure of the state spaces of possible processes, i.e., of the sets of physically possible trajectories through $\Gamma_{\mathbf{1.sys}}$ and $\Gamma_{\mathbf{2.sys}}$. These process state spaces $\Gamma_{\mathbf{1.proc}}$ and $\Gamma_{\mathbf{2.proc}}$ may be compared by means of morphisms $\mathcal{J}_{\mathbf{1.proc|2.proc}}$ and $\mathcal{J}_{\mathbf{2.proc|1.proc}}$.

In order to take into account that $\mathcal{T}_1$ and $\mathcal{T}_2$ are probabilistic theories, one may consider their sets of theoretically possible probability measures $\Gamma_{\mathbf{1.prob}}$ and $\Gamma_{\mathbf{2.prob}}$, which determine the statistical properties of ensembles of processes described by these theories. These probability state spaces $\Gamma_{\mathbf{1.prob}}$ and $\Gamma_{\mathbf{2.prob}}$ may be compared by means of morphisms $\mathcal{J}_{\mathbf{1.prob|2.prob}}$ and $\mathcal{J}_{\mathbf{2.prob|1.prob}}$.

The QM1 and QM2 of the above example are equivalent in the sense that there are isomorphisms between all their respective state spaces. There are other approaches in QM which do not lead to an equivalence in this sense, but this does not by itself entail that there must also be empirical differences. Empirical equivalence by itself does not imply the existence of isomorphisms between the state spaces of different theories, nor between the worlds represented by the individual states. There can be major conceptual differences between theories which do affect the structure of the state spaces and of the states, but which do not show at the empirical level. Such matters may be investigated by involving, besides the morphisms between the theoretical state space, also the morphisms between the theoretical and the empirical levels.

8.2 Nonclassical alternatives

Instead of interpreting QM in a classical logical framework by using contextual logic, one may consider the alternative of applying a nonclassical logic. Let us examine some problems to which this may lead.

8.2.1 Nondeterministic states

An approach which might seem natural with respect to the interpretation of QM, or at least I used to think so myself, is the introduction of a *probabilistic semantics*, in which the value of a magnitude is determined by the state of the system only in a probabilistic sense. In this way, one may even try to hold on to the Hilbert space $\mathcal{H}$, or the set of density operators $\mathcal{D}$, as providing the state space of QM. Magnitudes thus become stochastic functions, instead of ordinary functions, on the state space. Although this may seem an attractive approach at first sight, it does not solve the problem of finding a satisfactory interpretation of factual, nonprobabilistic sentences. For a given density operator W, one can assign a probability to a sentence "$X = x$", but how about truth? One may try to treat truth as unit probability, i.e., the truth of "$X = x$" is interpreted by means of $p(X = x) = 1$. If one wants to hold on to $\mathcal{H}$ or $\mathcal{D}$ as providing the state space of QM, however, this clashes with both subjective and objective interpretations of probability. Therefore, in as far as this is a viable approach at all, it at least requires a new interpretation of probability.

Subjective probabilities

A subjective probability represents a degree of rational belief, where various different subjective interpretations have different ideas about the contents of "rational". Within the logical framework used here, a natural way to formalize knowledge, or believe, is by introduction of a state Γ^*, representing what we know or belief about the system, i.e., Γ^* is the set of all elementary states that are possible according to our knowledge; if we know a sentence ψ to be true, this is expressed by $\Gamma^* \subseteq \gamma(\psi)$. A natural way to formalize degrees of belief is now to replace Γ^* by a probability space $< \Gamma, \mathcal{F}, p >$, where $\mathcal{F}$ is a σ-field of subsets of Γ and p is a probability measure on $\mathcal{F}$. Probabilistic statements like "$p(\phi) = a$" are now statements at the semantic level, not interpretable themselves by a state $\gamma(p(\phi) = a)$. It is indeed possible in this case to interpret ϕ by means of $p(\phi) = 1$. The sentence ϕ expresses $\Gamma^* \subseteq \gamma(\phi)$, and with Γ^* defined as

$$\Gamma^* = \inf\{\Lambda \in \mathcal{F} |\, p(\Lambda) = 1\} \tag{8.3}$$

this is indeed equivalent to $p(\phi) = 1$. If one wants to interpret the probabilities in QM in this way by means of a subjective theory of probability, however, then the Hilbert space $\mathcal{H}$ nor the set of density operators $\mathcal{D}$ are acceptable choices for the state space Γ. If we were to take $\mathcal{H}$ or $\mathcal{D}$, then the σ-field $\mathcal{F}$ would contain subsets of $\mathcal{H}$ or $\mathcal{D}$. In other words, the probabilities we would get would concern our belief about which state vector or density matrix is the right one; they would be probabilities of the type $p(W) = a$, and not of the type $p(X = x) = a$. If we want to have the latter kind of probabilities, then we must have a different state space, on which magnitudes are simply functions instead of stochastic functions.

Objective probabilities

Alternatively, let us consider an objective interpretation of probability. An objective probability is a physical quantity, such as the relative frequency of red balls in an urn or the average relative time spent by a physical system in a specific part of its phase space. Such a quantity may be called a probability if there is an associated type of measurements, leading to random sequences of outcomes with frequencies approaching the probability in numerical value[4]. Within our present logical frame-

[4]The terms "random" and "approach" are problematic concepts, relating to controversial issues in the interpretation of objective probability itself. This is not the proper place to discuss these

work the natural way to formalize objective probabilities is to define, apart from the state space Γ that serves to interpret nonprobabilistic sentences, a separate $\Gamma_{\mathbf{prob}}$ for interpreting the probabilistic ones. We then get $\Gamma' = \Gamma \times \Gamma_{\mathbf{prob}}$ as the total state space, on which sentences of both types can be interpreted. To obtain a suitable $\Gamma_{\mathbf{prob}}$, consider a σ-field $\mathcal{F}$ of subsets of Γ. If we now take for $\Gamma_{\mathbf{prob}}$ some set of probability measures μ on $\mathcal{F}$, then a probabilistic sentence like "$p(\phi) = a$" can get a corresponding meaning

$$\gamma_{\mathbf{prob}}(p(\phi) = a) = \{\mu \in \Gamma_{\mathbf{prob}} |\ \mu(\gamma(\phi)) = a\}. \tag{8.4}$$

Instead of such a set of measures, any suitable mathematical representation of it will also suffice, and this provides us with a way to regard $\mathcal{H}$ or $\mathcal{D}$ as acceptable state spaces for interpreting probabilistic sentences. A given experiment has outcomes in an index set $\mathbf{K}$. Let $\mathcal{F}$ be a σ-field of subsets of $\mathbf{K}$, let R be an EVM on $\mathcal{F}$, and let $K \in \mathcal{F}$ be the subset corresponding to the sentence ϕ. We may now define

$$\gamma_{\mathbf{prob}}(p(\phi) = a) = \{W \in \mathcal{D} |\ \mathrm{Tr}[R(K)W] = a\}. \tag{8.5}$$

This works very well, but we can no longer interpret "$X = x$" by means of $p(X = x) = 1$, as we did in the case of subjective probabilities. These two sentences are now both sentences in the object language, and they definitely have a different meaning under the present semantics; ϕ is interpreted on Γ, and "$p(\phi) = 1$" on $\Gamma_{\mathbf{prob}}$. Thinking of an urn with coloured balls, it is clear why this must be so. The sentence ϕ describes, for instance, that the ball just drawn is red. It is really nonsense to interpret this as being equivalent to "$p(\phi) = 1$", which would imply that all balls in the urn are red. Accordingly, although $\mathcal{H}$ or $\mathcal{D}$ can be accepted as part of the state space, for instance as a representation of $\Gamma_{\mathbf{prob}}$, they are not acceptable as representing the entire state space. Some part must be added to provide a meaning to the ordinary nonprobabilistic sentences.

Summarizing, we may conclude that, unless one finds another way to include probability theory, not heading under the subjective or objective approaches outlined above, stochastically defined magnitudes do not solve the problem of interpreting factual statements in QM and they are of no help in trying to save the Hilbert space as more or less the conceptual equivalent of a classical state space.

8.2.2 Quantum logic

The main attempt at finding an alternative logical framework which can make sense of the standard formalism of QM is based on the theory of nondistributive lattices and is known as quantum logic. The basis for this approach was laid in the work of Birkhoff and Von Neumann[5].

Quantum logic can be obtained as a generalization of the logic of semiinterpreted languages. A semiinterpreted language is a triple $< \mathcal{L}, \Gamma, \gamma >$ in which the meaning

matters, however, and our present analysis does not really require it.

[5] [28].

of a sentence $\phi \in \mathcal{L}$ is given by a subset $\gamma(\phi)$ of the state space Γ. Instead of $\gamma(\phi) \subseteq \Gamma$ we can of course write $\gamma(\phi) \in \wp(\Gamma)$, where $\wp(\Gamma)$ is the power set of Γ. Therefore, instead of the triple $< \mathcal{L}, \Gamma, \gamma >$ we might as well have chosen a triple $< \mathcal{L}, \Upsilon, \gamma >$, with $\Upsilon = \wp(\Gamma)$, to represent a semiinterpreted language. It would make no difference, apart from the fact that the latter kind of representation is readily generalized into a more general approach to semantics by admitting other choices for Υ. In particular, one can choose for Υ a complemented lattice, i.e., a partially ordered set with a maximal and a minimal element, a meet and a join, and a complement. The lattice ordering can then be taken to represent semantic entailment, the maximal and minimal element can represent truth and falsehood, and the complement represents negation.

This generalized semantics can be applied to QM by choosing for Υ the lattice of subspaces of the Hilbert space. Representing these subspaces by their associated projection operators, one can define:

$$\begin{aligned} P \sqsubseteq P' & \quad \text{iff } PP' = P \\ 1\!\!1 & \quad \text{is the maximal element of } \Upsilon \\ 0 & \quad \text{is the minimal element of } \Upsilon \\ P \sqcap P' & = \sup\{P''|\, P'' \sqsubseteq P, P'' \sqsubseteq P'\} \\ & = \lim_{n\to\infty} (PP')^n \\ P \sqcup P' & = \inf\{P''|\, P \sqsubseteq P'', P' \sqsubseteq P''\} \\ & = (P^\perp \sqcap P'^\perp)^\perp \\ P^\perp & = 1\!\!1 - P. \end{aligned} \tag{8.6}$$

A sentence $X = x$ in QM can now be interpreted by taking for $\gamma(X = x)$ the projection on the eigenspace of the operator X corresponding to the eigenvalue x.

This lattice of subspaces is a so–called *nondistibutive*, or also *nonboolean* lattice, because the above meet and join do not satify the distributive laws. Logic on a nonboolean lattice is conceptually more difficult to grasp than logic on a boolean lattice, because the relation with a state space is lost. If the lattice is boolean, one can find a set Γ such that the lattice is a subalgebra of the algebra of subsets of Γ. This makes it possible to hold on to the idea that the system one aims to describe is in some well–defined elementary state, given by an element of Γ. If we do not have a state space we do not have this picture. Along with the state space we also lose the idea that we are describing a set of possible worlds, of which the actual world is a particular one. Logic on a nonboolean lattice can no doubt provide a mechanism for ordering sentences, but whether it does more, whether it can also provide an intuitively satisfactory 'meaning' to these sentences, is questionable. It may be a matter of taste, but the approach has never appealed to me very much[6].

[6]What might be counted as attractive in quantum logic, however, by a sufficiently perverse mind, is that the meaning of many basic terms, such as determinism, completeness, locality, and reality, is so obscure within the new logac that they must, in fact, be entirely redefined. This opens the possibility of defining them such that the theory is deterministic, complete, and local after all.

8.2.3 Quantum probability

A funny thing about the probability measure $p_W(P) = \mathrm{Tr}[PW]$ of the standard formalism is that it is not a probability measure. Summation over even a small set of distinct projection operators P may easily yield results larger than 1. The expression behaves as a probability measure *only* if applied to some given set of *orthogonal* projection operators. We may therefore say that the expression in fact defines a family of measures, a *supermeasure*, rather than an ordinary measure, and the relation between a supermeasure and a measure is given by

supermeasure + context $\Rightarrow$ measure.

Realizing this, one may do several things. One possibility is to explicitly define the probability measures to be context dependent, as in contextual QM, where the context $\mathbf{u}$ appears as an index to p. A second possibility is to make implicit use of the context dependence, but without explicitly mentioning it. This is a bit sloppy. The third possibility is to make the transition to a generalized measure theory. This is what goes on in the *quantum probabilistic* approach[7], a branch of mathematical physics where supermeasures, rather than ordinary measures, are studied. Quantum probability should combine well with quantum logic, because the unorthodox state concept in quantum logic more or less requires an equally unorthodox probability theory.

From a mathematical point of view, both developments may be considered interesting. In mathematics it is very often successful to approach a subject from different points of view, because it may lead to theorems which would have been difficult to find otherwise. However, both in the case of quantum logic and in the case of quantum probability, one must be careful with drawing conclusions, because many of the common intuitions about logic and probability no longer function in this exotic context. Accordingly, even if the mathematics is correct, one may still be tempted to draw physical or philosophical conclusions that are debatable. As an example, I note that I hold the philosophical ideas of Accardi, a main proponent of quantum probability, to be untenable. The core of his approach is connected with Bayes' rule

$$p(A \cap B) = p(A|B)p(B) \tag{8.7}$$

In standard probability theory this follows directly from the definition of conditional probabilities, but for supermeasures the situation is less straightforward. Accardi takes Bayes' rule to be an *axiom* of probability theory, and he drops it in order to arrive at a more general type of measures. Mathematically this might be sound, if one is looking for a new type of mathematical objects, but he then proceeds by claiming that his approach can be combined with a naive realism where it can be maintained that in the two slit experiment particles pass through only one slit, where it is not necessary to introduce the notion of a "superposition of states", where Bell's inequalities do not lead to conclusions about nonlocality, and where, in short[8]: "all

[7] See, e.g., [4, 2, 1, 3].
[8] Accardi [3] p. 120.

the so–called paradoxes of quantum theory [...] are swept away". Accardi's position was criticized in 1990 in a paper of which I was one of the authors[9], one of the arguments being that basic problems in the interpretation QM can be formulated without mentioning conditional probabilities at all. Accardi has briefly responded to this criticism[10], but unfortunately without going into any of our arguments. He merely restates his position and concludes that we have misunderstood him. Apart from noting that our original criticism still stands, I have, at the moment, little to add to this discussion.

As I see it, the best way of making proper use of results obtained in quantum logic or quantum probability is via contextual interpretation. Contextual logic provides a very general logical scheme, and via some trick comparable to splitting magnitudes it should be possible to find contextual representations of even quite exotic types of nonclassical logic. A suitable representation theorem may then serve as a bridge by which interesting results in quantum logic, or quantum probability, may be transferred to safer ground, where such results are less easily misinterpreted.

8.3 Variations on the initial formalism

Of the contextual formalisms discussed so far, the initial formalism remains closest to the standard density operator formalism of QM. Nevertheless, at several points in the course of its development choices had to be made about the structure of its states and its dynamics. At such instants, I have chosen the alternatives which I myself regarded as the most natural and promising ones, but, of course, one is free to disagree, and another choice might have led to quite different results. After all, there are many different ideas about the meaning of the standard formalism, and these are likely to require different contextual formalizations. Three examples of this will be discussed below.

8.3.1 Statistical density operators

The initial formalism uses states $< \Gamma_{\mathbf{u}}, P, W >$, in which a density operators is included along with a contextual state represented by the projection P. One might consider as slightly overdone. Alternatively, one has the option of giving them only a statistical relevance, which may be attained by defining quantum states of the form $< \Gamma_{\mathbf{u}}, P >$. QM then becomes a theory about the statistics of finding a system in such a state under specified conditions, and the density operators can be regarded as representing the physically possible probability distributions of the theory. In fact, Gleason's theorem[11] can be understood to provide the basis for such a representation. The set $\mathcal{D}$ of density operators represents precisely the set of all families of probability measures satisfying the semiseparability condition (7.11).

[9] [24].
[10] [3] p. 123.
[11] See also § 7.3.

The dynamical equations of QM now provide a probabilistic theory about the possible histories of a system in terms of states $< \Gamma_{\mathbf{u}}(t), P(t) >$, i.e., the theory says that under specified conditions some histories are more likely to occur than others. The difference with the initial approach may perhaps seem subtle, but it is, in fact, considerable. Within the present approach the density operator is part of a statistical theory over a contextual state space, in the same sense in which an ensemble distribution in classical statistical mechanics is part of a statistical theory over a classical phase space. In the classical theory nobody would consider the idea of including the set of ensemble distributions within phase space itself, and, likewise, a quantum theory in which the density operator is included in the state space is conceptually quite different from one in which the density operator merely represents statistics.

Example: Consider two similar systems $\mathcal{S}^A$ and $\mathcal{S}^B$, with $\mathcal{H}^A = \mathcal{H}^B$, described in the initial approach by states $< \Gamma_{\mathbf{u}}, P, W^A >$ and $< \Gamma_{\mathbf{u}}, P, W^B >$, and assume that $\mathrm{Tr}[PW^A]$ is large and $\mathrm{Tr}[PW^B]$ small. Within the initial approach $\mathcal{S}^A$ and $\mathcal{S}^B$ are, indeed, in different states, and the fact that $p_{\mathbf{u},W^B}(P) = \mathrm{Tr}[PW^B]$ is small implies that systems in a state $< \Gamma_{\mathbf{u}}, P, W^B >$ are relatively rare. Within the present approach, however, $\mathcal{S}^A$ and $\mathcal{S}^B$ are in identical, i.e., both in the state $< \Gamma_{\mathbf{u}}, P >$. The density operator now represents information about the *environment*, rather than about the system itself, and the fact that $p_{\mathbf{u},W^B}(P)$ is small means that a system in this state is rarely found in conditions as represented by the density operator W^B.

This approach provides an alternative to the initial formalism, and, naturally, it runs into comparable difficulties. In particular, it requires generalization with respect to handling nonideal measurements and sequences of measurements. The former may be achieved by introducing EVMs[12]. The latter introduces some additional problems. In the first place, it turns out that the probability measures on the sets of states $< \Gamma_{\mathbf{u}}, P >$ do not have the Markov property, which is often understood as a sign that the state description is incomplete. Secondly, one should require the set of possible probability measures to be closed under conditionalization. Both problems may be simultaneously solved by making the transition to a theory in which operations[13] are used for describing quantum processes. Corresponding to a given experiment there is a contextual state space $\Gamma_{\mathbf{u}}$, being some specific set of operations, and the set of physically possible probability measures is generated by introducing pairs of positive operators W_{in} and W_{out}.

On thus obtains a version of process theory, where, however, W_{in} and W_{out} are interpreted as representing sets of possible probability measures. It seems possible to elaborate on this approach in an interesting way, and there should, in fact, exist a considerable amount of freedom for further variation in this direction.

Note: Some of the differences between a statistical and a more fundamental interpretation of the density operator may be elegantly discussed in terms of the

[12]See § 5.1.
[13]See 5.2.

contextual morphisms introduced earlier[14]. Assume that QM1 is the version of process theory as introduced in § 6.2, and QM2 is a version in which the density operators are interpreted statistically. This means that the process state space $\Gamma_{\mathbf{qm1.proc}}$ consists of a set of triples $< W_{\text{out}}, \phi, W_{\text{in}} >$, whereas $\Gamma_{\mathbf{qm2.proc}}$ contains merely operations ϕ. Evidently, there is no isomorphism between these state spaces. $\Gamma_{\mathbf{qm1.proc}}$ is much larger than $\Gamma_{\mathbf{qm2.proc}}$, and we have

$$\mathcal{J}_{\mathbf{qm1|qm2}} \circ \mathcal{J}_{\mathbf{qm2|qm1}}(\Lambda_{\mathbf{qm1}}) \supset \Lambda_{\mathbf{qm1}}, \text{ for most } \Lambda_{\mathbf{qm1}} \subseteq \Gamma_{\mathbf{qm1}}. \tag{8.8}$$

Empirical equivalence of the two theories is saved, however, by the fact that there is, nevertheless, an isomorphism between $\Gamma_{\mathbf{qm2.prob}}$ and $\Gamma_{\mathbf{qm1.prob}}$, which is possible because in QM2 the relation between process states and probabilistic states is much less strict. Suppressing, for the sake of simplicity, the dependence on the statistical weight function w, the probability measure in a context $\mathbf{u}$ of QM1 is uniquely determined by $p_{\mathbf{u}}(\Phi) = C_{\mathbf{u}} L(\Phi)$. In QM2, however, the probability of finding a state ϕ still depends on choosing a W_{in} and a W_{out}.

8.3.2 Complementarity versus modal interpretations

Both in the particle versions and in the process versions of contextual QM we have distinguished between, on the one hand, a system state space $\Gamma_{\mathbf{sys}}$ or a process state space $\Gamma_{\mathbf{proc}}$, and, on the other hand, a quantum state space $\Gamma_{\mathbf{qm}}$, containing pairs $< \Gamma_{\mathbf{u}}, \Theta >$ or $< \Gamma_{\mathbf{u}}, \Phi >$. One may take different points of view as to which type of states describe the actual physical systems. I would say that in a Bohr–like approach, with a heavy emphasis on complementarity, the quantum states are basic, because the contexts represent the quantum phenomena and physical magnitudes may aquire a meaning only within a specific phenomenon.

Alternatively, however, one may hold that, e.g. in a particle approach, a system is described by a a state Θ, whereas the specification of the contextual state space $\Gamma_{\mathbf{u}}$ refers to the experimental set–up, or in a wider sense to the environment of the system. In other words, contexts emerge from interactions, but they are not included in the state of the system itself. This possibility corresponds to the class of *modal interpretations* of QM, put forward by Bas van Fraassen[15] in a more quantum logic oriented framework. His state descriptions are based on pairs, consisting of a *value state* and a *dynamical state*. These may be regarded as corresponding to the contextual state and the density operator in, e.g., the initial formalism. Evidently, the modal interpretations remain empirically fully equivalent to the complementarity interpretations, because the assignment of probabilities must still take place within a definite context, determined by the experimental situation. As the difference between the approaches is subtle, one should expect also a close correspondence between their spectra of further interpretations. For example, what Van Fraassen introduces as the "Copenhagen variant" of the modal interpretation corresponds

[14] See § 8.1.

[15] See, e.g., [95].

to combining the modal interpretation with a version of the 'secret pure states' approach, discussed below in § 8.3.5.

8.3.3 Some more alternatives

The quantum states of the initial formalism are triples $< \Gamma_{\mathbf{u}}, P, W >$. The density operator W may be interpreted statistically, leaving system states $< \Gamma_{\mathbf{u}}, P >$. $\Gamma_{\mathbf{u}}$ may be interpreted as referring to the experimental set–up, leaving system states $< P, W >$. More generally, some of the variables in the initial quantum states may be interpreted as referring to physical properties other than those of the system itself. Let us make a summary of the possibilities.

(1) All elements in the system refer to the system, as in the initial formalism.

(2) $< \Gamma_{\mathbf{u}}, P >$ refers to the system, W gives a statistical distribution.

(3) $< P, W >$ refers to the system, $\Gamma_{\mathbf{u}}$ refers to the measurement set–up; the modal interpretations.

(4) P gives the state of the system, $\Gamma_{\mathbf{u}}$ refers to the measurement set–up, and W gives a statistical distribution.

(5) W gives the state of the system, $\Gamma_{\mathbf{u}}$ and P both refer to the conditions of the experiment.

(6) None of the elements of the triple describes the system. All these notions refer to relations between the system and its environment.

All these different interpretations of the quantum states in contextual QM may be maintained, and even appear natural from a certain point of view. Each of them may be a starting point for further interpretational endeavours. Presumably, the resulting spectra of interpretations are closely related, because, in a way, it does not really matter precisely what the various items in the states refer to. On the other hand, one must be careful with such a statement, because additional assumptions that seem natural from one point of view may seem absurd from another, and, in this way, the potential equivalence of the views may be broken.

8.3.4 Nondemolition properties

In the initial formalism the assumption was made[16] that the context applying to a dynamically isolated system is the trivial context $\mathbf{u}_{1\!\!1}$ with $\Gamma_{1\!\!1} = \{1\!\!1\}$. From some perspectives this is a very natural assumption, for instance, if the contextual states, projections or more generally operations, are assumed to refer to relations of a system with its environment. One may have very different intuitions about this, however, and it may then be natural to assume that the system does have

[16] See (6.8).

definite, nontrivial properties at all times, regardless of whether it does or does not interact with any other systems. At times when the systems is dynamically isolated, it is, by assumption, not observed, and, therefore, any consistent set of reasonable assumptions can be maintained in principle.

A more or less natural idea is that a system has those properties which can be predicted with certainty *if* a measurement of the appropriate observable *would* be performed, i.e., properties are assigned via a *counterfactual* argument. Taking the initial approach as a starting point this may be elaborated into a modification of the contextual part of the theory such that

> if at time $t = 0$ the contextual state of the system is given by a projection operator P, and the system is dynamically isolated up to time $t = t_1$, then the state $U(t_1)PU^+(t_1)$ applies at t_1.

In other words, suppose that a measurement at $t = 0$ produces a quantum state $Q(0) =< \Gamma_{\mathbf{u}}, P, W >$, with $PWP = W$, then at time t we get

$$\begin{aligned} Q(t) &= < \Gamma_{\mathbf{u}}(t), P(t), W(t) > \\ \text{with} \quad W(t) &= U(t)WU^+(t) \\ \Gamma_{\mathbf{u}}(t) &= \{U(t)PU^+(t) |\, P \in \Gamma_{\mathbf{u}}\} \\ P(t) &= U(t)PU^+(t). \end{aligned} \tag{8.9}$$

Now consider a magnitude X, definable in the context $\mathbf{u}$, with $X(P) = x$. Let X_t be a similar magnitude, definable in the context $\mathbf{u}(t)$, and with $X_t(P(t)) = x$. From the measurement result $X = x$ at $t = 0$, we can then predict with certainty that $X_t = x$, by virtue of the fact that $\mathrm{Tr}[P(t)W(t)P(t)] = \mathrm{Tr}[PWP] = 1$. Within the present formalism, we can not only say that this value would be found *if* a measurement were performed, we can actually say that the system *has* the property $X_t(P(t)) = x$.

Magnitudes like X_t, which can be defined as functions on the state space $\Gamma_{\mathbf{u}}(t)$, are precisely the magnitudes for which there is a so-called 'nondemolition measurement', i.e., they can in principle be measured at time t without a measurement disturbance[17]. Therefore, the property P can also be called a *nondemolition property*.

Generalizations

Evidently, this formalism too should exhibit difficulties of the kind encountered in the initial formalism, and, presumably, the use of OVMs and of the ABL measures should provide an appropriate way of overcoming these problems. The use of OVMs presents some difficulties in this approach, however, owing to the fact that the OVMs enable us to include interactions which do not lead to properties which can be predicted with certainty, because they cannot be represented by projection operators. A possible solution to this problem is to assign to a system the smallest projector P such that $P\phi_k(W)P = \phi_k(W)$, after an interaction with an associated operation ϕ_k.

[17]See also § 5.3.

This leads to a quantum state of the same form (8.9), but it has the consequence that, generally, the projectors $P \in \Gamma_{\mathbf{u}}$ no longer add up to $1\!\!1$. Instead one has

$$\sum_{P\in\Gamma_{\mathbf{u}}} P \geq 1\!\!1 . \tag{8.10}$$

This has further implications for the conceptual structure of the resulting theory, but we shall merely mention the problem without filling in the details.

There are also complications connected with the use of the ABL measures. This topic is of some interest because it has been discussed in recent literature[18], and is, moreover, related to a point discussed much earlier[19]. The problem is the following: suppose that a measurement of a magnitude X is performed at $t = 0$, with the result $X = x$, and a measurement of magnitude Y is performed at $t = t_1$, with the result $Y = y$. Let P_0 be the projector representing the result $X = x$ and P_1 the one representing $Y = y$. This leads to

$$P_0 W_{\text{in}} P_0 = W_{\text{in}}, \quad \text{and} \quad P_1 W_{\text{out}} P_1 = W_{\text{out}}. \tag{8.11}$$

For the period $0 < t < t_1$ one may then argue that $U(t)P_0U^+(t)$ is a counterfactual property, but the same can be argued for $U^+(t_1 - t)P_1U(t_1 - t)$. *If* an appropriate magnitude *were* measured at time t, *then* the value represented by $U^+(t_1-t)P_1U(t_1-t)$ could be predicted with certainty, by virtue of

$$\begin{aligned} &U^+(t_1 - t)P_1U(t_1 - t)W_{\text{out}}(t)U(t_1 - t)P_1U^+(t_1 - t) \\ &= U^+(t_1 - t)P_1W_{\text{out}}P_1U(t_1 - t) \\ &= W_{\text{out}}(t). \end{aligned} \tag{8.12}$$

This leads to difficulties if $U(t)P_0U^+(t)$ and $U^+(t_1-t)P_1U(t_1-t)$ are noncommuting projectors, such as a position eigenprojector and a momentum eigenprojector. One would then have to conclude that position and momentum simultaneously have a sharply defined value, a condition which is considered to be excluded in standard QM.

The most elegant solution to this problem may be to distinguish between in–properties and out–properties, and likewise between in–magnitudes and out–magnitudes, and between in–contexts and out–contexts. The resulting structure of the quantum states would then be

$$\begin{aligned} Q(t) &= \; < Q_{\text{out}}, Q_{\text{in}} > \\ &= \; < \Gamma_{\mathbf{u},\text{out}}(t), W_{\text{out}}(t), P_{\text{out}}(t), \Gamma_{\mathbf{v},\text{in}}(t), P_{\text{in}}(t), W_{\text{in}}(t) > . \end{aligned} \tag{8.13}$$

Within a formalism based on this kind of quantum states a system cannot have a simultaneously sharply defined in–position and in–momentum, nor a sharply defined out–position and out–momentum, but there is no objection against it having a

[18] Albert *et al.* [7], Bub *et al.* [37], Costa de Beauregard [50] p. 935.
[19] Heisenberg [114], ch. II § 2.

sharply defined in–position and out–momentum; and the same holds with respect to other pairs of incompatible observables.

The idea of assigning counterfactual properties to isolated systems leads to an approach which seems to be rather different from the approaches discussed earlier. I have not pursued this view any further, because it does not really appeal to me. Nevertheless, there may be considerable freedom for further elaborations.

8.3.5 Secret pure states

In the initial formalism the trivial context applies at times when the system is dynamically isolated, due to the fact that the contextual states are conceived as resulting from the interactions with the environment. Above, we considered a way of assigning nontrivial properties at all times, by means of a counterfactual argument. There are other way of doing this, however, and one of them is by assuming that, in fact, there is at all times a maximal quantum observable which applies to the system, even though we generally cannot know which one. Consequently, the contextual state is, at all times, a pure state, i.e., a one–dimensional projector, or, equivalently, a Hilbert space vector. This leads to a view where the quantum state is given by a triple $Q =< \Gamma_{\mathbf{u}}, \psi, W >$. At a time when the system is isolated one may, for instance, assume $\Gamma_{\mathbf{u}} = \{\psi\}$, and the system evolves according to $Q(t) =< \Gamma_{\mathbf{u}}(t), \psi(t), W(t) >$, with $W(t) = UWU^{+}$, $\psi(t) = U\psi$, and $\Gamma_{\mathbf{u}}(t) = \{\psi(t)\}$. However, during an interaction the evolution of the quantum state bifurcates. W changes, into a proper or improper mixture, or into a pure state if the interaction was a maximal measurement. One has a new set of possible vectors ψ_i, and $\Gamma_{\mathbf{u}}$ is the set of all possibilities. Within the context $\mathbf{u}$, the state ψ_i has a probability of $p_{\mathbf{u},W}(\psi_i) = \langle\psi_i|W|\psi_i\rangle$ of being realized, if the ψ_i are mutually orthogonal. If they are not, which may for instance happen during a nonideal measurement, then a normalization constant must be added.

A maximal measurement may, for instance, be analyzed in the following manner: the contextual state ψ is forced into becoming one of the eigenvectors ψ_i of the observable, with a probability $\langle\psi_i|W|\psi_i\rangle$, and after the measurement interaction W itself is transformed into $|\psi_i\rangle\langle\psi_i|$, if indeed a state reduction occurs. Nonideal and nonmaximal measurements, and interactions more generally, may be handled by using operations for the transformation of W, and by admitting nonorthogonal sets of ψs. Note that probabilities are determined by W, and not by ψ, and that ψ is not necessarily an eigenvector of W. In fact, W and ψ may be two different pure states.

Example: In Bohm's version of the EPR experiment one has two spin–1/2 in a state $2^{-1/2}(|+,-\rangle - |-,+\rangle)$. The two particles part, and let us follow the left one. Its density operator is given by $W = 1/2 \cdot \mathbb{1}$, and, according to the present view, it also has a contextual state ψ, being some vector in $\mathcal{H}$. One now has the problem that it is not clear what set of alternatives $\Gamma_{\mathbf{u}}$ to take. One approach is that some particular orthogonal basis, unknown to us, applies, but, alternatively, one might admit, for instance, the entire set $\mathcal{H}$. Ignoring $\Gamma_{\mathbf{u}}$, however, suppose that $|\psi\rangle = 2^{-1/2}(|+\rangle + |-\rangle)$. A spin measurement in the z–direction on the *right* particle may have the effect of changing W into $|+\rangle\langle+|$. There is no need, however, of assuming that this affects ψ also. Only: if a spin measurement z–direction is performed on the left particle too, then ψ is forced into either $|+\rangle$ or $|-\rangle$, but the latter possibility has probability 0, because of $W = |+\rangle\langle+|$.

One may develop this view in a time–symmetric and an asymmetric version. In the asymmetric version there are the usual problems associated with state reduction, and there is also the problem of finding the appropriate context $\Gamma_{\mathbf{u}}$. However, Dennis Dieks, who has a view of roughly this type, combined with a relative state approach[20], has published about some aspects of both these problems[21], using, in particular, the fact that the 'biorthogonal expansion' of the state of a composite system[22] is often unique.

In the time–symmetric case, an interesting possibility is to assume that ψ and W are both pure states, and identify them with the out–state and the in–state respectively. In this case one may take quantum states $Q =< \Gamma_{\mathbf{u}}, \langle\text{out}|, |\text{in}\rangle >=< \Gamma_{\mathbf{u}}, \Theta >$, with $p_{\mathbf{u}}(\Theta) = |\langle\text{out}|\text{in}\rangle|^2$. This leads to the problem, mentioned in § 6.3, that it is not immediately clear how to assign both an in–state and an out–state to a system in interaction with other systems, due to the simultaneous entanglement of both. This may be solved by making the transition to process theory, taking us back to views introduced earlier.

8.3.6 Hidden variables

There are many ways of elaborating on the theme of the two previous approaches, in which the assignment of properties was extended beyond what is minimally necessary. The general idea is that one may interpret the contextual states of the initial formalism as representing not *all* properties possessed by the system at a given moment, but merely those properties which are to the world at that moment exposed, via some interaction with other systems.

It may be noted that, by assigning unobservable properties to the system, we have in fact been adding hidden variables to the theory, although in a way which remains relatively close to standard QM. An important traditional motive for looking for hidden variable theories has been the restoration of determinism, which appears to be an important issue for at least a considerable minority of physicists. Interpreting determinism in the classical sense, it boils down to the condition that there is only one phase trajectory through each phase point. In our present terminology this means that there is only one elementary possible history $\lambda_{\mathbf{hv.proc}} \in \Gamma_{\mathbf{hv.proc}}$ for a given elementary system state $\lambda_{\mathbf{hv.sys}} \in \Gamma_{\mathbf{hv.sys}}$, which implies that there must be an isomorphism between $\Gamma_{\mathbf{hv.proc}}$ and $\Gamma_{\mathbf{hv.sys}}$. It is not necessary to assume that this isomorphism is computable, i.e., that the trajectory of a system can actually

[20]See (8.4.2).
[21]See, e.g., [63, 64, 66].
[22]See also § 8.5.2 below.

be computed from its present state. On the other hand, it would be necessary for a successful HV–theory to provide a solid account of how the probabilities of QM are reproduced. Presumably, the quantum probabilities would represent a special class of HV probability distributions corresponding to something like a 'thermodynamic equilibrium'. Evidently, HV–theory would become a lot more interesting if 'nonequilibrium' effects, violating QM, could be produced in a systematic way, i.e., one would like to find something like a subquantum Brownian motion. Thus far, however, QM seems to perform remarkably well.

Example: A simple example of a hidden variable theory is obtained by taking the initial formalism of contextual QM and adding the assumption that the system has a definite position at all times[23]. Adding a position variable $\mathbf{x}$ to the state description, we thus obtain a state space

$$\Gamma_{\mathbf{hv.sys}} = \{< \Gamma_{\mathbf{u}}, P, W, \mathbf{x} > \mid < \Gamma_{\mathbf{u}}, P, W > \in \Gamma_{\mathbf{qm.sys}}, \mathbf{x} \in I\!R^3\}. \tag{8.14}$$

Suppose that a position measurement is made, then the probability of finding the system in a volume V is given by $\int_{\mathbf{x}\in V} \mathrm{Tr}[|\mathbf{x}\rangle\langle\mathbf{x}|W]$. We are now free to assume that the probability density of the system to be at the position $\mathbf{x}$ is at all times given by

$$\rho_W(\mathbf{x}) = \mathrm{Tr}[|\mathbf{x}\rangle\langle\mathbf{x}|W] = \langle\mathbf{x}|W|\mathbf{x}\rangle \tag{8.15}$$

regardless of whether the position is actually measured at all. The above measures $\rho_W(\mathbf{x})$ must be interpreted in this case as representing some sort of subquantum thermodynamic equilibrium.

This way of introducing position as a hidden variable is entirely unproblematic, but if one wants to consider a theory $\mathcal{HV}$ of this type as a serious physical theory, then the significant step is in finding a dynamics which produces and conserves (8.15), and which can indeed give some explanation for the apparent absence of nonequilibrium effects. This includes finding a Chapman–Kolmogorov equation such that

$$\begin{aligned} \dot{\rho}(\mathbf{x}) &= \int \Big(T(\mathbf{x},\mathbf{x}')\rho(\mathbf{x}') - T(\mathbf{x}',\mathbf{x})\rho(\mathbf{x})\Big)d\mathbf{x}' \\ &= -i\langle\mathbf{x}|[H,W]|\mathbf{x}\rangle. \end{aligned} \tag{8.16}$$

From this equation an interesting conclusion can be inferred even without further knowledge about the dynamics and about the transition probabilities T. The probability density $\rho(\mathbf{x})$ is given by the diagonal terms of W in the position representation. On the other hand, from the second line of the above equation it follows that $\dot{\rho}(\mathbf{x})$ also depends on the off–diagonal terms. It then follows that the transition probabilities T must depend explicitly on W, i.e., $T(\mathbf{x},\mathbf{x}') = T_W(\mathbf{x},\mathbf{x}')$. This excludes the view, at least within a hidden variable theory of roughly this type, that W might in fact have only a statistical meaning. It must have a more fundamental physical relevance. This means that, within a HV–theory of

[23]See, e.g., Bell [13].

roughly the above type, W must be represented in a physical state space, either as referring to the state of the system itself or to the state of the environment. The latter alternative is taken in the De Broglie–Bohm hidden variable theory, where W, or rather the state vector ψ, is interpreted as the potential of a force field.

It is further interesting to note that, if W has physical significance other than as a statistical quantity, then the state reduction process also has a direct physical meaning, which must, somehow, be accounted for in the dynamics of the theory. Presumably, this should also involve the use of OVMs and ABL measures in some way or another, but this is a topic I have not pursued any further.

8.4 The Von Neumann chain

In standard QM, measurement theory has been a traditional source of disagreement. Especially the Von Neumann chain problem is central to many distinctions between the various interpretations of QM. As we may recall, the Von Neumann chain problem can be formulated by considering the expression (5.6):

$$\phi(W) = \mathrm{Tr}^A[(R^A \otimes 1\!\!1\,)U^C(W^A \otimes W)U^{C+}],$$

and noting that $\phi(W)$ depends on R^A. In a particle view, the operation ϕ describes the transition of the density operator W of a system $\mathcal{S}$, into a new one $W' = \phi(W)$, induced by the interaction of $\mathcal{S}$ with another system $\mathcal{S}^A$. The class of operations is sufficiently wide to explain a variety of transitions of W, ranging from unitary evolutions on the one hand to full state reductions on the other hand. Because ϕ depends on R^A, however, a state reduction of $\mathcal{S}$ can be explained only relative to the state reduction of $\mathcal{S}^A$, and this raises a problem of infinite regress. There are a number of different ways in which one might try to stop the regress and this has lead to a spectrum of alternative theories, harbouring major conceptual differences.

Within a time–asymmetric approach to process theory it may be expected that the Von Neumann chain problem generates a similar range of views, because the state reduction appearing here is quite similar to that of standard QM. On the other hand, in a time–symmetric approach, and especially in an event theory, the details of the Von Neumann chain problem are quite different. There is no state reduction here, as the operations are defined on the basis of a retrocausal view. This makes it interesting, however, to see whether perhaps some of the views generated in standard QM might, nevertheless, have representatives within time–symmetric process theory. We shall elaborate on this question in two stages. First, a brief summary is presented of some of the different views on state reduction. Subsequently, as a case study, some aspects are studied of a version of relative state theory within process theory.

8.4.1 Restricting the quantum domain

Although standard QM by itself cannot be coherently maintained within a classical logical framework, one might still consider the possibility of saving part of it by investigating its tenability for microscopic systems alone. In principle one might make a difference between two kinds of systems, microscopic quantum systems and macroscopic classical systems. Within such a view it is possible to regard the Hilbert space $\mathcal{H}$, or the set of density operators $\mathcal{D}$, as a classical state space. However, the standard physical observables are not magnitudes pertaining to the state of a microsystem in a classical sense, but they are functions on the state space of a macrosystem. Magnitudes in the classical sense may now be defined as functions on $\mathcal{H}$, but they are less interesting, because they are not the quantities which are measured on a quantum system. In fact, quantum measurements are not 'measurements' in a classical sense. They are stochastic physical processes, in which the state of the quantum system merely determines a probability measure on the set of final states of the measurement apparatus. Thus, observables are not the real magnitudes, but the real magnitudes cannot be measured.

Such a view may indeed be tenable, but evidently there are some questions remaining to be answered. Where does the boundary between microsystems and macrosystems lie? What is the precise form of the state space of a macroscopic system? What is the mechanism of the interaction between microsystems and macrosystems? etcetera.

Macroscopic magnitudes

One way of elaborating on such a view is obtained by assuming that macroscopic properties emerge from the structure of the Hilbert space of large quantum systems. Suppose that QM applies to all physical systems, but suppose, furthermore, that for a macroscopic object there is a specific set of compatible observables, its macroscopic observables, which pertain to the object under all physically possible conditions, i.e., these magnitudes are always wel–defined and have a definite value. A consequence of this assumption is that the set of contexts only contains contexts $\mathbf{u}$ in which all macroscopic magnitudes can be simultaneously well–defined, i.e., there are restrictions on the set Ψ of contexts admitted as physically relevant. This may also be formulated by saying that the superposition principle is not generally valid, because there are superselection rules. A picture of this kind is emerging in particular in the C^*-algebra approach to QM[24], but at least the main lines of the idea may be grasped also within a less sophisticated formalism[25].

Evidently, although the approach solves some problems, it also raises new ones, relating especially to the demarcation between macroscopic and microscopic observables. Some specification of which magnitudes are the macroscopic ones must be added to the theory. If there are macroscopic magnitudes, i.e., if there are superselection rules, this implies that there are, on the other hand, self–adjoint operators which cannot correspond to physical observables, i.e., they cannot be measured for fundamental reasons. Let us note that, for establishing a view of this type, it is not enough to show that the class of self–adjoint operators for which a measurement

[24]See, e.g., Primas [171].

[25]See, e.g., Wan [212].

procedure exists is restricted for practical reasons[26]. The basic idea is that the existence of superselection rules leads to restrictions on what quantities are measurable, but the implication does not automatically work the other way around. The mere circumstance that we are not able to measure certain quantities is by itself quite insufficient to produce a state reduction, or to produce a *fundamental* difference between microscopic and macroscopic magnitudes.

Other ways of restricting the validity of QM

What the above case in fact amounts to is that the Von Neumann regress is halted by reaching a level where QM is not strictly valid. There are a few more ways to elaborate on this option.

A well-known view holds that the regress only stops by the actual observation of the measurement outcome by a conscious being and that the collapse of the state represents a direct interaction between mind and matter[27].

Furthermore, there is a recent theory[28] which puts the blame already on the microscopic level. The idea is that quantum particles are submitted to spontaneous random collapses of the density operator at a very low rate. Owing to the entanglement of single particle states in larger bodies, however, the spontaneous collapse of a single particle can cause a collapse of the state of the entire system. As a consequence many macroscopic superpositions are very short-lived.

8.4.2 Open system views

A particular elegant view on the Von Neumann chain problem is to assume that the infinite regress is, actually, physically real. Reductions of the density operator can be interpreted to represent boundary conditions on a composite system, which determine the physical conditions in a subsystem, and in this way the subsystem derives its properties from those of the composite system. The problem of how the composite system obtains its properties can be answered by saying that it derives them from a larger system from which it is itself a subsystem. The problem of how the universe as a whole obtains its properties can be answered by saying that, at least within QM, there is no such thing as the universe as a whole. It can be consistently maintained that *every* physical system is a subsystem of a larger system, and that it is impossible to assign a quantum state to the totality of all systems. In this way, every system which can be described by QM at all is still an open system which can have definite properties which it derives from its interactions with the environment. Accordingly, the classical structure of the macroscopic world originates from the infinite complexity of the microworld. This view may provide a physically acceptable basis for the emergence of superselection rules, and C^*-algebra[29] may indeed provide an appropriate formalism for developing it.

[26]Such as in Daneri *et al.* [55].

[27]London and Bauer [153], Wigner [222].

[28]Ghirardi, Rimini, Weber [99, 98, 97].

[29]See, e.g., Primas [171].

Relative states

In the preceding 'open universe' approach, physical systems derive their properties from the existence of external boundary conditions, applying to every system for which a quantum description is possible. This means that the idea is not applicable within a theory in which the universe as a whole is conceived to be a closed system. There is another view, however, in which the key to solving the Von Neumann regress problem is also contained in the process of selecting subsystems, but which is applicable also in closed universes. The idea is that subsystems of an isolated system can have properties relative to each other. In Everett's words[30]:

> "...a constituent subsystem cannot be said to be in any single well–defined state, independent of the remainder of the composite system",

but Everett then proceeds to assign 'relative states' which do depend on the remainder of the composite system. The philosophical implications of this view are far–reaching, because the upshot is that empirically observed facts, for instance, become part of the 'relative reality' of a specific observer.

Making a balanced assessment of the relative state view on the basis of the original papers about the topic, those of Everett himself and of others such as De Witt, is a bit problematic. In the first place, because some complications are treated badly, and next, because in Everett's approach the relative state view appears in combination with other ideas which are, in fact, logically independent of it. We shall not discuss these problems at this particular spot. Within quantum process theory a view analogous to the relative state view can be introduced by considering the question of when a composite process can be split into definite subprocesses. This view will be discussed in the next section. After stating the problem of subprocess selection in more detail, we shall then have the opportunity to discuss some of the salient features and complications of the relative state view in a more systematic way.

8.5 Relative process states

Two main variants of process theory have been introduced, a time–symmetric and an asymmetric one. The asymmetry in the latter approach is due to a 'state reduction process', which is not only asymmetric in time but also generates a 'Von Neumann chain problem', leading to a range of different views in which the problem is overcome in different ways. Within the time–symmetric theory of quantum processes the Von Neumann chain problem as such does not exist, because there is no state reduction. However, there is a related topic concerning the definition of subprocesses, which, nevertheless, generates a more or less corresponding range of views.

[30][74] p.124.

Processes and subprocesses

Consider a physical process $\mathcal{P}^A$, for example an experiment in a laboratory. Evidently, $\mathcal{P}^A$ is part of a larger process $\mathcal{P}^{AB}$, involving, for instance, the history of a roomful of equipment during a certain period of time. In process theory we may now ask what the relation is between the description of the composite process $\mathcal{P}^{AB}$ and that of the process $\mathcal{P}^A$, and under what conditions it is possible at all to define a definite subprocess $\mathcal{P}^A$, starting from a specific process $\mathcal{P}^{AB}$. Clearly, this topic is highly relevant for many subjects, ranging from simple laboratory situations to cosmological problems. In cosmology it is, for instance, interesting to ask whether one can meaningfully speak of the entire history of the universe in terms of a physical process which is not a subprocess of some larger process.

Our considerations will be based on the formalism of pure process states given in § 6.4.1, with processes described, as in (6.72), by a triple $\Phi =< \langle \text{out}|, V, |\text{in}\rangle >$, where $\langle \text{out}| \in \mathcal{H}^+$, $|\text{in}\rangle \in \mathcal{H}$ and V is a linear operator $\mathcal{H} \to \mathcal{H}$.

In particular, consider an experiment described within process theory as a composite process $\mathcal{P}^{AB}$, taking place between times t_{in} and t_{out}, and with an associated Hilbert space $\mathcal{H}^{AB} = \mathcal{H}^A \otimes \mathcal{H}^B$. The state of this process is given by a triple

$$\Phi^{AB} =< \langle \text{out}^{AB}|, V^{AB}, |\text{in}^{AB}\rangle >, \tag{8.17}$$

being one of the elements of a contextual state space $\Gamma^{AB}_{\mathbf{u}}$, containing all states which may potentially be realized in this experiment. Process theory assigns to these states the probabilities (6.15), here equal to

$$p^{AB}_{\mathbf{u},w^{AB}}(\Phi) = C^{AB}_{\mathbf{u},w}\, L^{AB}(\Phi^{AB}) w^{AB}(\Phi^{AB}). \tag{8.18}$$

We shall investigate various special cases of this situation, with, in particular, the objective of defining corresponding subprocesses $\mathcal{P}^A$ and $\mathcal{P}^B$. This implies, in the first place, that there must be a way of defining subprocess states Φ^A and Φ^B. Therefore, let us assume that there is a function σ^A on $\Gamma_{\mathbf{proc}}$, such that

$$\Phi^A = \sigma^A(\Phi^{AB}), \tag{8.19}$$

and likewise for $\mathcal{P}^B$. Further, one may want to describe the entire experiment in terms of the process $\mathcal{P}^A$, and this implies that, corresponding to the state space $\Gamma^{AB}_{\mathbf{u}}$ one must find a state space $\Gamma^A_{\mathbf{u}}$, and, in order to compute the probabilities $p^A_{\mathbf{u},w^A}$, the statistical weight function w^{AB} must be transformed into a corresponding function w^A. In terms of the function σ^A, this may be achieved by defining

$$\begin{aligned} \Gamma^A_{\mathbf{u}} &= \{\sigma^A(\Phi^{AB}) |\, \Phi^{AB} \in \Gamma^{AB}_{\mathbf{u}}\} \\ w^A(\Phi^A) &= \sum_{\Phi^{AB},\, \sigma^A(\Phi^{AB})=\Phi^A} w^{AB}(\Phi^{AB}). \end{aligned} \tag{8.20}$$

With respect to the probabilities $p^A_{\mathbf{u},w^A}$, this should lead to

$$p^A_{\mathbf{u},w^A}(\Phi^A) = \sum_{\Phi^{AB},\, \sigma^A(\Phi^{AB})=\Phi^A} p^{AB}_{\mathbf{u},w^{AB}}(\Phi^{AB}). \tag{8.21}$$

Accordingly, the problem of defining subprocesses boils down to the question whether a suitable function σ^A can be found.

Scattering

In the first place, we consider a simple *scattering process*, for which we make the following assumptions:

$$\begin{aligned} \langle \mathrm{out}^{AB}| &= \langle \mathrm{out}^{A}| \otimes \langle \mathrm{out}^{B}| \\ |\mathrm{in}^{AB}\rangle &= |\mathrm{in}^{A}\rangle \otimes |\mathrm{in}^{B}\rangle. \end{aligned} \tag{8.22}$$

Suppose that, for the above scattering process, we want to find an associated subprocess $\mathcal{P}^A$. Defining

$$\begin{aligned} \Phi^A &= \sigma^A(\Phi^{AB}) \\ &:= \ < \langle \mathrm{out}^A|, \langle \mathrm{out}^B|V^{AB}|\mathrm{in}^B\rangle, |\mathrm{in}^A\rangle >, \end{aligned} \tag{8.23}$$

it is readily verified that one obtains the correct probabilities. For $L^A(\Phi^A)$ one finds

$$\begin{aligned} L^A(\Phi^A) &= \langle \mathrm{out}^A|\langle \mathrm{out}^B|V^{AB}|\mathrm{in}^B\rangle|\mathrm{in}^A\rangle \\ &= \langle \mathrm{out}^A|V^A|\mathrm{in}^A\rangle \\ &= L^{AB}(\Phi^{AB}), \end{aligned} \tag{8.24}$$

and therefore, using (8.20) and putting $C^A_{\mathbf{u},w^A} = C^{AB}_{\mathbf{u},w^{AB}}$:

$$\begin{aligned} p^A_{\mathbf{u},w^A}(\Phi^A) &= C^A_{\mathbf{u},w^A} L^A(\Phi^A) w^A(\Phi^A) \\ &= \sum_{\Phi^{AB},\sigma^A(\Phi^{AB})=\Phi^A} C^{AB}_{\mathbf{u},w^{AB}} L^{AB}(\Phi^{AB}) w^{AB}(\Phi^{AB}) \\ &= \sum_{\Phi^{AB},\sigma^A(\Phi^{AB})=\Phi^A} p^{AB}_{\mathbf{u},w^{AB}}(\Phi^{AB}). \end{aligned} \tag{8.25}$$

It thus appears that it is indeed possible, in the present case, to start from a given composite process $\mathcal{P}^{AB}$, and define subprocesses $\mathcal{P}^A$ and $\mathcal{P}^B$ to which definite process states can be assigned such that the probability measures match.

Note: We may note that, *vice versa*, the composite process $\mathcal{P}^{AB}$ cannot be reconstructed from its subprocesses $\mathcal{P}^A$ and $\mathcal{P}^B$, because the operator V^{AB} cannot be reconstructed from the submatrices $\langle \mathrm{out}^B|V^{AB}|\mathrm{in}^B\rangle$ and $\langle \mathrm{out}^A|V^{AB}|\mathrm{in}^A\rangle$. Accordingly, we lose information by describing a process in terms of its constituent subprocesses.

8.5.1 A chain–like problem

The above way of defining subprocesses depends crucially on the factorizability conditions (8.22) for the in–states and the out–states. This condition, however, introduces a problem which resembles the Von Neumann chain problem. Clearly we all witness many physical processes to which we assign definite properties. These processes, however, are subprocesses of a succession of larger processes, involving ourselves, the world around us, and, eventually, large fractions of the universe. By the conditions (8.22), the very possibility of assigning a definite state to a process $\mathcal{P}^A$ seems to depend on the properties of processes $\mathcal{P}^{AB}$ of which it is part. In its turn, however, this also holds for $\mathcal{P}^{AB}$ itself, so where can we start with assigning properties? Acceptable answers to this question must be able to accomodate the existing practice on which the empirical interpretation of QM is based, in particular the assignment of states to physical processes on the basis of observation. A minimal requirement for this is that we have well-defined properties ourselves, and this raises the question of *why* this is so.

Although this question is not entirely as malicious as the Von Neumann problem in the time-asymmetric theory, there are definite similarities both between the problems and the answers. One may construct a range of different views, corresponding more or less closely to existing views on solving the Von Neumann chain problem. For instance, one may hold the view that that we are sentient beings, or macroscopic objects, or otherwise systems which can be assumed to fall outside the domain of applicability of QM. Another type of answers is that there are no restrictions to the validity of QM and that we owe the quality of having nontrivial properties to being open systems ourselves. A first way to work out the latter option is by assuming that the in the history of the universe there are definite boundaries, and that there are a universal W_{in} and W_{out}, representing universal boundary conditions, which happen to satisfy appropriate factorizability conditions. An alternative to this view is that the the history of the universe has no boundaries, but cannot in its entirety be described as a quantum process, whereas every subprocess which *can* be described as a quantum process *does* have boundaries. In both cases, the core of the solution is the assumption that the processes from which we are built are sufficiently similar to scattering processes.

Note: Besides similarities there are also considerable differences between the present situation and the one discussed in the time-asymmetric theory. A significant difference is, in particular, that in time-symmetric process theory one can use *future* boundary conditions for assigning well-defined states to processes occurring *now*. For example, for a process $\mathcal{P}^{AB}$ with a state

$$\Phi^{AB} = \tag{8.26}$$
$$<< \langle \text{out}^A|, \langle \text{out}^B| >, < V^A(t_2,t_1) \otimes V^B(t_3,t_1), V^{AB}(t_1,t_0) >, |\text{in}^{AB}\rangle >,$$

where t_2 and t_1 are times in the neighbourhood of now, one may define a subprocess $\mathcal{P}^A$ with a state

$$\begin{aligned} \Phi^A &= < \langle \text{out}^A|, V^A(t_2,t_1), |\text{in}^A\rangle > \quad \text{with} \\ |\text{in}^A\rangle &= \langle \text{out}^B|V^B(t_3,t_1)V^{AB}(t_1,t_0)|\text{in}^{AB}\rangle. \end{aligned} \tag{8.27}$$

The in-state $|\text{in}^A\rangle$ at t_1 depends on the out-state $\langle \text{out}^B|$ at t_3, which may be in our very far future. In the time-asymmetric theory a corresponding procedure for assigning a state on the basis of future developments does not exist, and, accordingly, process theory has significantly more possibilities available for dealing with problems of the Von Neumann chain type.

8.5.2 Relative states

If one is dissatisfied by the idea of having to satisfy the factorizability conditions (8.22), one may try to find a way of defining subprocesses also in circumstances where these conditions are violated. We shall make an attempt at doing this in a way which corresponds to some extent to Everett's relative state view.

Suppose we were to assume that the properties of a process are, in fact, not monadic properties but relations with other subprocesses, i.e., processes have states *relative to each other.* This would make it possible for a subprocess to have properties which are not represented in the state of the composite process. Such a view would be applicable also to processes where (8.22) is violated. It would no longer be necessary to require the factorizability of either the in–state or the out–state of the composite process. Therefore, we shall proceed by exploring the structure of (8.17) under more general conditions, and more specifically we shall be interested in the following question, which, I will formulate in particle terminology for the sake of clarity: suppose that two systems interact repeatedly, but are dynamically isolated in intermediate periods. To what extent is it possible then to speak meaningfully about two separate systems having separate properties?

In order to investigate this question we consider a process $\mathcal{P}^{AB}$ with a corresponding Hilbert space $\mathcal{H}^{AB} = \mathcal{H}^A \otimes \mathcal{H}^B$ and a process state of type

$$\Phi^{AB} =< \langle \text{out}^{AB}|, V^A \otimes V^B, |\text{in}^{AB}\rangle >, \tag{8.28}$$

describing a process with entangled in–states and out–states, but with a factorizable Stinespring operator $V^{AB} = V^A \otimes V^B$.

Clearly, our earlier method for constructing subprocesses is no longer applicable because of the entanglement of the in–states and the out–states of the process $\mathcal{P}^{AB}$. Nevertheless, one might attempt to simply assume the existence of two subprocesses $\mathcal{P}^A$ and $\mathcal{P}^B$ of $\mathcal{P}^{AB}$, with well-defined separate process states, and see to what consequences it would lead. Therefore, we continue by assuming that these subprocesses exist, and in order to construct their states we pick one of them, say $\mathcal{P}^A$, and a time, say $t_{\text{in}} + \delta t$, and assume that there is a definite vector $\psi \in \mathcal{H}^A$, representing the in–state of $\mathcal{P}^A$: i.e.,

$$|\text{in}^A\rangle = |\psi\rangle. \tag{8.29}$$

The next step is assuming that, corresponding to this given ψ, we can assign to $\mathcal{P}^B$ a 'relative in–state'

$$|\text{in}^B(\psi)\rangle = \langle\psi|\text{in}^{AB}\rangle. \tag{8.30}$$

It then appears that the states of the two processes $\mathcal{P}^A$ and $\mathcal{P}^B$ can be computed by using the following 'trick': we note that $|\text{in}^{AB}\rangle$ and $\langle\text{out}^{AB}|$ are objects which are, in fact, quite similar to V^A and V^B. For instance, V^A is a linear operator $\mathcal{H}^A \to \mathcal{H}^A$, but at the same time it is a vector in $\mathcal{H}^A \otimes \mathcal{H}^{A+}$. Likewise, $|\text{in}^{AB}\rangle$ is a vector in $\mathcal{H}^A \otimes \mathcal{H}^B$, but it can also be regarded as a linear operator $\mathcal{H}^{A+} \to \mathcal{H}^B$, or $\mathcal{H}^{B+} \to \mathcal{H}^A$. Accordingly, vectors like $|\text{in}^{AB}\rangle$ and $\langle\text{out}^{AB}|$ and linear operators

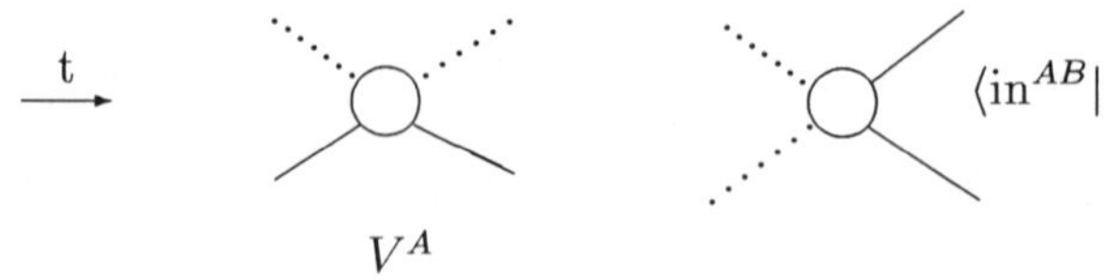

Figure 8.1: diagram 1 diagram 2

like V^A and V^B are similar, up to transformations $\mathcal{H} \to \mathcal{H}^+$ or $\mathcal{H}^+ \to \mathcal{H}$, which are associated with time–reversal. Speaking in terms of diagrams, V^A represents a diagram of type 1 in fig. 8.1, and $|\mathrm{in}^{AB}\rangle$ one of type 2, i.e., they are similar up to a reflection in spacetime.

Given a specific Φ^{AB} and a ψ, quantum states for both processes $\mathcal{P}^A$ and $\mathcal{P}^B$ can now be constructed by developing ψ along two different paths. To $\mathcal{P}^B$ we assign an in–state by using the relative state construction (8.30). This state is developed along V^B and $\langle \mathrm{out}^{AB}|$ to obtain the out–state $\langle \mathrm{out}^A(\Phi^{AB}, \psi)|$ of $\mathcal{P}^A$, i.e.:

$$\begin{aligned} \langle \mathrm{out}^A(\Phi^{AB}, \psi)| &= \langle \mathrm{out}^{AB}|V^B|\mathrm{in}^B\rangle \\ &= \langle \mathrm{out}^{AB}|V^B\langle\psi|\mathrm{in}^{AB}\rangle. \end{aligned} \tag{8.31}$$

We can thus assign to $\mathcal{P}^A$ a state:

$$\sigma^A(\Phi^{AB}, \psi) = < \langle \mathrm{out}^{AB}|V^B\langle\psi|\mathrm{in}^{AB}\rangle, V^A, |\psi\rangle >, \tag{8.32}$$

and via an entirely similar procedure we can define for $\mathcal{P}^B$ a state

$$\sigma^B(\Phi^{AB}, \psi) = < \langle \mathrm{out}^{AB}|V^A|\psi\rangle, V^B, \langle\psi|\mathrm{in}^{AB}\rangle > . \tag{8.33}$$

Cyclic processes

By treating the vectors $|\mathrm{in}^{AB}\rangle$ and $\langle \mathrm{out}^{AB}|$ as evolution phases, comparable to V^A and V^B, we have actually obtained a *cyclic process*, beginning with $|\psi\rangle$ and then proceeding via V^A to $\langle \mathrm{out}^{AB}|$ and then back in time via V^B to $|\mathrm{in}^{AB}\rangle$ and forwards in time again to arrive at $\langle\psi|$. The state of this cyclic process $\mathcal{P}^C$ can be represented by

$$\Phi^C = < \langle\psi|, \ < |\mathrm{in}^{AB}\rangle, V^B, \langle \mathrm{out}^{AB}|, V^A >, \ |\psi\rangle > , \tag{8.34}$$

where $|\mathrm{in}^{AB}\rangle, V^B, \langle \mathrm{out}^{AB}|$, and V^A are all regarded as phases in the evolution of $\mathcal{P}^C$. The corresponding evolution operator V^C is given by[31]

$$\begin{aligned} V^C|\psi\rangle &= < |\mathrm{in}^{AB}\rangle, V^B, \langle \mathrm{out}^{AB}|, V^A > \\ &= \langle \mathrm{out}^{AB}|V^A|\psi\rangle V^B|\mathrm{in}^{AB}\rangle. \end{aligned}$$

Let us proceed by observing that we could have chosen to arrange the phases of the process $\mathcal{P}^C$ in a different way, by beginning at a different point in the cycle. For

[31] See § 6.2.2.

instance, we could take the evolution phase V^A out of the evolution phase of $\mathcal{P}^C$ and group it together with $|\psi\rangle$ in the preparation phase to obtain a process $\mathcal{P}^D$ with a state

$$\Phi^D = < \langle\psi|, < |\text{in}^{AB}\rangle, V^B, \langle\text{out}^{AB}| > , < V^A, |\psi\rangle >>, \tag{8.35}$$

with an associated in–state $|\text{in}^D\rangle = V^A|\psi\rangle$ and an evolution operator

$$V^D|\phi\rangle = \langle\text{out}^{AB}|\phi\rangle V^B|\text{in}^{AB}\rangle. \tag{8.36}$$

It then becomes clear that the processes $\mathcal{P}^A$ and $\mathcal{P}^B$ are themselves simply different arrangements of the process $\mathcal{P}^C$, given by

$$\begin{aligned} \Phi^A &= << \langle\psi|, < |\text{in}^{AB}\rangle, V^B, \langle\text{out}^{AB}| > , V^A, |\psi\rangle >, \\ \Phi^B &= << \langle\text{out}^{AB}|, V^A, |\psi\rangle > , V^B, < \langle\psi|, |\text{in}^{AB}\rangle >> . \end{aligned} \tag{8.37}$$

The difference between such arrangements of the phases in a process has no physical significance, but is comparable to using different pictures, such as the Schrödinger or the Heisenberg picture, in standard QM. This also becomes apparent by computing the amplitudes and the L–values for these processes. The amplitudes associated with $\mathcal{P}^A$, $\mathcal{P}^B$, and $\mathcal{P}^C$, with states $\Phi^X = \sigma^X(\Phi^{AB}, \psi)$, are given by

$$\begin{aligned} A^A(\Phi^A) &= A^B(\Phi^B) = A^C(\Phi^C) \\ &= \langle\text{out}^C|V^C|\text{in}^C\rangle \\ &= \langle\text{out}^{AB}|V^B\langle\psi|\text{in}^{AB}\rangle V^A|\psi\rangle \\ &= \langle\text{out}^{AB}|\left(V^A \otimes V^B\right)\left(|\psi\rangle\langle\psi| \otimes 1\!\!1^B\right)|\text{in}^{AB}\rangle. \end{aligned} \tag{8.38}$$

We shall use $\mathcal{P}^C$ in our further analysis.

Probabilities

Defining a probability measure for the subprocess $\mathcal{P}^C$ requires, first of all, the construction of an appropriate context. This is slightly more complicated than in the case of scattering, because the state $\Phi^C = \sigma^C(\Phi^{AB}, \psi)$ depends not only on Φ^{AB}, but also on ψ. For this reason, (8.20) is not directly applicable. Actually, we must first decide what kind of probability we want, because there are several different types. In the first place, let us assume some fixed state Φ^{AB}. For this fixed state, we now want to find a probability of finding some given Φ^C. This problem, however, cannot be solved without determining some context $\mathbf{v}$, with a corresponding state space $\Gamma_{\mathbf{v}}$. At this point we encounter a problem which has also figured in the debate about the relative state view in standard QM. In this theory too, a context must be fixed, in the form of a special representation of the Hilbert space, in order to obtain probabilities[32].

[32]The necessity of fixing a specific Hilbert space representation in Everett's theory was pointed out in, e.g., d'Espagnat [73], Bell [13], Cooper [48], see also Ben–Dov [17, 16, 18].

Example: A common example in discussions about relative states concerns measurement interactions, in which the state of a measuring instrument becomes correlated with the state of a quantum system. An initial state $|\psi, \phi_0\rangle = \sum_i c_i|\alpha_i, \phi_0\rangle$ evolves into $\sum_i c_i|\alpha_i, \phi_i\rangle$, and the idea is that the different terms of the superposition have a probability $p(\phi_i) = |c_i|^2$ of being realized. However, this depends on choosing a particular decomposition of ψ into vectors α_i, which may be done in infinitely many diferent ways.

This problem may be solved in an *ad hoc* way by merely assuming that some particular context $\mathbf{v}$ is realized. In our own case[33] this boils down to assuming that, for a given state Φ^{AB} of the composite process $\mathcal{P}^{AB}$, there is some set $\Delta_\mathbf{v}$ of possible vectors ψ. For the process $\mathcal{P}^C$ we may then construct a state space

$$\Gamma^C_\mathbf{v} = \{\sigma^C(\Phi^{AB}, \psi) \mid \psi \in \Delta_\mathbf{v}\}, \tag{8.39}$$

leading to a probability measure

$$\begin{aligned} p^C_\mathbf{v}\left(\sigma^C(\Phi^{AB}, \psi)\right) &= C^C_\mathbf{v} L^C\left(\sigma^C(\Phi^{AB}, \psi)\right) \\ &= C^C_\mathbf{v}\left|\langle \mathrm{out}^{AB}|(V^A \otimes V^B)(|\psi\rangle\langle\psi| \otimes 1\!\!1^{\,B})|\mathrm{in}^{AB}\rangle\right|^2 \end{aligned} \tag{8.40}$$

Dropping the assumption of a fixed Φ^{AB} results in a second case, with a different context and different probabilities. With Φ^{AB} being itself an element of a set $\Gamma^{AB}_\mathbf{u}$ of possible states of $\mathcal{P}^{AB}$, one may, again, ask for the probability of the occurence of a specific Φ^C. Corresponding to each Φ^{AB} one may now have a different context $\mathbf{v}(\Phi^{AB})$, with a different set $\Delta_{\mathbf{v}(\Phi^{AB})}$ of vectors ψ. The set of all process states Φ^C which may be realized may then be collected in the state space $\Gamma^C_\mathbf{w}$ of a new context $\mathbf{w}$, with

$$\Gamma^C_\mathbf{w} = \bigcup_{\Phi^{AB} \in \Gamma^{AB}_\mathbf{u}} \Gamma^C_{\mathbf{v}(\Phi^{AB})}. \tag{8.41}$$

A corresponding probability measure is defined by

$$\begin{aligned} p^C_\mathbf{w}(\Phi^C) &= \sum_{\Phi^{AB},\, \psi,\, \sigma^C(\Phi^{AB}, \psi) = \Phi^C} p^C_{\mathbf{v}(\Phi^{AB})}(\Phi^C) p^{AB}_\mathbf{u}(\Phi^{AB}) \\ &= \sum_{\Phi^{AB},\, \psi,\, \sigma^C(\Phi^{AB}, \psi) = \Phi^C} C^C_\mathbf{w} w^{AB}_\mathbf{u}\left|\langle \mathrm{out}^{AB}|(V^A \otimes V^B)|\mathrm{in}^{AB}\rangle\right|^2 \times \\ &\qquad \left|\langle \mathrm{out}^{AB}|(V^A \otimes V^B)(|\psi\rangle\langle\psi| \otimes 1\!\!1^{\,B})|\mathrm{in}^{AB}\rangle\right|^2. \end{aligned} \tag{8.42}$$

Finally, one may wish to obtain the joint probability of the combination of a state Φ^C and a state Φ^{AB}. In this case, one should consider the combination of processes $\mathcal{P}^C$ and $\mathcal{P}^{AB}$. A suitable state space $\Gamma^{C,AB}_{\mathbf{w},\mathbf{u}}$ is defined by

$$\Gamma^{C,AB}_{\mathbf{w},\mathbf{u}} = \{< \Phi^C, \Phi^{AB} > \mid \Phi^C \in \Gamma^C_\mathbf{w},\ \Phi^{AB} \in \Gamma^{AB}_\mathbf{u}\}. \tag{8.43}$$

[33] i.e., in the pure state version of process theory. In the general version the amount of freedom is much larger still.

The corresponding joint probability is found from (8.42) by dropping the summation over Φ^{AB}:

$$p^C_{\mathbf{w},\mathbf{u}}(\Phi^C, \Phi^{AB}) = \sum_{\psi,\, \sigma^C(\Phi^{AB},\psi)=\Phi^C} p^C_{\mathbf{v}(\Phi^{AB})}(\Phi^C) p^{AB}_{\mathbf{u}}(\Phi^{AB}). \tag{8.44}$$

A pitfall

Looking at the amplitude (8.38) of $\mathcal{P}^C$ it is tempting to try defining a joint probablity for the combined process $\mathcal{P}^{C,AB}$ directly, in the following manner:

$$\begin{aligned} & p^{C,AB}_{\mathbf{wu}}\left(\sigma^C(\Phi^{AB},\psi), \Phi^{AB}\right) \\ = \; & C^{C,AB}_{\mathbf{wu}} \left| \langle \mathrm{out}^{AB} | \left(V^A \otimes V^B\right) \left(|\psi\rangle\langle\psi| \otimes 1\!\!1^{\,B}\right) |\mathrm{in}^{AB}\rangle \right|^2, \end{aligned} \tag{8.45}$$

where, for brevity, the statistical weight function w has been suppressed. This definition does not work, however. Obviously, one would have to require that the probabilities add up correctly and this would imply that

$$\sum_{\psi \in \Delta_{\mathbf{v}(\Phi^{AB})}} p^{C,AB}_{\mathbf{wu}}\left(\sigma^C(\Phi^{AB},\psi), \Phi^{AB}\right) = p_{\mathbf{u}}(\Phi^{AB}). \tag{8.46}$$

This condition is generally *not* satisfied.

Example: Suppose that we impose on $\Delta_{\mathbf{v}(\Phi^{AB})}$ the requirement that

$$\sum_{\psi\in\Delta} |\psi\rangle\langle\psi| = 1\!\!1^{\,A}, \tag{8.47}$$

then it is readily computed that

$$\begin{aligned} \sum_{\psi\in\Delta} \langle \mathrm{out}^A | V^A | \mathrm{in}^A\rangle &= \langle \mathrm{out}^{AB} | \left(V^A \otimes V^B\right) \left(|\psi\rangle\langle\psi| \otimes 1\!\!1^{\,B}\right) |\mathrm{in}^{AB}\rangle \\ &= \langle \mathrm{out}^{AB} | (V^A \otimes V^B) | \mathrm{in}^{AB}\rangle. \end{aligned} \tag{8.48}$$

Generally, $|\sum A(\Phi)|^2$ is not at all equal to $\sum |A(\Phi)|^2$. The two expressions differ by the occurrence of 'interference terms', which vanish only in very special circumstances. For arbitrary Φ^{AB} it is not generally possible to find a Δ which makes the interference terms disappear and therefore (8.46) will generally be violated.

The fact that joint probabilities for $\mathcal{P}^C$ and $\mathcal{P}^{AB}$ are given by (8.44) rather than (8.45) is closely related to what may be considered as the core of the relative state view. What it amounts to is the following: although the $\mathcal{P}^{AB}$ sets up the conditions for the existence of the subprocesses $\mathcal{P}^A$ and $\mathcal{P}^B$, the process $\mathcal{P}^{AB}$ and its subprocesses are for the rest independent of each other. Physical reality is split into different *levels* which are mutually inaccessible. Although $\mathcal{P}^{AB}$ creates the conditions for the existence $\mathcal{P}^C$, the two processes lead their own life. For an observer 'inside' $\mathcal{P}^C$, little can be inferred about $\mathcal{P}^{AB}$. *Vice versa*, probabilities at the level of the composite process do not depend on what happens at the level of the subprocesses. Probabilities at the level of $\mathcal{P}^{AB}$ are not affected if, at the level of $\mathcal{P}^C$, some specific vector ψ happens to be realized. An observer at the level of $\mathcal{P}^{AB}$ might decide to break into $\mathcal{P}^C$ by performing a measurement, but this implies, at the same time, changing the subprocess $\mathcal{P}^C$ into a different on. Therefore, if we restrict the analysis to the level of reality of $\mathcal{P}^{AB}$, then $\mathcal{P}^C$ can be regarded as a *virtual* process, even though it is quite real at the level of $\mathcal{P}^C$ itself. A *total reality* containing all such levels can be conceived to exist *in abstracto*, but no physical observer can obtain information about any other than the single level at which he himself is running as a physical process.

Biorthogonal expansions

With reference to the above analysis, there are a few points which seem to be in rather urgent need of further examination. In the first place, besides (8.28) there are various other cases which remain to be analyzed. It would be interesting, for instance, to see how the approach handles such situations as virtual pair creation. In the second place, the amount of arbitrariness in the model is rather disturbing. The construction of the state space $\Gamma^C_{\mathbf{v}}$ depends on the free parameter $\Delta_{\mathbf{v}}$. Our analysis has not thus far produced any reason for imposing restrictions on $\Delta_{\mathbf{v}}$, so in principle one might take any arbitrary subset of $\mathcal{H}^A$. This leaves us with a rather extreme amount of freedom, and there may or may not be plausible arguments for reducing the set of possibilities.

Example: Suppose that $|\mathrm{in}^{AB}\rangle$ factorizes into $|\mathrm{in}^A\rangle \otimes |\mathrm{in}^B\rangle$. In this case it seems natural to demand $|\psi\rangle = |\mathrm{in}^A\rangle$. A systematic way to achieve this would be to take the *biorthogonal expansion* of $|\mathrm{in}^{AB}\rangle$:

Regarding $|\mathrm{in}^{AB}\rangle$ as a linear operator $\mathcal{H}^A \to \mathcal{H}^{B+}$, it can be written in its polar decomposition[34] as the product of a Hermitian operator $R^{1/2} : \mathcal{H}^A \to \mathcal{H}^A$ and a unitary operator $|U\rangle : \mathcal{H}^A \to \mathcal{H}^{B+}$, where R satisfies

$$R|\psi\rangle = \langle \mathrm{in}^{AB}|\psi\rangle|\mathrm{in}^{AB}\rangle, \tag{8.49}$$

and $|U\rangle$ can be written in the form

$$|U\rangle = \sum_j e^{iu_j}|\phi^A_j, \chi^B_j\rangle, \tag{8.50}$$

with orthonormal sets of ϕ^A_j and χ^B_j. $|\mathrm{in}^{AB}\rangle$ can now be written as

$$|\mathrm{in}^{AB}\rangle = \sum_j \sqrt{r_j}|r^A_j, s^B_j\rangle, \tag{8.51}$$

where the vectors $|r^A_j\rangle$ are the eigenvectors of R, and the $|s^B_j\rangle$ are given by

$$|s^B_j\rangle = \langle r^A_j|U\rangle. \tag{8.52}$$

[34] See (5.20).

Because the $|r_j^A\rangle$ are orthogonal and $|U\rangle$ is unitary, the $|s_j^B\rangle$ are also orthogonal and for this reason the expansion (8.51) is called biorthogonal. The biorthogonal expansion of $|\text{in}^{AB}\rangle$ is unique if R is nondegenerate.

If we take for $\Delta_\mathbf{v}$ the set of vectors $|r_j^A\rangle$ from the biorthogonal expansion[35] of $|\text{in}^{AB}\rangle$, then in the case that $|\text{in}^{AB}\rangle = |\text{in}^A\rangle \otimes |\text{in}^B\rangle$, the vector $|\psi\rangle = |\text{in}^A\rangle$ is automatically the only remaining one which yields a nonzero probability. Accordingly, if we postulate that $\Delta_\mathbf{v}$ must be chosen in this way, the amount of freedom is reduced to generally uniqueness, unless R is degenerate, and in a seemingly quite plausible way. On the other hand, it must be noted that there are other choices which are plausible in precisely the same way. The operator$|\text{in}^{AB}\rangle$ is not the only one with a biorthogonal expansion. There are V^A, V^B, and $\langle\text{out}^{AB}|$ as well, and, furthermore, there is the operator V^C, which may actually be the nicest candidate. Corresponding to every one of the biorthogonal expansions of these operators one may construct a different $\Delta_\mathbf{v}$. Evidently, there is still some amount of thinking to be done with respect to this subject.

Relative states, universal states, and many worlds

The idea of relative states is an intriguing but very far–reaching concept. Evidently, we should be careful with drawing conclusions from the above analysis, which gives some considerations about the particular situation defined by (8.28), but which leaves many questions open and is in itself not sufficient as the basis for a view with such major implications. Nevertheless, a rough picture is emerging and the basic idea is that a composite process can contain virtual subprocesses which are real, but within their own level of reality, which is distinct from the level of reality of the composite process itself. The history of universe is thus divided into a hierarchy of processes and embedded, but independent, subprocesses. Because the definition of subprocesses no longer depends on the factorizability conditions (8.22), there is a possibility of describing the entire history of universe as a single quantum process. To this process, a definite process state may be assigned in principle, but this state is inaccessible to ourselves, as we are living at the level of some subprocess. However, one might indulge in the intriguing speculation that the universal process state is actually the vacuum state, making the world we know into a virtual subprocess of nothing.

There are a number of circumstances which complicate comparing the present presentation of the relative state approach with Everett's original ideas about the subject. In the first place, Everett's analysis is based entirely on the state concept of standard quantum mechanics, i.e., the state description by means of a Hilbert space vector. As is argued in the present work[36], this concept of state is seriously incomplete. Our own analysis above is based on one of the newly developed quantum state concepts, and it makes quite a difference in various respects. The above states, for instance, contain in–vectors and out–vectors, originating from the use of the ABL

[35] See, e.g., Zeh [226], Zurek [227], Dieks [63, 64].

[36] See § 7.4.

measures, which were not even available at the time of Everett's dissertation, and it is a state referring to processes, rather than to particles evolving in time as in Everett's case.

In the second place, in Everett's analysis the relative state concept is used in combination with a cluster of other ideas. In our own case we have seen that the relative state approach leads to a 'many–levels theory', i.e., total reality is divided into levels corresponding to subprocesses. This hierarchy of levels is not explicitly mentioned by Everett. In fact, the entire point of having to choose a specific representation of the Hilbert space $\mathcal{H}^A$, leading to a specific choice of $\Delta_{\mathbf{v}}$ was initially largely ignored both by Everett and by later advocates of his approach. Instead, the idea is introduced that different components of a Hilbert space vector can refer to different possible observation results, and that all these possibilities are 'equally real'. Later these possibilities were called "worlds", and the ideas of Everett and followers are often referred to as the 'many–worlds' interpretation of QM. What this part of the theory in fact amounts to is what may be called a many–worlds interpretation of *probability*, in which probability is assumed to belong to the structure of a gigantic reality consisting of many different branches, corresponding to different possible observation values. Beacause different possibilities are here depicted as equally real, this, in fact, amounts to introducing 'ontic modalities'[37]. This many–worlds structure is quite different, however, from the many–levels structure induced by the relative state idea itself, and, in fact, the relative state idea is entirely independent of any assumptions about a many–worlds interpretation of probability.

A main motive for Everett to put forward his idea of relative states was to obtain the possibility of assigning a state vector to the universe and set up a quantum cosmology. It is not entirely clear what remains of this motive now that the state description by means of a Hilbert space vector has turned out to be incomplete itself. Certainly we need some way to define subprocesses or to stop the Von Neumann infinite regress, certainly the relative state approach accomplishes this, and, contrary to some other approaches, it allows one to assign a quantum state to the whole universe. On the other hand, it is not clear that the relative state approach is the only one in which some version of a quantum state can be meaningfully assigned to the universe, nor that the assignment of such a state is a prerequisite for setting up a quantum cosmology.

Forming a tentative opinion about the relative state approach, I would say that, on the one hand, it seems far–fetched but, on the other hand, it has features which might be counted as attractive. In particular, the definition of subprocesses no longer depends on the factorizability conditions (8.22), and, as a consequence, there is in principle the possibility of describing the history of universe as a closed quantum process. There are also unsolved problems, however. In particular, one may question whether using a biorthogonal expansion for reducing the ambiguity in the choice of a context, i.e., of a set $\Delta_{\mathbf{v}}$, is satisfactory. Nevertheless, I think that at the present

[37]See also § 10.2.4 p. 244.

stage in the development of QM, with so many things remaining to be properly investigated, we are not in a position to discard the idea as absurd on any reasonable grounds.

8.6 Conclusions

Scientific theories are not isolated objects. They run against the background of a philosophical and logical framework, comparable to the way in which computer programs run within the framework of an operating system. In a developing science this background, too, is in need of maintainance and further development. Existing concepts may in the long run become obsolete or wear out, for instance because their use, and abuse, may generate slight changes in meaning, and new concepts, or revised versions, are required to replace them. This creates an ongoing need to design, debug, and redesign scientific concepts at many levels, including the philosophical background level.

Considering more specifically the ascent of QM in modern physics, it was soon recognized, or at least claimed, that this theory would require a thorough revision of some of the background ideas about, in particular, the concept of physical reality. For this reason a considerable part of the present work is dedicated to this subject. We shall try to briefly summarize the main conclusions and results.

Large fractions of the philosophical debate about the issue of realism are not particularly illuminating for our present purposes, because much of deals with the question of how much 'nature' resembles the theoretical pictures produced by science. It is not entirely evident that this question is open to meaningful examination, and our own interest concerns, in the first place, the practical sides of the concept of reality, i.e., what kind of role does it play in organizing science, what purposes does it serve, how does it work, and can we improve on it? Our basic idea of looking at these questions was to interpret reality as a 'glue concept', having an organizing role in integrating fragmentary pieces of knowledge into an intelligible whole. This idea may be worked out within different philosophical contexts. In chapter 1 three such contexts were discussed, those offered by the competing views of realism, empiricism, and pluralism.

Within a realistic context, scientific theories aim to contribute to the construction of a single scientific reality, conceived as ideally a literally true picture of nature. Within an empiristic context, scientific theories are artefacts which serve to organize our knowledge about an empirical reality which is supposed to be given by observation. Here we have opted for developing our views within a pluralistic environment, essentially because it best reflects the observation that wherever there are people there are different points of view, and we need the proper equipment to deal with this fact. Pluralism offers the possibility of a multireality environment in which different views can each have their own construction of reality.

Within such an context it is easy to be explicitly aware that the possibility of communication between different views is not self–evident. If communication is desired this requires the construction of a logical framework in which a meaningful exchange of information becomes possible. Therefore, it is important that the context includes facilities for building such interfaces. The chapters 2 and 3 contain an attempt to contribute to this field by means of some considerations about 'comparative metaphysics', leading to the construction of a 'contextual logic', a form of modal logic aimed especially at dealing with some of the problems in this area. Reality is now introduced by means of a specific type of construction within contextual logic. The resulting notion matches intuitive ideas about reality fairly well, at least those of the author himself. In fact, apart from the perhaps exceptionally liberal features which the concept owes to its construction within a pluralistic context, there may not be too much difference with common ideas about what "reality" is supposed to mean.

Having fixed the logical background, the attention was turned to QM. The main point of concern is the peculiar logical structure of QM, which seems to defy not only a realistic interpretation of the theory, but indeed the very possibility of its construal within a classical logical framework. In a theory of classical logic meaningful sentences are assumed to be either true or false, and their truth value is determined by reference to a 'model' or 'world' in which they are interpreted. For a given scientific theory one may consider its set of 'intended models', a set of 'possible worlds' in which the theoretical notions are meant to be interpreted. A 'state space' is a mathematical representation of such a set of possible worlds, a 'state' is a subset of possible worlds and a 'magnitude' is a parameterization of the state space, inducing a partition into equivalence classes of possible worlds. These concepts of state and of magnitude apply in all classical physics, but not in QM. This leaves us with two alternatives: either QM must be construed within a nonclassical logical framework, or QM has a fraudulent concept of state which must be replaced by something better. The latter alternative is regarded here as the most promising one, and is elaborated on.

The chapters 4 and 5 contain an introduction to QM, focussing especially on some known but badly assimilated results in measurement theory, in particular the application of so–called 'operation valued measures' and of the time–symmetric 'ABL measures'. In chapter 6 the logical problems of QM are addressed, and it then appears that contextual logic has some facilities which make it very suitable for formulating QM. More specifically, the concept of a 'context' turns out to be very well applicable for interpreting Bohr's notion of a 'quantum phenomenon', and 'complementarity' can then be understood in terms of the relations between contexts. By further developing this idea it turns out that, in fact, contextual logic may be applied in formulating quite a number of different 'interpretations' of QM, i.e., what we have achieved may be considered as a 'QM interpretation generator', with a menu of different options. Some of these options are introduced in chapter 6, and a rough survey of further possiblities is given in chapter 8. Together, these chapters should contain a concise guide to redrawing the basic conceptual structure of QM without loosing sight on how to preserve the empirical results already obtained.

This is where the main line of our argument ends. The two final chapters contain some more specialized elaborations. Chapter 9 contains an introduction to what I personally see as the most promising approach in the development of QM: a quantum event field theory, based on a rigorously time–symmetric approach, and containing some novelties about how to reconcile QM and relativity theory. Finally, chapter 10 gives some more details about contextual logic as a mathematical theory.

Literature: Accardi [4, 2, 1, 3], Albert *et al.* [7], Bell [13], Beltrametti and Cassinelli [15], Ben–Dov [17, 16, 18], Benioff [19, 20], Van den Berg et al. [24], Birkhoff and Von Neumann [28], Bub *et al.* [37], Cooper [48], Costa de Beauregard [50], Daneri *et al.* [55], De Witt and Graham [61], Dieks [63, 64, 66], D'Espagnat [73], Everett [74], Folse [85, 86], Van Fraassen [95, 94], Ghirardi *et al.* [97, 98, 99], Hawking [113], Heisenberg [114], Hooker [125], Jammer [129, 130], Joos [133], London and Bauer [153], Mielnik [161], Primas [171], Tipler [207], Wan [212], Wheeler and Zurek [219], Wigner [223, 222], Zeh [226, 225], Zurek [227].

Chapter 9

Quantum Event Theory

A Tetrode–Fokker version of Quantum Field Theory

Abstract: A theory of quantum events is presented and its relation to quantum field theory is discussed.

In the 1920s, H. Tetrode and A.D. Fokker gave a relativistic time–symmetric theory of electrodynamics[1]. It was an action–at–a–distance theory, without separate degrees of freedom for the electromagnetic field. In 1945 Wheeler and Feynman reintroduced the electromagnetic field and gave a theory of electromagnetism in which the fields are coupled to the sources by means of time–symmetric propagation. In 1964 Aharonov, Bergmann, and Lebowitz introduced a time–symmetric probability measure in quantum mechanics. In the early seventies, the Wheeler–Feynman approach was extended to quantum electrodynamics in the work of Hoyle and Narlikar and of Davies. In 1980 and afterwards, further work in this direction has appeared from Cramer. On various occasions, Costa de Beauregard has advocated time–symmetric causation in quantum mechanics, and there are quite a few other people who have been doing work relevant to this approach, such as Stueckelberg, Schwinger, Roberts, and Broyles.

The approach developed in this chapter fits into this tradition of time–symmetrical physics. It explores the possibility of taking a rigorous Tetrode–Fokker approach toward quantum field theory, such that *all* fields are deleted in favor of an action–at–a–distance picture. In quantum field theory, however, particles are merely manifestations of the fields and this introduces the obvious difficulty that, by deleting the fields, there are no particles left to interact with each other. Indeed, the natural interpretation for the resulting theory will be that it is about correlations between events. In this, and other respects, the theory links jointlessly with the time–symmetric approach in process theory, as developed in ch. 6.

[1]Tetrode [206], Fokker [84].

9.1 Quantum events

In quantum mechanics there is a long–standing problem concerning the fact that time is not treated in the same manner as the other spacetime coordinates. Whereas the space coordinates are represented by operators, time is merely a numerical parameter. In quantum field theory the problem is solved by the circumstance that, as a side effect of second quantization, the space coordinates too become numerical parameters. Every now and then, however, physicists have been fascinated by the alternative possibility of introducing an operator representing time[2]. The main difficulty in finding such an operator is the fact that the energy of a system is bounded from below. Ideally, one would expect the spacetime position and momentum operators to be canonical conjugates in the sense that

$$[x_\mu, p_\nu] = ig_{\mu\nu} \tag{9.1}$$

where $g_{\mu\nu}$ is the metric tensor, which we shall take to have the signature (1,-1,-1,-1). If we now identify p_0 with the Hamilton operator H, we are in immediate trouble because the energy has a lower bound and the desired commutation rule cannot be satisfied by an operator with a bounded spectrum. The deeper cause of the trouble is in the way in which we use a time–dependent Hilbert space vector $\psi(t)$ to describe the state of a dynamical system with an enduring existence, such as a particle. The difficulties vanish as soon as we decide to use Hilbert space vectors to represent events. In this case we can take the Hilbert space $L^2(I\!\!R^4)$, i.e., we act in the following way:

Let $x, y \in I\!\!R^4$ be spacetime positions. We now span a Hilbert space by means of an orthonormal set B of basisvectors[3]:

$$B = \{|x\rangle | x \in I\!\!R^4\} \tag{9.2}$$

with, accordingly, the inner product

$$\langle x|y\rangle = \delta_4(x-y). \tag{9.3}$$

Spacetime position operators are straightforwardly defined by

$$x^\mu = \int |x\rangle x^\mu \langle x| \, d^4x, \tag{9.4}$$

and this includes an operator x^0 representing time. Still, it remains true that an operator p_0 satisfying (9.1) cannot be given by the Hamiltonian of the system. Instead, however, we can define it as the operator conjugate to x^0 by simply requiring it to satisfy (9.1). We thus obtain

$$\langle x|p_\mu|\psi\rangle = i\partial_\mu \langle x|\psi\rangle \,, \quad \text{for all } \mu \in \{0,1,2,3\}. \tag{9.5}$$

[2]See, e.g., Stueckelberg [199, 200], Schwinger [190], Johnson [132], Rosenbaum [184], Broyles [36], Gotō *et al.* [103, 102] , and Horwitz [126].

[3]See, e.g., Broyles [36].

Instead of the problem of finding a time operator, what we are left with is 'merely' the problem of formulating quantum physics within this new context. This involves two things:

i formulating a theory which conserves the empirical content of quantum mechanics;

ii adjusting our physical world picture to fit the new theory.

In the present chapter we shall be largely concerned with (**i**). Our leading idea will come from the observation that the entire empirical content of quantum mechanics can be expressed in terms of Green functions. Accordingly, we are well on our way if we can find an operator which yields the Green functions as its matrix elements[4].

Example: Consider the case of the relativistic quantum theory for a free scalar 'particle' obeying the Klein–Gordon equation. The Green function, in this case, is equal up to a factor i to the free particle propagator, which is found as the solution of

$$(-\partial^2 - m^2)K(x,y) = \delta_4(x-y). \tag{9.6}$$

We assume that $K(x,y)$ is given as the matrix element of an operator K,

$$K(x,y) = \langle x|K|y\rangle. \tag{9.7}$$

Inserting this in (9.6), we find

$$\begin{aligned}(-\partial^2 - m^2)\langle x|K|y\rangle &= \langle x|(p^2-m^2)K|y\rangle \\ &= \delta_4(x-y) = \langle x|y\rangle\end{aligned} \tag{9.8}$$

which has as its general solution

$$K = \frac{1}{p^2 - m^2 + i\epsilon} + A(p)\delta(p^2 - m^2). \tag{9.9}$$

The constants of integration $A(p)$ raise the question of fixing boundary conditions, which we shall for the present simply ignore by giving no motivation for choosing $A(p) = 0$ at every instant where, in the sequel, we encounter integrational freedom of this type. We thus find

$$\begin{aligned}K(x,y) &= \langle x|\frac{1}{p^2-m^2+i\epsilon}|y\rangle \\ &= \int \frac{e^{-ip(x-y)}}{(2\pi)^4(p^2-m^2+i\epsilon)}d^4p.\end{aligned} \tag{9.10}$$

---

[4]See also Schwinger [190].

Note: The topic of fixing boundary conditions relates directly to the discussion about the cosmological condition of "complete absorption", which has played an important role in the entire debate about the Wheeler–Feynman approach[5]. It is not evident, however, that conclusions reached by other authors can be straightforwardly transposed to the present situation, especially because the proposed empirical interpretation of event theory[6] contains some subtleties which may be relevant in this respect. We shall not give a detailed discussion of the subject at this point, but some tentative ideas have been put forward in § 7.6.1.

9.2 Event fields

In order to formulate a more general theory about the correlations between larger numbers of events, we first apply second quantization, as in quantum field theory. For simplicity, we continue to restrict our analysis to scalar events. In the cases (spin $\frac{1}{2}$ and spin 1) I have to some extent checked, however, the introduction of spin degrees of freedom appears to be straightforward.

The Hilbert space is constructed by introducing fields of annihilation operators $a(x)$ and $b(x)$, and creation operators $a^+(x)$ and $b^+(x)$, satisfying the commutation relations:

$$\begin{aligned} [a(x), a(y)] &= [a(x), b(y)] = 0 \\ [a(x), a^+(y)] &= [b(x), b^+(y)] = \delta_4(x-y) \\ [a(x), b^+(y)] &= [b(x), a^+(x)] = 0 \end{aligned} \tag{9.11}$$

Note: The occurrence of the $\delta_4(x-y)$ in the second line expresses the fact that we are dealing with events. It is unusual from the standard field theoretic point of view, but, indeed, one should not compare the above expressions to commutator relations between standard fields, but to expressions like $[\phi(x), \partial_0\phi^+(y)]\delta(x^0 - y^0) = i\delta_4(x-y)$. See also (9.33) and (9.36).

States are now defined, as in standard quantum field theory, by applying combinations of these operators to a vacuum state $|0\rangle$, here representing an 'empty' spacetime. The resulting states span an event Fock space $\mathcal{F}$. Let us consider two such states $\Psi, \Phi \in \mathcal{F}$, for instance with Φ representing, what would in standard field theory be regarded as, a 'particle creation' at point y, and Ψ 'its annihilation' at point x. In standard field theory we would then be interested in the corresponding Green function, defined as the vacuum expectation of the time–ordered product $\langle 0|T[\phi(x)\phi^+(y)]|0\rangle$ of the fields $\phi^+(y)$ and $\phi(x)$, which, for a free field, yields i times the bare Feynman propagator. More complex states Φ, Ψ correspond to more complex Green functions with time–ordered products of larger numbers of fields. In event theory we want to obtain these quantities as the matrix elements $\langle\Psi|G|\Phi\rangle$ of an operator G.

[5]See, e.g., [221, 59, 128].
[6]See § 9.5.

In order to achieve this, we shall postulate event field equations. First, however, we shall slightly simplify our notation by imposing two equations which are linked to the CPT invariance of the theory:

$$\begin{aligned} Ga^+(x) &= b(x)G \\ Gb^+(x) &= a(x)G \end{aligned} \tag{9.12}$$

Note: Note that for $\tilde{a}(p) := \int \langle p|x\rangle a(x)d^4x$ this leads to

$$\begin{aligned} G\tilde{a}^+(p) &= \int \langle x|p\rangle Ga^+(x)d^4x & (9.13) \\ &= \int \langle x|p\rangle b(x)Gd^4x \\ &= \tilde{b}(-p)G, & (9.14) \end{aligned}$$

i.e., these equations express the condition that, in a particle picture, an incoming particle may be alternatively interpreted as its outgoing CPT–transformed antiparticle.

Note: In standard quantum field theory, time reversal is obtained by an antiunitary operator, which may be decomposed into a unitary operator and an antilinear complex conjugation operator. A nice feature of event theory is that time reversal, like parity change, is represented by a unitary operator, whereas antilinearity is associated only with the transformation between ingoing and outgoing 'particles'. This latter transformation, however, is purely formal, expressing merely a change of view without any physical meaning.

Starting with $\langle\Psi|G|\Phi\rangle$, we can use these equations to bring Φ to the left, with the result

$$\langle\Psi|G|\Phi\rangle = \langle\Psi, \bar{\Phi}|G|0\rangle, \tag{9.15}$$

where $\bar{\Phi}$ is the 'anti–state' of Φ, its reinterpretation as an outgoing rather than incoming state. It is obtained by writing $|\Phi\rangle$ as a complex of creation operators working on $|0\rangle$, applying a transformation $a^+(x) \to b(x)$, or $\tilde{a}^+(p) \to \tilde{b}(-p)$, on this complex of operators, and letting the result work to the left on $\langle 0|$, or in the above case on $\langle\Psi|$. If we now define

$$|G\rangle = G|0\rangle \tag{9.16}$$

then any Green function can be written in the form

$$G(\Psi) = \langle\Psi|G\rangle. \tag{9.17}$$

Accordingly, we are now looking for a theory about this vector $|G\rangle$. Rather than trying to give an account of the, rather unsystematic, way in which I arrived at such a theory, I shall merely write down the fundamental equations.

9.3 Field equations

We assume that, corresponding to a given spacetime volume τ, there is an 'action' $S(\tau)$ which is an expression in terms of the fields $a(x)$ and $b(x)$ and their derivatives. Our field equations will be of the form

$$\delta S(\tau)/\delta a(x)|G\rangle = ia^{+}(x)|G\rangle \ , \ \text{with } x \in \tau, \tag{9.18}$$

and similarly for $b(x)$, and for any other field. Furthermore, we add the equation:

$$\langle 0|G\rangle = 1. \tag{9.19}$$

Example: Let us consider a free scalar field for which we postulate an action of the form

$$\begin{aligned} S(\tau) &= \int_\tau \{\partial_\mu a(x)\partial^\mu b(x) - m^2 a(x)b(x)\}d^4x \\ &= \int_\tau b(x)(-\partial^2 - m^2)a(x)d^4x + \oint_{\partial\tau} b(x)\partial_\mu a(x)dV^\mu. \end{aligned} \tag{9.20}$$

This yields, for $x \in \tau$

$$\delta S/\delta b(x) = (-\partial^2 - m^2)a(x). \tag{9.21}$$

For a two–event Green function $G(x,\bar{y}) = \langle x,\bar{y}|G\rangle = \langle x|G|y\rangle$, we now find that it satisfies the following equation:

$$\begin{aligned} (-\partial^2 - m^2)G(x,\bar{y}) &= \langle\bar{y}|(-\partial^2 - m^2)a(x)|G\rangle \\ &= \langle\bar{y}|\delta S/\delta b(x)|G\rangle \\ &= \langle\bar{y}|ib^{+}(x)|G\rangle \\ &= i\delta_4(x-y)\langle 0|G\rangle \\ &= i\delta_4(x-y) \end{aligned} \tag{9.22}$$

with the solution

$$G(x,\bar{y}) = iK(x,y). \tag{9.23}$$

Example: For a second example, consider an action of the form

$$\begin{aligned} S &= \int_\tau \{b(x)(-\partial^2 - m^2)a(x) + \frac{1}{2}c(x)(-\partial^2 - M^2)c(x) \\ &\quad + ga(x)b(x)c(x)\}d^4x + O(\partial\tau) \\ &= S_0 + S_I \end{aligned} \tag{9.24}$$

where S_I is the interaction term. This yields:

$$\begin{aligned} \delta S/\delta a(x) &= (-\partial^2 - m^2)b(x) + gb(x)c(x), \\ \delta S/\delta b(x) &= (-\partial^2 - m^2)a(x) + ga(x)c(x), \\ \delta S/\delta c(x) &= (-\partial^2 - M^2)c(x) + ga(x)b(x). \end{aligned} \tag{9.25}$$

Considering, again, the two–event Green function $G(x,\bar{y})$, we find

$$\begin{aligned}(-\partial^2 - m^2)G(x,\bar{y}) &= \langle\bar{y}|\{\delta S/\delta b(x) - \delta S_I/\delta b(x)\}|G\rangle \\ &= i\delta_4(x-y) + i\langle\bar{y}|iga(x)c(x)|G\rangle\end{aligned} \tag{9.26}$$

with the solution

$$G(x,\bar{y}) = iK(x,y) + \int iK(x,z)\langle\bar{y}|iga(z)c(z)|G\rangle d^4z. \tag{9.27}$$

Apart from the free field propagator we now find an additional $O(g)$ term, involving the three–event Green function $\langle 0|b(y)a(z)c(z)|G\rangle$. This term can be further analyzed by picking any of the fields and dealing with it in the above manner. We have, for instance,

$$(-\partial^2 - M^2)c(z)|G\rangle = \{\delta S/\delta c(z) - \delta S_I/\delta c(z)\}|G\rangle \tag{9.28}$$

and, therefore:

$$c(z)|G\rangle = \int i\Delta(z,u)\{c^+(u) + iga(u)b(u)\}|G\rangle d^4u, \tag{9.29}$$

where $\Delta(z,u)$ is the c–field propagator. By virtue of $\langle 0|c^+(u) = 0$, we obtain

$$\begin{aligned}G(x,\bar{y}) &= iK(x,y) \\ &\quad - g^2\int iK(x,z)i\Delta(z,u)\langle 0|b(y)a(z)a(u)b(u)|G\rangle d^4z d^4u.\end{aligned} \tag{9.30}$$

Proceeding in this way, we find:

$$\begin{aligned}G(x,\bar{y}) &= iK(x,y) \\ &\quad - g^2\int iK(x,z)i\Delta(z,u)iK(z,u)iK(u,y)d^4z d^4u \\ &\quad - g^2\int iK(x,z)i\Delta(z,u)iK(u,u)iK(z,y)d^4z d^4u \\ &\quad + O(g^3),\end{aligned} \tag{9.31}$$

where the $O(g^3)$ terms contain more Green functions, which can be further analyzed to any desired order.

9.4 The correspondence between field theory and event theory

The relation between ordinary quantum field theory and event theory seems clear, at least from a computational point of view. Starting within field theory, we may take the expression for the action, replace all particle field operators by corresponding event annihilation operators, and perform our calculations within event theory. For low–order diagrams it is readily verified by hand that the results agree with standard theory, including the correct combinatorial factors. For a more general proof of a correct correspondence between the two theories, we may observe that in both cases the Green functions can be obtained via the same iteration procedure, as we shall now proceed to demonstrate.

In event theory, let us assume the action to be of the general form

$$\begin{aligned} S &= S_0 + S_I \\ S_0 &= \frac{1}{2}\int \sum_\alpha \{\partial_\mu a_\alpha(x)\partial^\mu b_\alpha(x) - m_\alpha^2 a_\alpha(x) b_\alpha(x)\} d^4x, \end{aligned} \tag{9.32}$$

where the factor $\frac{1}{2}$ appears because summation runs over *all* fields a_α and, accordingly, there is for every α a β such that $a_\alpha = b_\beta$. Every field thus appears twice, except in the case of a field which is its own anti–field, but then the factor $\frac{1}{2}$ comes in handy because $\delta S/\delta a$ produces a factor 2 which must be canceled.

We assume the following commutation relations:

$$\begin{aligned} [a_\alpha(x), a_\beta(x)] &= 0 \\ [a_\alpha(x), a_\beta^+(y)] &= \delta_{\alpha\beta}\delta_4(x-y). \end{aligned} \tag{9.33}$$

By virtue of (9.19), (9.32), and (9.33), and by the methods of the previous section applied to the term $a_\alpha(x)$, the Green functions satisfy

$$\begin{aligned} &\langle 0|a_{\alpha_1}(x_1)...a_{\alpha_N}(x_N)a_\alpha(x)|G\rangle = \\ &\int iK_\alpha(x,y)\Big\{\langle 0|\sum_{i=1}^N a_{\alpha_1}(x_1)...[a_{\alpha_i}(x_i), b_\alpha^+(y)]...a_{\alpha_N}(x_N)|G\rangle \\ &+i\langle 0|a_{\alpha_1}(x_1)...a_{\alpha_N}(x_N)\delta S_I/\delta b_\alpha(y)|G\rangle\Big\}d^4y. \end{aligned} \tag{9.34}$$

If the coupling constants are sufficiently small and divergences can be renormalized, or be dealt with otherwise, this equation enables us to approximate the Green functions by means of iteration. The next step is to show that the same procedure works within standard field theory.

In standard field theory the action is expressed in terms of fields $\phi_\alpha(x)$ and their derivatives. Corresponding to (9.32), we take

$$\begin{aligned} S &= S_0 + S_I \\ S_0 &= \int \tfrac{1}{2}\sum_\alpha \{\partial_\mu\phi_\alpha(x)\partial^\mu\phi_\alpha^+(x) - m_\alpha^2\phi_\alpha(x)\phi_\alpha^+(x)\}d^4x. \end{aligned} \tag{9.35}$$

Assuming, for the moment, the absence of time derivatives in S_I, we have that the fields satisfy the following commutation relations:

$$[\phi_\alpha(x), \partial_0\phi_\beta^+(y)]\delta(x^0 - y^0) = i\delta_{\alpha\beta}\delta_4(x-y). \tag{9.36}$$

The Green functions are defined as the vacuum expectation values of time–ordered products of fields, i.e.,

$$G(\alpha_1, x_1; ...; \alpha_N, x_N; \alpha, x) = \langle 0|T[\phi_{\alpha_1}(x_1)...\phi_{\alpha_N}(x_N)\phi_\alpha(x)]|0\rangle. \tag{9.37}$$

The field equations, the Euler–Lagrange equations for the fields, now have the form $\delta S/\delta\psi(x) = 0$, and taking $\psi = \phi_\alpha(x)$, and $m = m_\alpha$, we have

$$(-\partial^2 - m^2)\psi(x) \;=\; -\delta S_I/\delta\psi^+(x). \tag{9.38}$$

If we calculate the expression

$$(-\partial^2 - m^2)T[\phi_{\alpha_1}(x_1)...\phi_{\alpha_N}(x_N)\psi(x)], \tag{9.39}$$

while inserting (9.38), we find

$$\begin{aligned}&(-\partial^2 - m^2)T[\phi_{\alpha_1}(x_1)...\phi_{\alpha_N}(x_N)\psi(x)] = \\ &= R - T[\phi_{\alpha_1}(x)...\phi_{\alpha_N}(x_N)\delta S_I/\delta\psi^+(x)],\end{aligned} \tag{9.40}$$

where the term R originates from the action of ∂_0 on the step functions $\theta(x_i^0 - x^0)$ and $\theta(x^0 - x_i^0)$, which are included in the definition of the time–ordering. By a slightly tedious but straightforward computation, it can be shown that R equals

$$R \;=\; \sum_{n=1}^{N} T\Big[\,[\phi_{\alpha_n}(x_n), \partial_0\psi(x)]\delta(x^0 - x_n^0) \prod_{i=1;i\neq n}^{N} \phi_{\alpha_i}(x_i)\Big]. \tag{9.41}$$

Proof: A proof is simple in the case of a 2–point Green function. Using the field equations (9.38), we can write

$$\begin{aligned}&(-\partial^2 - m^2)T[\psi(x)\phi(y)] \\ =\;& (-\partial^2 - m^2)\{\theta(x^0 - y^0)\psi(x)\phi(y) + \theta(y^0 - x^0)\phi(y)\psi(x)\} \\ =\;& R + T[(-\partial^2 - m^2)\psi(x)\phi(y)] \\ =\;& R - T[(\delta S_I/\delta\psi^+(x))\phi(y)]\end{aligned} \tag{9.42}$$

where

$$\begin{aligned}R \;=\;& -\partial_0^2\theta(x^0 - y^0)\psi(x)\phi(y) - 2\partial_0\theta(x^0 - y^0)\partial_0\psi(x)\phi(y) \\ & -\partial_0^2\theta(y^0 - x^0)\phi(y)\psi(x) - 2\partial_0\theta(y^0 - x^0)\phi(y)\partial_0\psi(x) \\ =\;& -\partial_0\delta(x^0 - y^0)\psi(x)\phi(y) - 2\delta(x^0 - y^0)\partial_0\psi(x)\phi(y) \\ & +\partial_0\delta(x^0 - y^0)\phi(y)\psi(x) + 2\delta(x^0 - y^0)\phi(y)\partial_0\psi(x) \\ =\;& -\delta(x^0 - y^0)\partial_0\psi(x)\phi(y) + \delta(x^0 - y^0)\phi(y)\partial_0\psi(x) \\ =\;& \delta(x^0 - y^0)[\phi(y), \partial_0\psi(x)],\end{aligned} \tag{9.43}$$

where we used $\partial_0\theta(x^0 - y^0) = \delta(x^0 - y^0)$

and $\partial_0\delta(x^0)f(x^0) = -\delta(x^0)\partial_0 f(x^0)$.

The general case, for an arbitrary n–point Green function, is more complicated only by the abundant reshuffling of terms, as is clear from the following computation.

Adopting a more frugal notation to avoid drowning in indices, we consider the expression

$$\begin{aligned}(-\partial^2 - m^2)T[\psi \prod \phi] &= R + T[(-\partial^2 - m^2)\psi \prod \phi] \\ &= R - T[(\delta S_I/\delta\psi^+) \prod \phi],\end{aligned} \tag{9.44}$$

which corresponds to (9.42). In order to calculate R we write out T in terms of strings of fields and step functions θ. Defining

$$\begin{aligned}\theta_i &= \theta(t_i - t_{i-1}) & \delta_i &= \delta(t_i - t_{i-1}) \\ \theta_{\psi,n} &= \theta(t - t_n) & \delta_{\psi,n} &= \delta(t - t_n) \\ \theta_{n,\psi} &= \theta(t_n - n) & \delta_{n,\psi} &= \delta(t_n - n)\end{aligned} \tag{9.45}$$

we get

$$T\left[\psi \prod_{i=1}^{N} \phi_{\alpha_i}\right] = \sum_P \sum_{n=0}^{N} (\prod_{i=1}^{n} \theta_i^P \phi_{\alpha_i}^P)(\theta_{\psi,n}^P \psi \theta_{n+1,\psi}^P)(\prod_{i=n+1}^{N} \theta_i^P \phi_{\alpha_i}^P), \tag{9.46}$$

where the index P runs over all permutations of the fields ϕ_{α_i}.

All contributions to R arise from the action of ∂_0 on the step functions $\theta_{\psi,n}^P$ and $\theta_{n,\psi}^P$, and, more specifically, we have

$$R = \sum_P \sum_{n=0}^{N} (\prod_{i=1}^{n} \theta_i^P \phi_{\alpha_i}^P) F_n^P (\prod_{i=n+1}^{N} \theta_i^P \phi_{\alpha_i}^P), \tag{9.47}$$

with

$$\begin{aligned}F_n &= \partial_0^2(\theta_{\psi,n}\psi\theta_{n+1,\psi}) - \theta_{\psi,n}(\partial_0^2\psi)\theta_{n+1,\psi} \\ &= (\delta_{\psi,n}\partial_0\psi - \delta_{n+1,\psi}\partial_0\psi).\end{aligned} \tag{9.48}$$

We thus obtain

$$\begin{aligned}&\sum_{n=0}^{N}(\prod_{1}^{n}\theta\phi)F_n(\prod_{n+1}^{N}\theta\phi) \\ &\quad = \sum_{n=0}^{N}(\prod_{1}^{n-1}\theta\phi)(\theta_n\theta_{n+1}\delta_{\psi,n}\phi_{\alpha_n}\partial_0\psi)(\prod_{n+1}^{N}\theta\phi) \\ &\quad - \sum_{n=0}^{N}(\prod_{1}^{n}\theta\phi)(\theta_{n+1}\theta_{n+2}\delta_{\psi,n+1}\partial_0\psi\phi_{\alpha_{n+1}})(\prod_{n+2}^{N}\theta\phi).\end{aligned} \tag{9.49}$$

Relabelling the summation index, $n \to n-1$, in the second term, can be done without changing the summation boundaries, by virtue of $\delta_{\psi,0} = \delta_{N+1,\psi} = 0$. Therefore, we get

$$\begin{aligned}&\sum_{n=0}^{N}(\prod_{1}^{n}\theta\phi)F_n(\prod_{n+1}^{N}\theta\phi) \\ &= \sum_{n=0}^{N}(\prod_{1}^{n-1}\theta\phi)\theta_n\theta_{n+1}\delta_{\psi,n}(\phi_{\alpha_n}\partial_0\psi - \partial_0\psi\phi_{\alpha_n})(\prod_{n+1}^{N}\theta\phi) \\ &= \sum_{n=0}^{N}(\prod_{1}^{n-1}\theta\phi)\theta_n\theta_{n+1}[\phi_{\alpha_n},\partial_0\psi]\delta_{\psi,n}(\prod_{n+1}^{N}\theta\phi).\end{aligned} \tag{9.50}$$

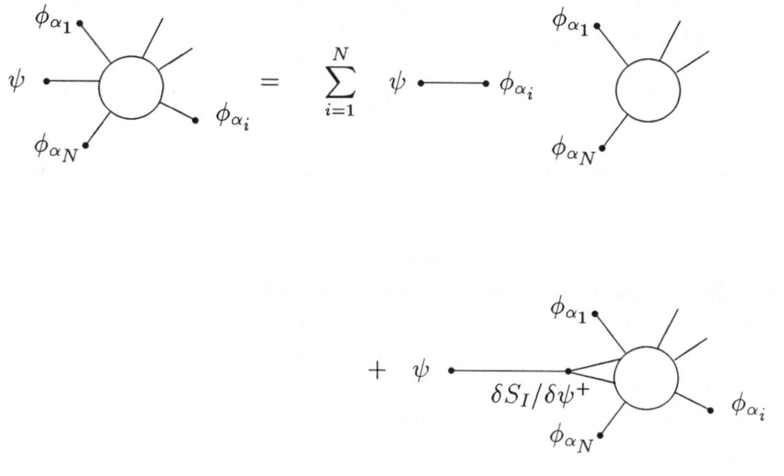

Figure 9.1: *An iterative equality for Green functions*

Using this result, we can write R as

$$\begin{aligned} R &= \sum_P \sum_{n=1}^{N} \Big(\prod_{1}^{n-1} \theta^P \phi^P\Big)\theta_n^P \theta_{n+1}^P [\phi_{\alpha_n}^P, \partial_0 \psi]\delta_{\psi,n}^P \Big(\prod_{n+1}^{N} \theta^P \phi^P\Big) \qquad (9.51) \\ &= T\Big[\sum_{n=1}^{N} [\phi_{\alpha_n}, \partial_0\psi]\delta_{\psi,n} \prod_{i\neq n} \phi\Big]. \end{aligned}$$

□

We may insert (9.41) into (9.40) and solve the equation by means of Fourier analysis. As a result, we find

$$\begin{aligned} &\langle 0|T[\phi_{\alpha_1}(x_1)...\phi_{\alpha_N}(x_N)\psi(x)]|0\rangle \qquad (9.52) \\ &= \int iK(x,y)\langle 0|\Big\{T[\sum_{i=1}^{N} \phi_{\alpha_1}(x_1)...[\phi_{\alpha_i}(x_i), \partial_0\psi(y)]\delta(x_i^0 - y^0)...\phi_{\alpha_N}(x_N)] \\ &\quad + iT[\phi_{\alpha_1}(x_1)...\phi_{\alpha_N}(x_N)\delta S_I/\delta\psi^+(y)]\Big\}|0\rangle d^4y. \end{aligned}$$

Recalling that $\partial_0\psi(y)$ is precisely the canonical conjugate of $\psi^+(y)$, and using the commutation relation (9.36), we find a precise correspondence between (9.52) and (9.34). In terms of diagrams these equations are expressed by fig. 9.1.

Note: Two things should be noted about the graphs as used here. In the first place, the combinatorial factors are not included in the definition of the vertices. They are obtained by iterating up to the desired order and collecting identical diagrams. In the second place, as long as the combinatorial factors have not been collected, vertices which look the same in a diagram may, nevertheless, be prefixed with different factors. As an example, consider

$$S_I = \int ga(x)b(x)c^2(x)d^4x. \tag{9.53}$$

In this case $\delta S_I/\delta a(x)$ will generate a factor $(2\pi)^4 ig$ but $\delta S_I/\delta c(x)$ will produce $2(2\pi)^4 ig$.

Finally, we may recall that correspondence between field theory and event theory was shown only on the condition that no time derivatives occur in S_I. If, for instance, $\partial_0\phi$ does occur in S_I, then $\partial_0\phi^+$ is no longer the canonical conjugate of ϕ. In quantum field theory this gives rise to 'normal dependent' terms that are not manifestly covariant[7]. Special arguments are now needed to show that, on the proper premises, such terms cancel. In event theory such complications do not arise because normal dependent terms do not arise. An occurrence of $\partial_0 a$ in S_I simply produces a straightforward factor $-ip_0$.

9.5 Probabilities reconsidered

A quantum probability measure in event theory can be obtained by a suitable adjustment of (6.15):

$$\begin{aligned} p_{\mathbf{u},w}(\Phi_{\partial\tau}) &= C_{\mathbf{u},w}\, L(\Phi_{\partial\tau})w(\Phi_{\partial\tau}) \\ &= C_{\mathbf{u},w}\, |\langle \text{out}|V|\text{in}\rangle|^2 w(\Phi_{\partial\tau}), \end{aligned}$$

which, in its turn, connects to more standard formulations.

The triple $\Phi_{\partial\tau} =< \langle\text{out}|, V, |\text{in}\rangle >$ was conceived as an expression referring to some sort of flux through the hypersurface $\partial\tau$ of a spacetime volume τ, but, in this particular form, its use is restricted to a specific Lorentz frame and to a τ which has the form of a space–time box with two moments t_{in} and t_{out} as boundaries. In order to find a more general expression, we shall start by considering spacial distributions of events and then work our way toward an expression in terms of fluxes via a procedure which involves 'cutting' the propagators. More specifically, let an elementary process in event theory be described by a vector $\Phi \in \mathcal{F}$. As in process theory, an experimental setup is assumed to determine a set $\Gamma_{\mathbf{u}}$ of alternative processes, and choosing a particular set $\Gamma_{\mathbf{u}}$ amounts to defining, in Bohr's terminology, a quantum phenomenon. Probabilities are obtained by using (6.15), with Φ substituted for $\Phi_{\partial\tau}$. We now postulate the following expression for the function $L(\Phi)$:

$$L(\Phi) = |\langle\Phi|G\rangle|^2. \tag{9.54}$$

[7]See, e.g., Rohrlich [182].

Nonrelativistic event theory

For clarity, we shall first proceed to establish the connection with (6.15) within a nonrelativistic version of event theory, based on a field equation which is first order, rather than second order in time, like the Schrödinger equation in standard QM. Moreover, we begin by considering the case of an isolated process, with V given by

$$V = e^{-iE(t_{\mathrm{out}}-t_{\mathrm{in}})}, \quad \text{with } t_{\mathrm{out}} > t_{\mathrm{in}}. \tag{9.55}$$

What we are now looking for is an equation equivalent to (6.15), but formulated in terms of Green functions and vectors on the new Hilbert space $L^2(\mathbb{R}^4)$. This is, in fact, quite straightforward; we can take

$$\langle \mathrm{out}|V|\mathrm{in}\rangle = \langle \mathrm{out}, t_{\mathrm{out}}|G|\mathrm{in}, t_{\mathrm{in}}\rangle, \tag{9.56}$$

with G given by

$$G = \int |y\rangle e^{-iE(y^0-x^0)}\theta(y^0 - x^0)\langle x|d^4y d^4x. \tag{9.57}$$

Note: Operators G of the above form are readily obtained in a nonrelativistic version of event theory, based on the following field equations:

$$\begin{aligned}
\langle 0|G|0\rangle &= 1 \\
\delta S/\delta a^+(x)G &= iGa(x) \\
G\delta S/\delta a(x) &= ia^+(x)G \\
S &= S_0 + S_I \\
\text{with} S_0 &= \sum_\alpha \int a_\alpha^+(x)(i\partial_0 - E)a_\alpha(x)d^4x \\
\text{and, e.g.,} S_I &= \sum_{\alpha\beta} \int A_{\alpha\beta}(x,y)a_\alpha^+(x)a_\beta^+(y)a_\alpha(x)a_\beta(y)d^4xd^4y.
\end{aligned} \tag{9.58}$$

It must be noted here that the 'CPT-trick' (9.12) doesn't work, and, therefore, $\langle\Psi|G|\Phi\rangle$ cannot be written in the form $\langle\Psi, \bar{\Phi}|G\rangle$.

In the case of a free 'particle', the operator V on $L^2(\mathbb{R}^3)$ may now be given in terms of the operator G on $L^2(\mathbb{R}^4)$ by means of

$$V = \int |\mathbf{y}\rangle\langle \mathbf{y}, t_{\mathrm{out}}|G|\mathbf{x}, t_{\mathrm{in}}\rangle\langle \mathbf{x}|d^3\mathbf{x}d^3\mathbf{y}, \tag{9.59}$$

where boldfacing denotes a 3–vector. In the more general case where the 'particle' has interacted with some other system between t_{in} and t_{out}, this may be replaced by

$$V = \int |\mathbf{y}\rangle\langle \mathbf{y}, t_{\mathrm{out}}|\mathrm{A}(\Psi_1)G\mathrm{A}^+(\Psi_0)|\mathbf{x}, t_{\mathrm{in}}\rangle\langle \mathbf{x}|d^3\mathbf{x}d^3\mathbf{y}, \tag{9.60}$$

where $\mathrm{A}^+(\Psi_0)$ and $\mathrm{A}(\Psi_1)$ are the operators creating the incoming and outgoing states of the other system:

$$\begin{aligned}
|\Psi_0\rangle &= \mathrm{A}^+(\Psi_0)|0\rangle \\
\langle\Psi_1| &= \langle 0|\mathrm{A}(\Psi_1).
\end{aligned} \tag{9.61}$$

If, moreover, $\langle\mathrm{out}|$ and $|\mathrm{in}\rangle$ are 'many–particle' states, then all corresponding propagators must be treated in this way.

A cutting theorem

The relativistic case is more complicated because the propagators have more structure. Related to this is the fact that in standard quantum field theory the probabilities are not derived directly from the Green function, but from the S–matrix. In order to obtain the correct probabilities we must take into account the proper relations between the Green functions and the S–matrix elements, and for this purpose we first prove the following cutting theorem for propagators:

Theorem: Let τ be a spacetime volume such that $x \notin \tau$ and $y \in \tau$. We then have

$$\begin{aligned} iK(x,y) &= \oint_{\partial\tau} iK(x,z) i \overleftrightarrow{\partial_\mu} (z) iK(z,y) dV^\mu(z) \\ &= \langle x|iK \left\{ \oint_{\partial\tau} \left(p_\mu|z\rangle\langle z| + |z\rangle\langle z|p_\mu\right) dV^\mu(z) \right\} iK|y\rangle \end{aligned} \tag{9.62}$$

where $\overleftrightarrow{\partial_\mu}=\overrightarrow{\partial_\mu} - \overleftarrow{\partial_\mu}$.

Proof: From Gauss' theorem, we have

$$\begin{aligned} &\oint_{\partial\tau} iK(x,z) i \overleftrightarrow{\partial_\mu} (z) iK(z,y) dV^\mu(z) = \\ &= \int_\tau \{iK(x,z) i \overrightarrow{\partial^2} (z) iK(z,y) - iK(x,z) i \overleftarrow{\partial^2} (z) iK(z,y)\} d^4z \\ &= \int_\tau \{iK(x,z)(-\partial^2(z) - m^2)K(z,y) - \\ &\qquad ((-\partial^2(z) - m^2)K(x,z)) iK(z,y)\} d^4z \\ &= \int_\tau \{iK(x,z)\delta_4(z-y) - \delta_4(x-z) iK(z,y)\} d^4z \\ &= iK(x,y). \end{aligned} \tag{9.63}$$

□

Let us select a particular inertial frame and two points x, y with $x^0 > y^0$, and consider a matrix element $\langle x|iK|y\rangle$. Let τ_A be a simultaneity hyperplane, given by the points z with $z^0 = t_A$, for some time t_A such that $y^0 < t_A < x^0$, and let τ_B be a simultaneity hyperplane given by $z^0 = t_B$, for a time t_B such that $t_B < y^0$. The matrix element $\langle x|iK|y\rangle$ may now be written as

$$\begin{aligned} \langle x|iK|y\rangle &= \langle x|iK \left\{ \int_{\partial\tau_A} (p_0|z\rangle\langle z| + |z\rangle\langle z|p_0)\, d^3\mathbf{z} \right\} iK|y\rangle \\ &\quad - \langle x|iK \left\{ \int_{\partial\tau_B} (p_0|z\rangle\langle z| + |z\rangle\langle z|p_0)\, d^3\mathbf{z} \right\} iK|y\rangle. \end{aligned} \tag{9.64}$$

We may recall that K is given by

$$\begin{aligned} K &= (p^2 - m^2 + i\epsilon)^{-1} \\ &= (2\pi)^{-3} \int \left\{ \frac{\delta(p_0 - E)\theta(x_0 - y_0)}{p_0 + E} + \frac{\delta(p_0 + E)\theta(y_0 - x_0)}{-p_0 + E} \right\} \times \\ &\quad e^{ip(x-y)} |x\rangle\langle y|\, d^4x d^4y d^4p, \end{aligned} \tag{9.65}$$

where $E = E(\mathbf{p}) = (m^2 + \mathbf{p}^2)^{1/2}$. Inserted in the *first term* of the r.h.s. of (9.64) only the first term of K, with $\delta(p_0 - E)$, remains, in both occurrences of K. Therefore the factors p_0 in (9.64) may both be replaced by the operator

$$\begin{aligned} H &= \int |p_0\rangle\langle p_0| dp_0 \otimes E \\ &= \int |t\rangle\langle t| dt \otimes E \end{aligned} \tag{9.66}$$

where E is the standard '1-particle' Hamilton operator:

$$E = \int E(\mathbf{p})|\mathbf{p}\rangle\langle\mathbf{p}|d^3\mathbf{p}. \tag{9.67}$$

H is therefore the equivalent of the Hamiltonian on the new Hilbert space $L^2(\mathbb{R}^4)$. We now obtain

$$\begin{aligned} &\langle x|iK\left\{\int_{\partial\tau_A}(p_0|z\rangle\langle z| + |z\rangle\langle z|p_0)\, d^3\mathbf{z}\right\} iK|y\rangle \\ &= \langle x|iK\left\{\int_{\partial\tau_A}(H|z\rangle\langle z| + |z\rangle\langle z|H) d^3\mathbf{z}\right\} iK|y\rangle \\ &= \langle x|iK(2H)^{1/2}\left\{\int_{\partial\tau_A}|z\rangle\langle z|d^3\mathbf{z}\right\}(2H)^{1/2} iK|y\rangle, \end{aligned} \tag{9.68}$$

where the last line follows by virtue of

$$[H, \int_{\partial\tau}|z\rangle\langle z|d^3\mathbf{z}] = 0, \tag{9.69}$$

which is valid for any arbitrary simultaneity hyperplane $\partial\tau$. The *second term* on the r.h.s. of (9.64) can be treated in an entirely similar manner. In this case, however, only the $\delta(p_0 - E)$ term remains of the first propagator, but only the $\delta(p_0 + E)$ term remains of the second one. Accordingly, the first occurrence of p_0 can be replaced by H, but the second occurrence gives $-H$, and taken together this leads to a factor

$$\int_{\partial\tau_B}(H|z\rangle\langle z| - |z\rangle\langle z|H)d^3\mathbf{z},$$

which is equal to zero, however, because of (9.69). Therefore, the second term of the r.h.s. of (9.64) vanishes, and we obtain

$$\langle x|iK|y\rangle = \langle x|iK(2H)^{1/2}\left\{\int_{\partial\tau_A}|z\rangle\langle z|d^3\mathbf{z}\right\}(2H)^{1/2} iK|y\rangle. \tag{9.70}$$

The operators $(2H)^{1/2}$ in this formula account for the factors $(2E)^{\pm 1/2}$ that are encountered in standard field theory in the definition of the fields and in the normalization of the sources.

In– and out–states

Let us consider again the case of a single free 'particle'. Let Φ_{in} be a 'particle creation', taking place some time before t_{in}, and Φ_{out} its 'annihilation', taking place some time after t_{out}. The relation between (6.15) and (9.54) can now be established by means of the following construction:

$$\begin{aligned} |\text{in}\rangle &= \langle t_{\text{in}}|(2H)^{1/2} iK|\Phi_{\text{in}}\rangle \\ \langle\text{out}| &= \langle\Phi_{\text{out}}|iK(2H)^{1/2}|t_{\text{out}}\rangle \\ V &= \langle t_{\text{out}}|(2H)^{1/2}G(2H)^{1/2}|t_{\text{in}}\rangle. \end{aligned} \tag{9.71}$$

In the more general case where the 'particle' has interacted with some other system between t_{in} and t_{out}, the last line must be replaced by

$$V = \int |\mathbf{y}\rangle\langle\mathbf{y}, t_{\text{out}}|(2H)^{1/2}\text{A}(\Psi)G(2H)^{1/2}|\mathbf{x}, t_{\text{in}}\rangle\langle\mathbf{x}|d^3\mathbf{x}d^3\mathbf{y}. \tag{9.72}$$

If, moreover, $\langle\text{out}|$ and $|\text{in}\rangle$ are multi–event states, either because Φ_{in} and Φ_{out} are themselves multi–event states or because of creation and annihilation processes occurring before t_{in} or after t_{out}, then all propagators must be treated in this way, which effectively amounts to replacing Green functions by S–matrix elements. Evidently, this construction applies only if $\partial\tau$ consists of two hyperplanes of constant time, defined by the moments t_{in} and t_{out}. It may be generalized for arbitrary shapes of τ and arbitrary Lorentz frames, however, by using the 'flux' operator

$$\oint_{\partial\tau} |x\rangle i \overleftrightarrow{\partial}_\mu \langle x|\, dV^\mu(x).$$

The relation between (9.54) and (6.15) may now be summarized as follows: let Φ be a configuration of events such that

$$\langle\Phi| = \langle 0|\text{A}(\Phi_{\text{out}})\text{A}(\Psi)\text{A}(\bar{\Phi}_{\text{in}}), \tag{9.73}$$

with Φ_{in} taking place before t_{in} and Φ_{out} after t_{out}. By the above construction, the complex number $\langle\Phi|G\rangle$ may then be rewritten as

$$\langle\Phi|G\rangle = \langle\text{out}|V|\text{in}\rangle. \tag{9.74}$$

Finally, it may be noted that probabilities in event theory may be elegantly expressed in terms of process valued measures[8]. Corresponding to a subset $K \subseteq \Gamma_{\mathbf{u}}$, one may define its associated statistical state $\bar{\Phi}_{\mathbf{u},w}(K)$ by means of

$$\bar{\Phi}_{\mathbf{u},w}(K) = C_{\mathbf{u},w} \sum_{\Phi\in K} w(\Phi)|\Phi\rangle\langle\Phi|. \tag{9.75}$$

The function $\bar{\Phi}_{\mathbf{u},w}$ is a PVM, and has a corresponding probability measure

$$p_{\mathbf{u},w}(K) = \langle G|\bar{\Phi}_{\mathbf{u},w}(K)|G\rangle. \tag{9.76}$$

[8]See § 6.2.3.

Literature: Broyles [36], Costa de Beauregard [49, 50], Cramer [51, 52], Davies [58, 59, 60], Fokker [84], Gotō *et al.* [103, 102], Horwitz [126], Hoyle *et al.* [128, 127], Johnson [132], Roberts [181], Rohrlich [182], Rosenbaum [184], Schwinger [190], Stueckelberg [199, 200], Tetrode [206], Weinberg [215], Wheeler [221, 220].

Chapter 10

Contextual logic

Abstract: In ch. 3 an introduction to contextual logic was given. We shall elaborate on this by giving the outlines of a more mathematical set–up and some applications.

Contextual logic was designed for the specific purpose of investigating the conceptual structure of quantum mechanics within a classical logical framework. It is, however, a heuristically powerful multipurpose logic; links to interesting developments in various other branches of logic, cognitive science, artificial intelligence, and philosophy of science general are readily imagined. In earlier chapters an introduction to contextual logic was given in relation to some general ideas about theory structure. In the present chapter some more details and further applications will be developed.

10.1 The general structure of contextual logic

The basic concept in contextual logic is that of a *logical labyrinth*, consisting of a set of *contexts*, where a context is conceived as a linguistic environment with a fixed syntax and a fixed primary, 'local', semantics. Logical argumentation within the labyrinth may proceed in two different ways: in the form of a *local argument* within a fixed context, or as a *connective argument*, running from one context to another. Contextual logic may be viewed as being the theory of transportation of logical truth through a labyrinth, both locally and connective.

Contexts

Some salient features of the approach may be illustrated by elementary considerations about manipulations with the simplest form of context, that of a 'semiinterpreted language', consisting of a syntax $\mathcal{L}$, a state space Γ, and a satisfaction function γ, mapping sentences of $\mathcal{L}$ on subsets of Γ. Consider an extremely simple context $\mathbf{e}$, with a syntax $\mathcal{L}_{\mathrm{e}}$, given by

$$\mathcal{L}_{\mathrm{e}} = \{\text{The light switch is on, The light switch is off}\}, \tag{10.1}$$

a state space

$$\Gamma_{\mathbf{e}} = \{a, b\}, \tag{10.2}$$

and a satisfaction function defined by

$$\begin{aligned}\gamma_{\mathbf{e}}(\text{The light switch is on}) &= \{a\} \\ \gamma_{\mathbf{e}}(\text{The light switch is off}) &= \{b\},\end{aligned} \tag{10.3}$$

i.e., the local semantics contains two states, and the local language merely consists of two sentences referring to these states. An equally simple context **f** is created by defining

$$\begin{aligned}&\mathcal{L}_{\mathbf{f}} = \{\text{The light bulb is bright, The light bulb is dark}\} \\ &\Gamma_{\mathbf{f}} = \{c, d\} \\ &\gamma_{\mathbf{f}}(\text{The light bulb is bright}) = \{c\} \\ &\gamma_{\mathbf{f}}(\text{The light bulb is dark}) = \{d\}.\end{aligned} \tag{10.4}$$

Please note that this has been purely abstract formal logic thus far; meaningful English sentences have been selected for $\mathcal{L}_{\mathbf{e}}$ and $\mathcal{L}_{\mathbf{f}}$ merely for making the example more vivid. We may, however, choose to assign to these sentences a factual meaning by connecting the contexts **e** and **f** to empirical contexts in which, for instance, the observed brightness of a certain lightbulb and the position of a certain switch are expressed. Via such a connection the sentences receive a secundary, in this case empirical, interpretation.

Connective contexts

Choosing the right switch to the right bulb we may note an empirical regularity, correlating the position of the switch and the brightness of the bulb, and we may want to linguistically express this newly discovered relation. To this end we introduce a *connective context* **c** over **e** and **f**, containing sufficient structure to do the job. Such connective contexts are the vital building blocks in contextual logic, and we shall take some time to explain the machinery by which they perform their work as connecting units.

To begin with, let us consider the language $\mathcal{L}_{\mathbf{c}}$. We shall use **c** for connecting **e**-sentences and **f**-sentences, so we must be able to refer to these sentences in **c**. To this end we introduce within **c** sentences of the form $[\mathbf{e} : \psi]$ and $[\mathbf{f} : \phi]$, with $\psi \in \mathcal{L}_{\mathbf{e}}, \phi \in \mathcal{L}_{\mathbf{f}}$. More explicitly, this gives us

$$\begin{aligned}\mathcal{L}_{\mathbf{c}} \supseteq \Big\{&[\mathbf{e} : \text{The light switch is on}], \\ &[\mathbf{e} : \text{The light switch is off}], \\ &[\mathbf{f} : \text{The light bulb is bright}], \\ &[\mathbf{f} : \text{The light bulb is dark}]\Big\}.\end{aligned} \tag{10.5}$$

To this set we now add all sentences which can be composed of them by use of the standard truthfunctional connectives $\neg, \vee, \wedge, \Rightarrow$ and $\equiv$. The operators $[\mathbf{e} : \_]$ and

$[\mathbf{f} : \_]$ will be called *connective operators*. Such operators serve to import, within a connective context, sentences and other terms from its member contexts.

For the state space of **c** we shall simply take

$$\begin{aligned} \Gamma_{\mathbf{c}} &= \Gamma_{\mathbf{e}} \times \Gamma_{\mathbf{f}} \\ &= \{<a,c>, <a,d>, <b,c>, <b,d>\}, \end{aligned} \tag{10.6}$$

to begin with, and the satisfaction function $\gamma_{\mathbf{c}}$ is constructed in the obvious way, with, e.g.,

$$\begin{aligned} \gamma_{\mathbf{c}}([\mathbf{e} : \text{The light switch is on}]) &= \{<a,c>, <a,d>\} \\ \gamma_{\mathbf{c}}(\neg\psi) &= \gamma_{\mathbf{c}}^{-}(\psi) \\ \gamma_{\mathbf{c}}(\psi \vee \phi) &= \gamma_{\mathbf{c}}(\psi) \cup \gamma_{\mathbf{c}}(\phi), \\ &\text{etcetera.} \end{aligned} \tag{10.7}$$

Whereas a simple semiinterpreted language is a triple $< \mathcal{L}_{\mathbf{u}}, \Gamma_{\mathbf{u}}, \gamma_{\mathbf{u}} >$, a connective context is at least a quadruple $< \mathcal{L}_{\mathbf{u}}, \Psi_{\mathbf{u}}, \Gamma_{\mathbf{u}}, \gamma_{\mathbf{u}} >$, where $\Psi_{\mathbf{u}} \subseteq \Psi$ is the set of contexts 'visible' from **u**.

Natural modalities

Now let us return to the empirical observation that the position of the light switch is correlated with the brightness of the bulb. This may be expressed by saying that the states $<a,d>$ and $<b,c>$ do not occur, and we may react to this by deleting them from the state space, giving

$$\Gamma_{\mathbf{c}} = \{<a,c>, <b,d>\}. \tag{10.8}$$

This semantical manoeuvre may be matched syntactically within $\mathcal{L}_{\mathbf{c}}$ by introducing a modal necessity operator $\Box$ by means of:

$$\begin{aligned} \gamma(\Box\phi) &= \Gamma \text{ iff } \gamma(\phi) = \Gamma \\ &= \emptyset \text{ otherwise.} \end{aligned} \tag{10.9}$$

This enables us to obtain:

$$\Box\Big([\mathbf{e} : \text{The light switch is on}] \equiv [\mathbf{f} : \text{The light bulb is bright}]\Big). \tag{10.10}$$

This formula says that, within the context **c**, it is *necessary* that the position of the switch is correlated with the brightness of the bulb. 'Necessity' may seem a big word for spending it on such a modest observation, especially since it is very well conceivable that the observed regularity is violated, for instance by malfunctioning of the bulb. The objective of the above example, however, is to explain a certain logical construction. The status we attach to a given notion of necessity is determined by the context in which it is defined. A notion of physical necessity originating in a context of a fundamental theory of physics may be more impressive than the above one, but this does not mean that its logical structure is different.

Simple modalities of the above type, which can be defined in terms of truth in all states, are also called "natural modalities". Richer modal systems are generally interpreted by means of a 'many–worlds' semantics, where extra structure is added by means of an accessibility relation between the worlds. Within the contextual approach a richer system will be obtained in a different way, via the relations between contexts.

Interpretation functions

A language $\mathcal{L}$ with a semantics defined in terms of a state space Γ and a satisfaction function γ is called *semiinterpreted* because only *sentences*, also called Boolean terms, are interpreted. Regardless of the syntactical structure of $\mathcal{L}$ this puts the semantics at the level of a propositional logic. A richer semantics, interpreting also other, nonboolean terms, such as individual constants, predicates and functions, is obtained by taking for the state space Γ a set of structures, or 'possible worlds'. In this case Γ is also called the set of 'intended models' of $\mathcal{L}$. If $\lambda \in \Gamma$ is such a world, then an interpretation of $\mathcal{L}$ in λ is a function I_λ, assigning objects in λ to terms of $\mathcal{L}$. We now define an *interpretation function* η as a function which assigns to terms $\psi \in \mathcal{L}$ a set of pairs $< \lambda, I_\lambda(\psi) >$, giving all objects in all possible worlds in which ψ has an interpretation.

The interpretation function $\eta_\mathbf{u}$ will be admitted to be partial, i.e., it is not necessarily defined for all expressions in $\mathcal{L}_\mathbf{u}$. If the domain of $\eta_\mathbf{u}$ is empty, or if it is restricted to sentences only, then $\mathbf{u}$ is effectively a semiinterpreted language. If $\eta_\mathbf{u}$ interprets a sentence ϕ at all, then $I_\lambda(\phi)$ takes a value in the set $\{0,1\}$ of truth values, and the relation between $\eta_\mathbf{u}(\phi)$ and $\gamma_\mathbf{u}(\phi)$ is given by

$$\gamma_\mathbf{u}(\phi) = \{\lambda |\, I_\lambda(\phi) = 1\}. \tag{10.11}$$

The definition of the connective operators $[\mathbf{u} : \psi]$ in a connective language $\mathcal{L}_\mathbf{c}$ is readily extended from sentences to nonboolean terms, i.e., $[\mathbf{u} : \psi]$ is an expression in $\mathcal{L}_\mathbf{c}$ for an object called ψ in the context $\mathbf{u}$.

In a representation of contexts in terms of quintuples $< \mathcal{L}_\mathbf{u}, \Psi_\mathbf{u}, \Gamma_\mathbf{u}, \gamma_\mathbf{u}, \eta_\mathbf{u} >$, nonconnective contexts may be included in this format by admitting $\Psi_\mathbf{u} = \emptyset$, and semiinterpreted languages may be obtained either by taking the domain of $\eta_\mathbf{u}$ to be empty or by restricting it to sentences only.

Intensional operators

A connective language like the above $\mathcal{L}_\mathbf{c}$, based entirely on manipulations with connective sentences of the type $[\mathbf{u} : \psi]$, is relatively primitive, and it is often convenient to give it a much richer structure. In this case one may also need a new type of operators, for constructing a contextual reference for terms in $\mathcal{L}_\mathbf{c}$. We shall call such operators *intensional operators*, and they will be written in the form $[\psi]\mathbf{u}$. For instance, consider a language $\mathcal{L}_\mathbf{c}$ suitable for describing various systems of classical mechanics, and an empirical context $\mathbf{u}$ in which a particular set of measurement results on a given billiard ball are represented in terms of positions and velocities. The sentence ψ = "The ball has a kinetic energy of 10 Joules" may be well-defined in $\mathcal{L}_\mathbf{c}$, and $[\psi]\mathbf{u}$ then means that the given ball has this amount of kinetic energy. The operators $[\ \_\ ]\mathbf{u}$ are closely related to the connective operators $[\mathbf{u} : \ \_\ ]$, but whereas in $[\mathbf{u} : \psi]$ the term ψ must be well-defined in the syntax of $\mathcal{L}_\mathbf{u}$, in $[\psi]\mathbf{u}$ it must be well-defined in $\mathcal{L}_\mathbf{c}$.

Example: Frege's paradigm example of the difference between extensional and intensional meaning concerned the terms "morning star" and "evening star". Both terms have the same extension, they are both thought to denote the planet Venus, but they have a different intension. This is clear from the fact that the two terms are not interchangeable, *salva veritate*, in arbitrary sentences. Within our present framework, we may analyze the situation by making a distinction between two different contexts. One of these, $\mathbf{u}$, relates to our staring at the sky at various times of the day and witnessing the presence of bright points. The other one, $\mathbf{v}$, relates to our ideas about the structure of the solar system. The terms "morning star" and "evening star" belong primarily to the first context, whereas "Venus" is a term from the second one. In a connective context $\mathbf{c}$ over $\mathbf{u}$ and $\mathbf{v}$ we may use all three terms and formulate their mutual difference in the following manner:

$$
\begin{array}{l}
[\text{morning star} = \text{Venus}]\mathbf{v} \\
[\text{morning star} = \text{evening star}]\mathbf{v} \\
[\text{morning star} \neq \text{evening star}]\mathbf{u} \\
[\text{Venus} = \text{morning star}]\mathbf{u} \rightarrow \text{Venus is visible at dawn.}
\end{array}
$$

"Extension" can now be used either as a contextually bound concept or as a concept referring to a predominant context. The latter meaning is used if we say that "morning star" and "evening star" have the same extension. "Intension", on the other hand, can be analyzed as multicontextual extension, i.e., two terms have a different intension if there is some context in which they have a different extension.

10.1.1 Summary

The general structure of contextual logic may be summarized in the following scheme:

Labyrinths:

A labyrinth Ψ is a set of contexts.

Contexts:

A context in Ψ is a quintuple $< \mathcal{L}_\mathbf{u}, \Psi_\mathbf{u}, \Gamma_\mathbf{u}, \gamma_\mathbf{u}, \eta_\mathbf{u} >$, where

$\mathcal{L}_{\mathbf{u}}$ is a syntax.

$\Psi_{\mathbf{u}}$ is a subset of Ψ.

$\Gamma_{\mathbf{u}}$ is a state space, represented as a structured set of structures.

$\gamma_{\mathbf{u}}$ is a satisfaction function mapping sentences from the set $\mathcal{L}_{\mathbf{u}}^{sent}$ of sentences of $\mathcal{L}_{\mathbf{u}}$ to the power set $\wp(\Gamma_{\mathbf{u}})$ of $\Gamma_{\mathbf{u}}$.

$\eta_{\mathbf{u}}$ is an interpretation function, assigning to expressions $\phi \in \mathcal{L}_{\mathbf{u}}$ pairs $< \lambda, I_\lambda(\phi) >$, where $\lambda \in \Gamma_{\mathbf{u}}$ is a structure, also called a "possible world", and $I_\lambda(\phi)$ is an element of λ.

If $\Psi_{\mathbf{u}} \supset \emptyset$, then **u** is called a connective context. The interpretation function $\eta_{\mathbf{u}}$ is admitted to be partial, i.e., it is not necessarily defined for all expressions in $\mathcal{L}_{\mathbf{u}}$. If the domain of $\eta_{\mathbf{u}}$ is empty, or if it is restricted to sentences only, then **u** is called a semiinterpreted language.

Local logic:

Local logic concerns the internal structure of a context without its connective machinery. Syntax may differ per context, but for our convenience we will assume the presence of the standard truthfunctional connectives, $\neg \vee \wedge \Rightarrow \equiv$, and of the nontruthfunctional natural modality operator $\Box$.

Typed languages, variables, quantifiers, and an identity relation may be introduced via standard procedures wherever desired. We shall use such features freely, withou discussing the details of their introduction.

Connective logic:

A connective context **c** can be compared to a corridor, and the contexts $\mathbf{u} \in \Psi_{\mathbf{c}}$ are the rooms to which the corridor **c** gives access. These rooms may be corridors themselves, in their turn giving access to other rooms and corridors. We can thus build labyrinths of arbitrary complexity. Connective logical structure is introduced into the labyrinth for the purpose of being able to take information from one context to another. The basic tools for achieving this are two types of contextual operators:

connective operators $[\mathbf{u} : \_\,] : \mathcal{L}_{\mathbf{u}} \rightarrow \mathcal{L}_{\mathbf{c}}$;

intensional operators $[\,\_\,]\mathbf{u} : \mathcal{L}_{\mathbf{c}} \rightarrow \mathcal{L}_{\mathbf{c}}$.

To some exent, the meaning of these operators is explained above; further details will be provided below.

The above scheme should provide a concise survey of the main features of contextual logic. We shall now proceed by filling in more detail.

10.1.2 Contextual semantics

The problem of finding a suitable semantics for a connective context boils down to the problem of how to construct its state space. For brevity we shall restrict our analysis to the case of semiinterpreted languages, i.e., the interpretation of expressions other than sentences will not be discussed. Even so, depending on the situation, there are various possibilities of how the state space of a connective context may relate to that of its member contexts.

Example: Suppose, for instance, that we start with a single context $\mathbf{a} = <\mathcal{L}_{\mathbf{a}}, \Gamma_{\mathbf{a}}, \gamma_{\mathbf{a}}>$. A set Ψ of contexts may then be formed by selecting a set of *subcontexts*,

$$\mathbf{u} \subseteq \mathbf{a} \ \text{ with } \ \mathcal{L}_{\mathbf{u}} \subseteq \mathcal{L}_{\mathbf{a}}, \Gamma_{\mathbf{u}} \subseteq \Gamma_{\mathbf{a}}, \gamma_{\mathbf{u}} \subseteq \gamma_{\mathbf{a}}. \tag{10.12}$$

We may now construct a connective context $\mathbf{c}$ with

$$\Gamma_{\mathbf{c}} = \bigcup_{\mathbf{u} \in \Psi} \Gamma_{\mathbf{u}} = \Gamma_{\mathbf{a}}. \tag{10.13}$$

In other situations, for instance in the case of contextual QM, it can be natural to take for $\Gamma_{\mathbf{c}}$ the disjoint union of the state spaces $\Gamma_{\mathbf{u}}$, i.e.,

$$\Gamma_{\mathbf{c}} = \{<\mathbf{u}, \lambda> | \mathbf{u} \in \Psi, \lambda \in \Gamma_{\mathbf{u}}\} = \bigcup_{\mathbf{u} \in \Psi} (\{\mathbf{u}\} \times \Gamma_{\mathbf{u}}). \tag{10.14}$$

If all $\Gamma_{\mathbf{u}}$ are equal here, say to some $\Gamma_{\mathbf{a}}$, we obtain

$$\Gamma_{\mathbf{c}} = \Psi \times \Gamma_{\mathbf{a}}. \tag{10.15}$$

Furthermore, for composite systems the Cartesian product of the state spaces of the subsystems is a natural choice, i.e.:

$$\Gamma_{\mathbf{c}} = \prod_{\mathbf{u} \in \Psi} \Gamma_{\mathbf{u}}. \tag{10.16}$$

If in this case all $\Gamma_{\mathbf{u}}$ are equal, $\Gamma_{\mathbf{u}}=\Gamma_{\mathbf{a}}$, we obtain

$$\Gamma_{\mathbf{c}} = \Gamma_{\mathbf{a}}^{\Psi} = \Psi \rightarrow \Gamma_{\mathbf{a}}, \tag{10.17}$$

i.e., the set of functions of Ψ to $\Gamma_{\mathbf{a}}$.

State spaces in connective contexts

The above structures are merely examples of possible choices for $\Gamma_{\mathbf{c}}$, or in fact for its connective part, because $\mathbf{c}$ may also have 'private' parts, not originating from its member contexts. In order to avoid having to deal with many separate cases, we shall first look for a general framework which can accommodate them all. Our first step in doing this is to define to each context $\mathbf{u}$ its *trivial complete extension* $\bar{\mathbf{u}}$ with

$$\begin{aligned} \mathcal{L}_{\bar{\mathbf{u}}} &= \mathcal{L}_{\mathbf{u}} \cup \{T\mathbf{u}, \neg T\mathbf{u}\}, \\ \Psi_{\bar{\mathbf{u}}} &= \Psi_{\mathbf{u}}, \\ \Gamma_{\bar{\mathbf{u}}} &= \Gamma_{\mathbf{u}} \cup \{\dagger_{\mathbf{u}}\} \\ \gamma_{\bar{\mathbf{u}}}(T\mathbf{u}) &= \Gamma_{\mathbf{u}}, \ \gamma_{\bar{\mathbf{u}}}(\neg T\mathbf{u}) = \{\dagger_{\mathbf{u}}\}. \end{aligned} \tag{10.18}$$

The point of this is the following: within the context $\mathbf{u}$ we aim to describe some system. What is expressed, in the context $\bar{\mathbf{u}}$, by describing a system by the sentence "$\neg T\mathbf{u}$" is that it is not a system which can be described within the context $\mathbf{u}$, i.e., $\dagger_{\mathbf{u}}$ represents the state that $\mathbf{u}$ does not apply. Accordingly, in a trivial sense, $\bar{\mathbf{u}}$ is a context which applies to any arbitrary system.

Our next step is now to define a state space $\Gamma_{\mathbf{c.con}}$, by means of

$$\Gamma_{\mathbf{c.con}} = \prod_{\mathbf{u} \in \Psi_{\mathbf{c}}} \Gamma_{\bar{\mathbf{u}}}. \tag{10.19}$$

By choosing for the connective part of $\Gamma_{\mathbf{c}}$ a suitable subset of $\Gamma_{\mathbf{c.con}}$ we can accommodate all cases mentioned in the above example.

Example: Consider a set of contexts $\Psi_{\mathbf{c}} = \{\mathbf{u1}, \mathbf{u2}, \mathbf{u3}\}$ and let $\Gamma_{\mathbf{c}}$, for instance, be the disjoint union of $\Gamma_{\mathbf{u1}}$, $\Gamma_{\mathbf{u2}}$, and $\Gamma_{\mathbf{u3}}$, as in (10.14). Instead of this particular $\Gamma_{\mathbf{c}}$ we can also take a state space $\Gamma_{\mathbf{c}}'$, given by

$$\begin{aligned} \Gamma_{\mathbf{c}}' \;=\; & (\Gamma_{\mathbf{u1}} \times \{\dagger_{\mathbf{u2}}\} \times \{\dagger_{\mathbf{u3}}\}) \\ & \cup(\{\dagger_{\mathbf{u1}}\} \times \Gamma_{\mathbf{u2}} \times \{\dagger_{\mathbf{u3}}\}) \\ & \cup(\{\dagger_{\mathbf{u1}}\} \times \{\dagger_{\mathbf{u2}}\} \times \Gamma_{\mathbf{u3}}). \end{aligned} \tag{10.20}$$

A $\Gamma_{\mathbf{c}}$-state like $< \mathbf{u1}, \lambda >$ then translates into the $\Gamma_{\mathbf{c}}'$-state $< \lambda, \dagger_{\mathbf{u2}}, \dagger_{\mathbf{u3}} >$. The state space $\Gamma_{\mathbf{c}}'$ is indeed a subset of $\Gamma_{\mathbf{c.con}}$.

Besides its connective part, the state space $\Gamma_{\mathbf{c}}$ may have a private part, which we shall call $\Gamma_{\mathbf{c.priv}}$. A combined state space $\Gamma_{\mathbf{c.comb}}$ will now be defined by:

$$\Gamma_{\mathbf{c.comb}} = \Gamma_{\mathbf{c.priv}} \times \Gamma_{\mathbf{c.con}}. \tag{10.21}$$

We shall from now on assume $\Gamma_{\mathbf{c}}$ to be of the form

$$\Gamma_{\mathbf{c}} \subseteq \Gamma_{\mathbf{c.comb}}. \tag{10.22}$$

With reference to any of the state spaces considered thus far we lose no generality by confining ourselves to this case. In particular cases there will often be a simpler and more elegant choice, but then we can safely assume that our present considerations are readily transposed to deal with it.

The meaning of connective and intensional operators

Let us now return to the issue of connective semantics. In order to define the meaning of sentences like $[\mathbf{u} : \phi]$ we first construct a type of operators

$$\cap_{\mathbf{c}|\mathbf{u}} : \wp(\Gamma_{\mathbf{u}}) \times \wp(\Gamma_{\mathbf{c}}) \to \wp(\Gamma_{\mathbf{c}}),$$

by means of

$$\Lambda_{\mathbf{c}} \cap_{\mathbf{c}|\mathbf{u}} \Lambda_{\mathbf{u}} = \{\lambda_{\mathbf{c}} \in \Lambda_{\mathbf{c}} | \, \pi_{\bar{\mathbf{u}}}(\lambda_{\mathbf{c}}) \in \Lambda_{\mathbf{u}}\}, \tag{10.23}$$

where $\pi_{\bar{\mathbf{u}}}(\lambda_{\mathbf{c}})$ gives the $\bar{\mathbf{u}}$-component of $\lambda_{\mathbf{c}}$. For **c**=**u** this is just the standard intersection, whereas more generally 'intersections' of this type generate cylinders in $\Gamma_{\mathbf{c}}$, and the natural interpretation of the sentence $[\mathbf{u} : \phi]$ within $\Gamma_{\mathbf{c.con}}$ is precisely the cylinder generated by the interpretation $\gamma_{\mathbf{u}}(\phi)$ of ϕ in **u**, i.e.:

$$\gamma_{\mathbf{c}}([\mathbf{u} : \phi]) = \Gamma_{\mathbf{c}} \cap_{\mathbf{c}|\mathbf{u}} \gamma_{\mathbf{u}}(\phi). \tag{10.24}$$

For a sentence of the form $[\phi]\mathbf{u}$ we require, if it is meaningful at all, the existence of a similar interpretation. In other words, a sentence $[\phi]\mathbf{u}$ obtains a meaning by specifying, in some way, a satisfaction function $\gamma_{\mathbf{u};\mathbf{c}}(\phi) \subseteq \Gamma_{\mathbf{u}}$ such that

$$\gamma_{\mathbf{c}}([\phi]\mathbf{u}) = \Gamma_{\mathbf{c}} \cap_{\mathbf{c}|\mathbf{u}} \gamma_{\mathbf{u};\mathbf{c}}(\phi). \tag{10.25}$$

This can be done, for instance, by defining the meaning of $[\phi]\mathbf{u}$ in terms of a sentence $[\mathbf{u} : \psi]$ by means of a rule of the form

$$\Box([\phi]\mathbf{u} \equiv [\mathbf{u} : \psi]), \tag{10.26}$$

which precisely forces the equality

$$\gamma_{\mathbf{c}}([\phi]\mathbf{u}) = \gamma_{\mathbf{c}}([\mathbf{u} : \psi]). \tag{10.27}$$

Now we must still fix the details of the definition of $\gamma_{\mathbf{c}}([\phi]\mathbf{u})$ in cases when it is not, or only partially meaningful. This raises some special problems.

The 'meaningful'–predicate

Example: Let **c** be a connective context over a set $\Psi_{\mathbf{c}}$ of contexts associated with various fairy tales. Let $\mathbf{u} \in \Psi_{\mathbf{c}}$ be the context associated with the tale of Cinderella, then the sentence "Snowwhite is asleep"$\in \mathcal{L}_{\mathbf{c}}$ is not meaningful with respect to this context, and we might define $\gamma_{\mathbf{u};\mathbf{c}}(\phi)$ by means of

$$\gamma_{\mathbf{u};\mathbf{c}}(\text{Snowwhite is asleep}) = \Gamma_{\mathbf{u}}, \tag{10.28}$$

because nothing in the tale of Cinderella gives us any information whatsoever about Snowwhite being asleep or not. By the same argument, however, we would also have

$$\begin{aligned} \gamma_{\mathbf{u};\mathbf{c}}(\text{Snowwhite is not asleep}) &= \Gamma_{\mathbf{u}} \\ &= \gamma_{\mathbf{u};\mathbf{c}}(\text{Snowwhite is asleep}), \end{aligned} \tag{10.29}$$

which can be regarded as peculiar, because it leads to

$$[\text{Snowwhite is asleep}]\mathbf{u} \wedge [\text{Snowwhite is not asleep}]\mathbf{u}. \tag{10.30}$$

Problems of this type will be dealt with by admitting

$$\gamma_{\mathbf{c}}([\phi]\mathbf{u} \wedge [\neg\phi]\mathbf{u}) \supset \emptyset. \tag{10.31}$$

while *denying* the general validity of

$$[\neg\phi]\mathbf{u} \equiv \neg[\phi]\mathbf{u} \tag{10.32}$$

With respect to the statements $[\phi]\mathbf{u}$ and $[\neg\phi]\mathbf{u}$, the state space $\Gamma_{\mathbf{c}}$ may now be divided into four different regions, where any one of the following statements holds:

$[\phi]\mathbf{u} \wedge [\neg\phi]\mathbf{u}$
: corresponds to the case that ϕ does not make sense with respect to $\mathbf{u}$.

$\neg[\phi]\mathbf{u} \wedge \neg[\neg\phi]\mathbf{u}$
: implies that the context $\mathbf{u}$ does not apply at all. Within $\mathcal{L}_{\mathbf{c}}$ we shall refer to this as[1] $\neg T\mathbf{u}$.

$[\phi]\mathbf{u} \wedge \neg[\neg\phi]\mathbf{u}$
: ϕ is true with respect to $\mathbf{u}$.

$\neg[\phi]\mathbf{u} \wedge [\neg\phi]\mathbf{u}$
: $\neg\phi$ is true with respect to $\mathbf{u}$.

The latter two states correspond to the case that (10.32) does hold and $[\phi]\mathbf{u}$ is meaningful in a standard sense. We shall further express this within $\mathcal{L}_{\mathbf{c}}$ by means of a predicate $M_{\mathbf{u}}$ defined by

$$M_{\mathbf{u}}(\phi) \equiv ([\neg\phi]\mathbf{u} \equiv \neg[\phi]\mathbf{u}). \tag{10.33}$$

With respect to the use of the other connectives we further postulate that

$$(M_{\mathbf{u}}(\phi) \wedge M_{\mathbf{u}}(\psi)) \;\Rightarrow\; M_{\mathbf{u}}(\phi \vee \psi) \tag{10.34}$$

$$(M_{\mathbf{u}}(\phi) \wedge M_{\mathbf{u}}(\psi)) \;\Rightarrow\; (([\phi]\mathbf{u} \vee [\psi]\mathbf{u}) \equiv [\phi \vee \psi]\mathbf{u}). \tag{10.35}$$

As the other connectives can be defined in terms of $\neg$ and $\vee$, it is readily verified that from $(M_{\mathbf{u}}(\phi) \wedge M_{\mathbf{u}}(\psi))$ one also obtains $M_{\mathbf{u}}(\phi \wedge \psi)$, $M_{\mathbf{u}}(\phi \Rightarrow \psi)$, $[\phi \wedge \psi]\mathbf{u} \equiv ([\phi]\mathbf{u} \wedge [\psi]\mathbf{u})$ etcetara. Further, the use of quantifiers can be accommodated by generalizing (10.34) and (10.35) to

$$\forall x \in A\, M_{\mathbf{u}}(\phi(x)) \;\Rightarrow\; M_{\mathbf{u}}(\exists x \in A\, \phi(x)) \tag{10.36}$$

$$\forall x \in A\, M_{\mathbf{u}}(\phi(x)) \;\Rightarrow\; \big(\exists x \in A\, [\phi(x)]\mathbf{u} \equiv [\exists x \in A\, \phi(x)]\mathbf{u}\big). \tag{10.37}$$

---

[1] See also (10.18).

Looped labyrinths

The membership relation induces an ordering of the contexts within a logical labyrinth. This ordering is either strict, arranging the contexts in a treelike hierarchy, or not strict, arranging the contexts in some more general kind of graph. In this latter case the labyrinth contains loops. A simple example is the case that $\mathbf{u} \in \Psi_{\mathbf{u}}$. In this case we shall say that the labyrinth has a loop of length 0. Likewise, if $\mathbf{u} \in \Psi_{\mathbf{v}}$ for some $\mathbf{v} \in \Psi_{\mathbf{u}}$, it has a loop of length 1, etcetera. If we consider the construction (10.22) for the state space of a context $\mathbf{c}$ for which $\Psi_{\mathbf{c}}$ contains loops, then we have a problem. Suppose that $\mathbf{c} \in \Psi_{\mathbf{c}}$. From (10.21) and (10.19) we obtain

$$\Gamma_{\mathbf{c}} \subseteq \Gamma_{\mathbf{c.priv}} \times \prod_{\mathbf{u} \in \Psi_{\mathbf{c}}} \Gamma_{\bar{\mathbf{u}}}. \tag{10.38}$$

This expression contains $\Gamma_{\mathbf{c}}$ itself on the r.h.s., i.e., it is of the form $\Gamma_{\mathbf{c}} \subseteq \Gamma_{\mathbf{c}} \times A$, and the only way to satisfy it would be to consider an infinite product of A's and take $\Gamma_{\mathbf{c}} \subseteq A \times A \times \cdots$. This, however, is a rather unwieldy solution, especially because entirely satisfactory results are achieved with merely $\Gamma_{\mathbf{c}} \subseteq A$. Therefore, we take

$$\Gamma_{\mathbf{c}} \subseteq \Gamma_{\mathbf{c.priv}} \times \prod_{\mathbf{u} \in (\Psi_{\mathbf{c}} - \{\mathbf{c}\})} \Gamma_{\bar{\mathbf{u}}}, \tag{10.39}$$

and sentences like $[\mathbf{c} : \phi]$ and $[\phi]\mathbf{c}$ are treated simply by regarding them as equivalent to ϕ itself, i.e.:

$$\gamma_{\mathbf{c}}([\mathbf{c} : \phi]) = \gamma_{\mathbf{c}}([\phi]\mathbf{c}) = \gamma_{\mathbf{c}}(\phi). \tag{10.40}$$

Loops of greater lengths lead to similar problems, which may be solved by similar means.

Contextual morphisms

We may view the state space $\Gamma_{\mathbf{c}}$ as being obtained from $\Gamma_{\mathbf{c.comb}}$ by deleting all superfluous, impossible states. Truth, if considered as necessary truth, is thus carved into the structure of the state space. This is one way to think of $\Gamma_{\mathbf{c}}$. Another way to explicate its structure is by defining interpretative morphisms between the contexts in Ψ. Let us, therefore, proceed to show that $\Gamma_{\mathbf{c}}$ can alternatively be specified by means of a mapping

$$\mathcal{J} : \wp(\Gamma_{\mathbf{c.comb}}) \to \wp(\Gamma_{\mathbf{c.comb}}), \tag{10.41}$$

or, equivalently, by means of a matrix of mappings

$$\mathcal{J}_{\bar{\mathbf{v}}|\bar{\mathbf{u}}} : \wp(\Gamma_{\bar{\mathbf{u}}}) \to \wp(\Gamma_{\bar{\mathbf{v}}}), \tag{10.42}$$

where $\mathbf{u}, \mathbf{v} \in \Psi_{\mathbf{c}} \cup \{\mathbf{c.priv}\}$. With $\Gamma_{\mathbf{c}}$ given, the mapping $\mathcal{J}$ is defined by

$$\mathcal{J}(\Lambda_{\mathbf{c}}) = \Lambda_{\mathbf{c}} \cap \Gamma_{\mathbf{c}}. \tag{10.43}$$

It is clear, then, that $\mathcal{J}$ and $\Gamma_{\mathbf{c}}$ determine each other completely. With $\mathcal{J}$ given, we can obtain $\Gamma_{\mathbf{c}}$ from

$$\Gamma_{\mathbf{c}} = \mathcal{J}(\Gamma_{\mathbf{c.comb}}). \tag{10.44}$$

The matrix of mappings $\mathcal{J}_{\bar{\mathbf{v}}|\bar{\mathbf{u}}}$ is defined similarly by

$$\mathcal{J}_{\bar{\mathbf{v}}|\bar{\mathbf{u}}}(\Lambda_{\bar{\mathbf{u}}}) = \{\pi_{\bar{\mathbf{v}}}(\lambda_{\mathbf{c}}) |\, \pi_{\bar{\mathbf{u}}}(\lambda_{\mathbf{c}}) \in \Lambda_{\bar{\mathbf{u}}}, \lambda_{\mathbf{c}} \in \Gamma_{\mathbf{c}}\}, \tag{10.45}$$

where $\pi_{\bar{\mathbf{v}}}(\lambda_{\mathbf{c}})$ gives the $\bar{\mathbf{v}}$-component of $\lambda_{\mathbf{c}}$. Given the whole matrix $\mathcal{J}_{\mathbf{v}|\mathbf{u}}$, the mapping $\mathcal{J}$ is reconstructed from

$$\mathcal{J}(\Lambda) = \bigcup_{\lambda_{\mathbf{c}}\in\Lambda_{\mathbf{c}}} \prod_{\mathbf{v}\in\{\mathbf{c.priv}\}\cup\Psi_{\mathbf{c}}} \bigcap_{\mathbf{u}\in\{\mathbf{c.priv}\}\cup\Psi_{\mathbf{c}}} \mathcal{J}_{\bar{\mathbf{v}}|\bar{\mathbf{u}}}(\lambda_{\bar{\mathbf{u}}}). \tag{10.46}$$

Using these results, we see that indeed a connective context of the form

$$\mathbf{c} = < \mathcal{L}_{\mathbf{c}}, \Psi_{\mathbf{c}}, \Gamma_{\mathbf{c}}, \gamma_{\mathbf{c}} >$$

can alternatively be specified in the form

$$\mathbf{c} =< \mathcal{L}_{\mathbf{c}}, \Psi_{\mathbf{c}}, \Gamma_{\mathbf{c.comb}}, \mathcal{J}, \gamma_{\mathbf{c}} >,$$

where $\mathcal{J}$ is either a mapping or a matrix of mappings. Various useful properties of the interpretative mappings follow directly from their definitions (10.43) and (10.45). It is clear, for instance, that they are monotone and that they are linear under union, but not necessarily under intersection.

10.1.3 Contextual syntax

Truth, logical consequence, logical inference

By the definition of $\gamma_{\mathbf{u}}$, a sentence $\phi \in \mathcal{L}_{\mathbf{u}}$ is true in a given state $\Lambda_{\mathbf{u}} \subseteq \Gamma_{\mathbf{u}}$ iff (if and only if) $\Lambda_{\mathbf{u}} \subseteq \gamma_{\mathbf{u}}(\phi)$. This is expressed, in standard notation, by

$$\Lambda_{\mathbf{u}} \models \phi \;\text{ iff }\; \Lambda_{\mathbf{u}} \subseteq \gamma_{\mathbf{u}}(\phi). \tag{10.47}$$

Instead of $\gamma_{\mathbf{u}}(\psi) \models \phi$ we shall also use the notation $\psi \models \phi$. If this holds, then ϕ is called a logical consequence of ψ. Further, if $\Gamma_{\mathbf{u}} \models \phi$ then ϕ is called necessarily true, and in agreement with (10.9) this holds iff $\Gamma_{\mathbf{u}} \models \Box\phi$.

If we want to know whether a sentence ϕ is indeed a logical consequence of ψ, then one way of trying to find out is by semantic analysis: we try to construct $\gamma_{\mathbf{u}}(\phi)$ and $\gamma_{\mathbf{u}}(\psi)$ and find out whether $\gamma_{\mathbf{u}}(\psi) \subseteq \gamma_{\mathbf{u}}(\phi)$. This means that the basic work is done within an *external*, in this case set theoretic environment, in which we talk *about* the sentences of $\mathcal{L}_{\mathbf{u}}$, and deductions are made by means of the usual logical and mathematical rules that we have learned to apply in set theory. Alternatively, we may want to obtain the same result by means of an *internal* reasoning within $\mathcal{L}_{\mathbf{c}}$. This means that the structure of $\mathcal{L}_{\mathbf{u}}$ must be extended to the effect that logical deduction becomes possible in it, i.e., a logical calculus must be added to its syntax. A *logical calculus* is some set of axioms and inference rules defining a *deducibility relation* $\vdash$, where

$$\psi_1 \cdots \psi_n \vdash \phi \;\text{ iff there is a deduction from }\psi_1 \cdots \psi_n \text{ to } \phi. \tag{10.48}$$

A currently very popular calculus is the system of *natural deduction*, based on a set of inference rules for the introduction or elemination of logical constants.

Example: One of the rules of natural deduction is that we may introduce a premise, which will be one of the ψ_i occurring on the left of the $\vdash$ in (10.48), unless it is later 'discharged'. A premise may be discharged by applying a rule $I \Rightarrow$, for "implication introduction".

$$\begin{array}{l} \psi \\ \lfloor \phi \\ \hline \psi \Rightarrow \phi, \quad I \Rightarrow \end{array}$$

This rule is traditionally called "conditional proof". There is a similar rule, $E \Rightarrow$, for implication elimination, which is traditionally called "modus ponens":

$$\begin{array}{c} \psi \\ \psi \Rightarrow \phi \\ \hline \phi, \quad E \Rightarrow \end{array}$$

There are similar rules for the introduction and elimination of the other connectives, and for quantifiers and identity in as far as defined within $\mathcal{L}_{\mathbf{u}}$. Further, for the modal operator $\Box$ we adopt an elimination rule "$E\Box$" saying that from $\Box\phi$ one can infer ϕ. The corresponding introduction rule "$I\Box$" is traditionally called "necessitation", and says that from ϕ one can infer $\Box\phi$ *if* the deduction of ϕ does not contain any 'charged' premises, i.e., premises which have not been discharged.

Connective arguments

Within the framework of contextual logic one may introduce the idea of a connective deduction, a multilingual argument which runs from one context to another, proceeding in each context via the rules of the local syntax. In order to make this possible, we need new tools, and we shall introduce two new types of operators, *shift operators* of the form "[u :", and corresponding *return operators* "]". These operators will be used for shifting from one context to another in the course of the deduction, in a way which is somewhat similar to a 'change directory' command in a computer operating system. Let us discuss some tentative ideas about how a connective calculus of this type might be made to work properly.

In the first place, we demand that a deduction begins by opening it in a certain context, i.e., the first line in a proof must consist of a shift operator [$\mathbf{u}$:, opening the argument in some context $\mathbf{u} \in \Psi$. The next lines may then, as usual, contain sentences $\phi \in \mathcal{L}_{\mathbf{u}}$, obtained by the rules of the calculus of $\mathcal{L}_{\mathbf{u}}$. Alternatively, however, they may contain a shift [$\mathbf{v}$: to a new context $\mathbf{v} \in \Psi_{\mathbf{u}}$, or a return operator]. In the first case the argument proceeds with sentences in $\mathcal{L}_{\mathbf{v}}$, obtained by the rules of *its* calculus. In the second case the return operator either closes the argument, if no preceding context is available, or it produces a shift back to the preceding context, *allowing the continuation of the argument as it was proceeding there.*

Note: Imagine a labyrinth Ψ, containing a loop, such that $\mathbf{u} \in \Psi_{\mathbf{v}}$ and $\mathbf{v} \in \Psi_{\mathbf{u}}$. Following a shift from $\mathbf{u}$ to $\mathbf{v}$ there are now two possible ways of returning to $\mathbf{u}$, by means of the return operator "]", or by a further shift with the operator "[**u** :". The action of the return operator differs, however, from that of a further shift by means of "[**u** :". In the latter case the 'fresh version' of **u** must be considered as lying on a 'deeper level' of the argument than the original **u**. In an argument, this means that conclusions reached in the original version of **u**, may be used in the new version only if they have been transported there, from the original **u**, through the other context **v**, by means of rules which we shall discuss shortly. Moreover, we shall impose the rule that a premise can be discharged only at the level at which it was introduced.

Example:

$$\begin{array}{ccc} \begin{array}{l} [\mathbf{u} : \\ \phi \\ [\mathbf{v} : \\ ] \\ \hline \phi \end{array} & \qquad & \begin{array}{l} [\mathbf{u} : \\ \phi \\ [\mathbf{v} : \\ [\mathbf{u} : \\ \hline \phi \end{array} \\ \\ \text{Argument 1} & & \text{Argument 2} \end{array}$$

Argument 1 is a valid argument, with $\phi \in \mathcal{L}_{\mathbf{u}}$ and $\mathbf{v} \in \Psi_{\mathbf{u}}$, but argument 2 is not.

The arguments of the deducibility relation now have the following form: on the l.h.s. it has a sequence of context shifts and premises made in the various contexts, and on the r.h.s. it has a conclusion, which must be a sentence in the language open by virtue of the argument on the l.h.s., e.g.,

$$[\mathbf{u} :, \ \phi \wedge \psi, \ [\mathbf{v} :, \ \chi, \ ] \vdash \phi.$$

The point of having a connective calculus is to obtain the ability to have premises in one context and draw conclusions in another. To this end we introduce a scheme of connective inference rules, and in the style of natural deduction we formulate them as rules for the introduction and elimination of the connective operators $[\mathbf{u} : \_]$. The introduction rules are given by

$$\begin{array}{l} [\mathbf{v} : \\ [\mathbf{u} : \\ \phi \\ ] \\ \hline [\mathbf{u} : \phi], \quad I[\mathbf{u} : \_] \end{array} \tag{10.49}$$

with $\mathbf{u} \in \Psi_{\mathbf{v}}$, $\phi \in \mathcal{L}_{\mathbf{u}}$, and $[\mathbf{u} : \phi] \in \mathcal{L}_{\mathbf{v}}$. There is a matching scheme of elimination rules $E[\mathbf{u} : \_]$:

$$\begin{array}{l} [\mathbf{v} : \\ [\mathbf{u} : \phi] \\ [\mathbf{u} : \\ \hline \phi, \qquad E[\mathbf{u} : \_] \end{array} \tag{10.50}$$

These rules enable us to argue from one context to another by syntactic movements. Let us proceed by considering the semantic counterpart of these movements.

Actualizations

The semantic counterpart of introducing a premise is the confinement to a corresponding part of the state space, i.e., introducing a premise ϕ confines us to the state $\gamma(\phi)$. Let us define $\Gamma^*_{\mathbf{u}}$ to be the actually remaining part of the state space $\Gamma_{\mathbf{u}}$ at a given stage of an argument in **u**. We may want to represent it explicitly by defining a context of the form

$$\mathbf{u}(\Gamma^*_{\mathbf{u}}) = < \mathcal{L}_{\mathbf{u}}, \Psi_{\mathbf{u}}, \Gamma_{\mathbf{u}}, \Gamma^*_{\mathbf{u}}, \gamma_{\mathbf{u}}, \eta_{\mathbf{u}} >. \tag{10.51}$$

Such a context will be called an *actualization* of **u**. An argument in **u** can now be regarded as a walk through a sequence of actualizations $\mathbf{u}(\Gamma^*_{\mathbf{u}})$.

Fixed and floating shift operators

In terms of going from one actualization to another, the action of the shift and the return operators can now be interpreted in the following manner. By using a shift operator [**v** : in an argument which is running in **u** at that moment, we step from an actualization $\mathbf{u}(\Gamma^*_{\mathbf{u}})$ of **u** to an actualization $\mathbf{v}(\Gamma^*_{\mathbf{v}})$ with $\Gamma^*_{\mathbf{v}}$ given by

$$\Gamma^*_{\mathbf{v}} = \mathcal{J}_{\mathbf{v}|\mathbf{u}}(\Gamma^*_{\mathbf{u}}). \tag{10.52}$$

This also makes clear why, in a looped labyrinth with $\mathbf{u} \in \Psi_{\mathbf{v}}$ and $\mathbf{v} \in \Psi_{\mathbf{u}}$, there is a difference between a shift back from a context **v** to a former context **u** by using the return operator "]", and a further shift forward to the same context **u** by means of "[**u** :". The action of the return operator will be interpreted to restore the original $\Gamma^*_{\mathbf{u}}$, whereas application of [**u** : would take us to

$$\mathcal{J}_{\mathbf{u}|\mathbf{v}}(\Gamma^*_{\mathbf{v}}) = \mathcal{J}_{\mathbf{u}|\mathbf{v}} \circ \mathcal{J}_{\mathbf{v}|\mathbf{u}}(\Gamma^*_{\mathbf{u}}),$$

which is not necessarily the same.

Occasionally it will be desirable to shift to a context while dropping all previous premises, i.e., we may want to move from **u** to **v** while making $\Gamma^*_{\mathbf{v}}$ equal to $\Gamma_{\mathbf{v}}$ rather than $\mathcal{J}_{\mathbf{v}|\mathbf{u}}(\Gamma^*_{\mathbf{u}})$. To this end we shall introduce a new type of shift operators, *fixed shift operators*, of the form "[**v**| :", whereas the ones introduced earlier will further be called *floating shift operators*. The action of the shift operators, applied within **u** and for a $\mathbf{v} \in \Psi_{\mathbf{u}}$, may thus be summarized as follows:

[**v** :	shift to **v**, with $\Gamma^*_{\mathbf{v}} = \mathcal{J}_{\mathbf{v}\|\mathbf{u}}(\Gamma^*_{\mathbf{u}})$	(10.53)
[**v**| :	shift to **v**, with $\Gamma^*_{\mathbf{v}} = \Gamma_{\mathbf{v}}$	
]	shift to the previous context while restoring the original Γ^*.	

10.1.4 Soundness and completeness

What we require of a logical calculus is that the deducibility relation $\vdash$ should give a good representation of the relation $\models$ of logical consequence, and this leads to two

criteria for the adequacy of the calculus,
Soundness:

$$\text{if } \psi \vdash \phi \text{ then } \psi \models \phi, \tag{10.54}$$

and
Completeness:

$$\text{if } \psi \models \phi \text{ then } \psi \vdash \phi. \tag{10.55}$$

Both criteria require further reflection. In the first place, let us give a brief comment on the soundness criterion. In as far as sentences have a well–defined interpretation there is little room for compromising with soundness, because it is not acceptable that true premises in an argument would lead to false conclusions. Admitting partial interpretations, however, one can also consider the case in which there are types of sentences which remain *hidden*, i.e., which are not interpreted. For such sentences there are no objections against their being deducible in spite of the fact that they are not true under the particular semantics at hand.

With respect to completeness it must be noted that it is a delicate subject on which there is extensive literature, and we shall restrict ourselves to making a few summarizing remarks.

To begin with, we may observe that in principle completeness can be trivially produced by admitting sufficiently many axioms and/or inference rules. We can simply postulate that every true statement is an axiom. The reason for considering this as unacceptable is that we impose comprehensibility requirements on deducibility. These comprehensibility criteria are available in varying degrees of strictness. A rather strict requirement is decidability of the set of deducible statements, i.e., there should be a halting, algorithmic procedure giving us, for every statement in $\mathcal{L}$, an answer to the question whether this statement is deducible or not. This amounts to saying that the set of deducible statements must be *recursive*. Within propositional logic, complete calculi meeting this requirement exist. Within a quantifier logic such as predicate logic, however, the criterion is generally too strict. In order to obtain a complete calculus in first order predicate logic, decidability of the set of deducible statements must be given up. A weaker comprehensibility requirement can still be maintained, however. The set of deducible statements is still *recursively enumerable*: there are procedures which systematically generate the entire set. In a second order predicate logic, if one imposes a standard interpretation, even this fails and one must either drop completeness or comprehensibility of the calculus, at least in this form.

The issue of completeness thus relates to, on the one hand, the comprehensibility of a language and, on the other hand, its ability to describe its subject matter accurately and adequately. Bluntly said, simple systems are easy to describe but complex systems may not admit an accurate and comprehensible description at all. The completeness issue is about the intermediate cases between these extremes and about the exchange between degrees of comprehensibility versus degrees of adequacy of the language. On the syntactic side, comprehensibility is one of the subjects studied in recursion theory. On the semantic side, the complexity of a system relates to topological properties of the state space, such as separability and compactness. Furthermore, the adequacy of a language in describing the system relates to the properties of the algebra of subspaces that is induced by the language on the state space by the interpretation function γ.

With respect to contextual logic, if all local logics within a given labyrinth are complete, one might like to have a theorem like

$$\text{if } [\mathbf{u} : \psi, \cdots [\mathbf{v} : \models \phi \text{ then } [\mathbf{u} : \psi, \cdots [\mathbf{v} : \vdash \phi. \tag{10.56}$$

saying that logical consequence within a connective argument can be represented by a comprehensible deducibility relation. Presumably, however, such a theorem is not generally valid, independent of the details of the labyrinth structure.

Contextual completeness

If a calculus is not complete, in the sense of (10.55), we can look at this from two different points of view. Either the calculus has not enough rules to make all valid statements deducible, or the state space lacks some of the states needed to satisfy all consistent sentences. Taking the latter view, completeness can be formulated as the condition that there are 'enough states'[2]. In this form the concept of completeness is readily generalized into a relation between contexts, and as such the concept has interesting applications for instance for expressing the empirical claims of a theory[3]. We shall distinguish various different types of completeness. In the first place, a context $\mathbf{v}$ will be called weakly complete relative to a context $\mathbf{u}$ iff for every nonempty state in $\mathbf{u}$ there is a corresponding nonempty state in $\mathbf{v}$, i.e.,

Weak completeness:

$$\mathcal{J}_{\mathbf{v}|\mathbf{u}}(\Lambda_{\mathbf{u}}) = \emptyset \quad \text{iff} \quad \Lambda_{\mathbf{u}} = \emptyset, \tag{10.57}$$

where, similar to (10.45):

$$\mathcal{J}_{\mathbf{v}|\mathbf{u}}(\Lambda_{\mathbf{u}}) = \{\pi_{\bar{\mathbf{v}}}(\lambda_{\mathbf{c}}) \in \Gamma_{\mathbf{v}} |\ \pi_{\bar{\mathbf{u}}}(\lambda_{\mathbf{c}}) \in \Lambda_{\mathbf{u}}, \lambda_{\mathbf{c}} \in \Gamma_{\mathbf{c}}\}. \tag{10.58}$$

Weak completeness is rather easily satisfied, for instance if two contexts are entirely independent, in the sense that every state of either context is compatible with every state of the other one. A much stronger condition is obtained by demanding, furthermore, that disjoint states in $\mathbf{u}$ have disjoint representations in $\mathbf{v}$. This form of completeness can be formulated as

Strong completeness:

$$\mathcal{J}_{\mathbf{v}|\mathbf{u}}(\Lambda_{\mathbf{u}}) \cap \mathcal{J}_{\mathbf{v}|\mathbf{u}}(\Lambda_{\mathbf{u}}') = \emptyset \quad \text{iff} \quad \Lambda_{\mathbf{u}} \cap \Lambda_{\mathbf{u}}' = \emptyset. \tag{10.59}$$

Strong completeness is a combination of weak completeness and a condition that we shall call accuracy, and that is expressed by each of the following, equivalent,

[2]Cf. Fourman and Grayson [87].

[3]See §3.3.1.

formulations.
Accuracy: For all $\Lambda_{\mathbf{u}}, \Lambda_{\mathbf{u}}' \subseteq \Gamma_{\mathbf{u}}$:

$$\begin{aligned} \mathcal{J}_{\mathbf{v}|\mathbf{u}}(\Lambda_{\mathbf{u}}) \cap \mathcal{J}_{\mathbf{v}|\mathbf{u}}(\Lambda_{\mathbf{u}}^{-}) &= \emptyset & (10.60)\\ \mathcal{J}_{\mathbf{v}|\mathbf{u}}(\Lambda_{\mathbf{u}} \cap \Lambda_{\mathbf{u}}') &= \mathcal{J}_{\mathbf{v}|\mathbf{u}}(\Lambda_{\mathbf{u}}) \cap \mathcal{J}_{\mathbf{v}|\mathbf{u}}(\Lambda_{\mathbf{u}}') & (10.61)\\ \mathcal{J}_{\mathbf{v}|\mathbf{u}}(\Lambda_{\mathbf{u}}^{-}) &= \mathcal{J}_{\mathbf{v}|\mathbf{u}}^{-}(\Lambda_{\mathbf{u}}) \cap \mathcal{J}_{\mathbf{v}|\mathbf{u}}(\Gamma_{\mathbf{u}}). & (10.62) \end{aligned}$$

Proof: Showing the equivalence of these equations is an elementary exercise in Boolean algebra. From the definition (10.45) of the interpretative mappings we directly obtain:

$$\mathcal{J}_{\mathbf{v}|\mathbf{u}}(\Lambda_{\mathbf{u}} \cup \Lambda_{\mathbf{u}}') = \mathcal{J}_{\mathbf{v}|\mathbf{u}}(\Lambda_{\mathbf{u}}) \cup \mathcal{J}_{\mathbf{v}|\mathbf{u}}(\Lambda_{\mathbf{u}}'). \tag{10.63}$$

We now proceed step by step.
(10.60)⇒(10.61): Using (10.63) and distributivity we have

$$\mathcal{J}_{\mathbf{v}|\mathbf{u}}(\Lambda_{\mathbf{u}}) \cap \mathcal{J}_{\mathbf{v}|\mathbf{u}}(\Lambda_{\mathbf{u}}') = \mathcal{J}_{\mathbf{v}|\mathbf{u}}(\Lambda_{\mathbf{u}} \cap \Lambda_{\mathbf{u}}') \cup \Big(\mathcal{J}_{\mathbf{v}|\mathbf{u}}(\Lambda_{\mathbf{u}} \cap \Lambda_{\mathbf{u}}'^{-}) \cap \mathcal{J}_{\mathbf{v}|\mathbf{u}}(\Lambda_{\mathbf{u}}^{-} \cap \Lambda_{\mathbf{u}}')\Big). \tag{10.64}$$

Because $\Lambda_{\mathbf{u}} \cap \Lambda_{\mathbf{u}}'^{-} \subseteq (\Lambda_{\mathbf{u}}^{-} \cap \Lambda_{\mathbf{u}}')^{-}$ we may apply (10.60). Therefore, the second term on the r.h.s. is equal to $\emptyset$ and (10.61) follows directly.
(10.61)⇒(10.62): Using distributivity and (10.61) we get

$$\begin{aligned} \mathcal{J}_{\mathbf{v}|\mathbf{u}}(\Lambda_{\mathbf{u}}^{-}) &= \Big(\mathcal{J}_{\mathbf{v}|\mathbf{u}}(\Lambda_{\mathbf{u}}) \cup \mathcal{J}_{\mathbf{v}|\mathbf{u}}^{-}(\Lambda_{\mathbf{u}})\Big) \cap \mathcal{J}_{\mathbf{v}|\mathbf{u}}(\Lambda_{\mathbf{u}}^{-}) & (10.65)\\ &= \mathcal{J}_{\mathbf{v}|\mathbf{u}}^{-}(\Lambda_{\mathbf{u}}) \cap \mathcal{J}_{\mathbf{v}|\mathbf{u}}(\Lambda_{\mathbf{u}}^{-})\\ &= \mathcal{J}_{\mathbf{v}|\mathbf{u}}^{-}(\Lambda_{\mathbf{u}}) \cap \Big(\mathcal{J}_{\mathbf{v}|\mathbf{u}}(\Lambda_{\mathbf{u}}) \cup \mathcal{J}_{\mathbf{v}|\mathbf{u}}(\Lambda_{\mathbf{u}}^{-})\Big)\\ &= \mathcal{J}_{\mathbf{v}|\mathbf{u}}^{-}(\Lambda_{\mathbf{u}}) \cap \mathcal{J}_{\mathbf{v}|\mathbf{u}}(\Gamma_{\mathbf{u}}). \end{aligned}$$

(10.62)⇒(10.60): Trivial. □

Furthermore, a proof that strong completeness as defined above is equivalent to weak completeness + accuracy is also immediate.

Even if completeness of **v** relative to **u** holds, in either form, it is still possible that **v** does not apply while **u** does. Applying a nonsensical theory, for instance, can still produce a context that is complete relative to **u**. Completeness merely means that to each state in **u** there is a possible state in **v**, not that there is a necessary correspondence between these states. If we strengthen the completeness concept to this effect, we obtain a concept that we shall call "coverage". There are again two versions, corresponding to weak and strong completeness. Weak coverage can be formulated by:
Weak coverage:

$$\mathcal{J}_{\bar{\mathbf{v}}|\mathbf{u}}(\Lambda_{\mathbf{u}}) = \mathcal{J}_{\mathbf{v}|\mathbf{u}}(\Lambda_{\mathbf{u}}). \tag{10.66}$$

Note that for all $\Lambda_{\mathbf{u}} \supset \emptyset$, $\mathcal{J}_{\bar{\mathbf{v}}|\mathbf{u}}(\Lambda_{\mathbf{u}}) \supset \emptyset$ by definition. Weak coverage is equivalent to weak completeness combined with the demand that **v** applies whenever **u** does, i.e.

$$\{\dagger_{\mathbf{v}}\} \cap \mathcal{J}_{\bar{\mathbf{v}}|\mathbf{u}}(\Gamma_{\mathbf{u}}) = \emptyset. \tag{10.67}$$

If, moreover, accuracy applies, this yields
Strong coverage:

$$\mathcal{J}_{\bar{\mathbf{v}}|\mathbf{u}}(\Lambda_{\mathbf{u}}) \cap \mathcal{J}_{\bar{\mathbf{v}}|\mathbf{u}}(\Lambda_{\mathbf{u}}') = \mathcal{J}_{\mathbf{v}|\mathbf{u}}(\Lambda_{\mathbf{u}} \cap \Lambda_{\mathbf{u}}'). \tag{10.68}$$

If this holds, we shall say that **v** strongly covers **u**, or that **u** is reducible to **v**. This is true, for instance, if **u** is a subcontext of **v**.

Note: Completeness of a theory relative to the class of phenomena it aims to describe is not a quality one would demand from just any theory about an arbitrary subject. Weak completeness is achieved rather easily, for instance by the trivial construction (10.18). However, if such a construction is not applied, then even weak completeness may amount to a rather strong claim. Assuming coverage indicates either a strong belief in the applicability of a theory, or a situation in which the theoretical context is *by definition* applicable to the phenomena which can be described in a given empirical context. Accuracy is a quality which may be very difficult to test if a class of phenomena is sufficiently complex. Generally, however, for a good and serious theory one will aim, on the one hand, to maximize the class of phenomena to which it applies, but, on the other hand, to demarcate the intended class of phenomena in such a way that the theory describes this particular class of phenomena in a complete and accurate way. Indeed, if one has a good theory, arousing sufficient enthusiasm, then one will try to maintain the claim that the theory strongly covers its intended class of phenomena. If this fails, then it certainly counts as an argument against the theory, giving it a tentative, rather than fundamental, status.

With respect to the incompleteness of the standard formalism of QM, however, there is more at stake than the fundamental status of the theory. By maintaining that a quantum system is at all times completely described by a Hilbert space vector, simple facts concerning relevant empirical observations cannot be adequately represented in the theoretical context. For a fundamental physical theory this is unacceptable, but, furthermore, it has the consequence that the standard formalism can account for only a small fragment of scientific practice, where, of course, the empirical observations continue to play a key role. What makes the theory run is, to a large extent, that physicists use all kinds of intuitive ideas which are not represented in the formalism.

10.2 Some applications

We now proceed by indicating some fields where the logical framework developed above may be applied. A first example concerns such matters as formulating notions like "empirical adequacy", "theoretical claims" and other concepts relating to the empirical interpretation of theories. A second example shows how the more standard approaches to modal logic in terms of a many–worlds semantics can be transferred into contextual terminology. A third example outlines some relations with epistemic and constructive logic. Finally, the fourth example very briefly sketches some ideas about a contextual tense logic.

10.2.1 Theoretical claims and predictions

Consider the situation that, through previous work, we have obtained a logical labyrinth that we now want to use for interpreting present experiences and for deriving expectations about future ones. How do we proceed to get along with this? Let us suppose that we have a context **emp** in which we describe our direct observations[4]. Clearly, if we are dealing with sequences of observations and with expectations about future observations, this involves a temporal aspect. Our changing knowledge can be interpreted as a dynamical process taking place within the labyrinth, and we may account for this by considering families of actualizations[5] $\mathbf{emp}(t)$, where t is a time variable [6]:

$$\mathbf{emp}(t) = < \mathcal{L}_{\mathbf{emp}}, \Gamma_{\mathbf{emp}}, \Gamma^*_{\mathbf{emp}}(t), \gamma_{\mathbf{emp}} > . \tag{10.69}$$

This sequence of actualizations describes our increasing knowledge and assuming that, in assembling our observations, we make no mistakes that must later be corrected, we have for all $t \geq t'$:

$$\Gamma^*_{\mathbf{emp}}(t') \subseteq \Gamma^*_{\mathbf{emp}}(t). \tag{10.70}$$

Our knowledge about the observable features of a system is complete if there is only a single elementary state left in $\Gamma^*_{\mathbf{emp}}(t)$.

Let us assume that, apart from the empirical context **emp**, the labyrinth contains a theoretical context **th**, and on the basis of the observations we want a theoretical interpretation. This involves translating the observation results into the theoretical context and computing the corresponding theoretical state. Subsequently, we then want to derive expectations about future observations, and this involves translating back from the theoretical into the empirical context. How does this technically work? How can it be, for instance, that such a twofold translation results in expectations? Where do the presumed restrictions on our future observations originate? Syntactically, the point is that the translation from **emp** to **th** and back involves applying the shift operators "[**th**: " and "[**emp**:", and this can change the actual state. What this amounts to semantically is that by merely translating to the theoretical context and back we throw away states. We shift from the context **emp**

---

[4]or at least what we regard as direct observations at the level we have chosen. Actually, there may be lots of theory involved in assembling them.

[5]See (10.51).

[6]We must take care here not to confuse the dynamics of changing knowledge with knowledge about a dynamical process. These two cases require an entirely different treatment.

to a context **th** that is not complete relative to **emp**; the latter context contains states that have no theoretical representation. In principle such a move could lead to trouble, but we may, nevertheless, admit it by using a theory's claim to empirical adequacy. In order to formulate this claim we introduce a sequence of contexts $\mathbf{pred}(t) \subseteq \mathbf{emp}(t)$, with a state space given by

$$\begin{aligned} \Gamma_{\mathbf{pred}} &= \mathcal{J}_{\mathbf{emp}|\mathbf{th}}(\Gamma_{\mathbf{th}}) \\ \Gamma^*_{\mathbf{pred}}(t) &= \Gamma^*_{\mathbf{emp}}(t) \cap \mathcal{J}_{\mathbf{emp}|\mathbf{th}}(\mathcal{J}_{\mathbf{th}|\mathbf{emp}}(\Gamma^*_{\mathbf{emp}}(t))). \end{aligned} \tag{10.71}$$

Note: On a set of contexts Ψ one may define an algebraic operation "$|$", by means of

$$\begin{aligned} &\mathbf{u}(\Gamma^*_{\mathbf{u}})|\mathbf{v}(\Gamma^*_{\mathbf{v}}) = \\ &\quad < \mathcal{L}_{\mathbf{u}}, \mathcal{J}_{\mathbf{u}|\mathbf{v}}(\Gamma_{\mathbf{v}}), \Gamma^*_{\mathbf{u}} \cap \mathcal{J}_{\mathbf{u}|\mathbf{v}}(\Gamma^*_{\mathbf{v}}), \gamma_{\mathbf{u}} \cap \mathcal{J}_{\mathbf{u}|\mathbf{v}}(\Gamma_{\mathbf{v}}) > . \end{aligned} \tag{10.72}$$

$\mathbf{pred}(t)$ is then defined as $\mathbf{emp}(t)|(\mathbf{th}|\mathbf{emp}(t))$.

Empirical adequacy can now be formulated as the condition that we shall never encounter observations that make $\Gamma^*_{\mathbf{pred}}$ vanish, i.e.,
Empirical adequacy:

$$\Gamma^*_{\mathbf{pred}}(t) \supset \emptyset \quad \text{for all } t. \tag{10.73}$$

Furthermore, the empirical content of a theory, relative to a given body of observations, can be defined as the difference of $\Gamma^*_{\mathbf{emp}}(t)$ and $\Gamma^*_{\mathbf{pred}}(t)$. Accordingly, a theory has empirical content if we have

$$\Gamma^*_{\mathbf{pred}}(t) \subset \Gamma^*_{\mathbf{emp}}(t). \tag{10.74}$$

What empirical adequacy amounts to is that the context **pred**, and *ipso facto* also the context **th**, is complete relative to the set of states that we shall ever observe, i.e., an observed state should always be theoretically possible. Therefore, if the theory has empirical content **emp** contains states that are, in this sense, redundant.

10.2.2 Embedding many–worlds semantics into contextual semantics

Consider a modal language $\mathcal{L}$, with a necessity operator "$\triangle$". A common way to interpret such a language is by means of a many–worlds semantics[7], sometimes called a Kripke semantics, consisting of a set of worlds W with an accessibility relation A defined between them, and an interpretation $\mathcal{I}_w$ for each world w, i.e., we have a tuple

$$< \mathcal{L}, W, A, \mathcal{I} > .$$

Necessity in a world w is now defined as truth in all worlds accessible from w, i.e., for all $\phi \in \mathcal{L}$

$$\mathcal{I}_w(\triangle\phi) = 1 \quad \text{iff} \quad \mathcal{I}_{w'}(\phi) = 1 \quad \text{for all } w' \text{ such that } \ wAw'. \tag{10.75}$$

[7]See, e.g., Kripke [138], Hintikka [119], Lewis [150], Bowen [31].

Given such a semantics for $\mathcal{L}$, one may ask how, and under what conditions, one can find a matching contextual semantics. It is clear that this is not generally possible if the Kripke necessity operators $\triangle$ are translated into contextual necessity operators $\Box$, because for the latter we have, in any context:

$$\Box\phi \Rightarrow \phi \quad \text{and} \quad \Box\phi \Rightarrow \Box\Box\phi, \tag{10.76}$$

as is readily verified. This corresponds to requiring reflexivity and transitivity of the accessibility relation and, therefore, Kripke necessity operators cannot be generally represented by contextual necessity operators.

There turns out to be a much better representation of Kripke necessity within a contextual framework, however, and through its application Kripke semantics can always be embedded in a contextual semantics. This embedding is achieved by means of the following construction of a logical labyrinth: we take a global language $\mathcal{L}_{\mathbf{c}}$, applicable in all contexts, containing $\mathcal{L}$ as a fragment and complemented with the contextual notions introduced earlier. To each world w we define a single state connective context $\mathbf{e}(w)$ with

$$\begin{aligned} \Gamma_{\mathbf{e}(w)} &= \{w\} \\ \Psi_{\mathbf{e}(w)} &= \{\mathbf{e}(w') |\, wAw'\}. \end{aligned} \tag{10.77}$$

An interpretation function $\eta_{\mathbf{e}(w)}$ is constructed by

$$\eta_{\mathbf{e}(w)}(\phi) = \{< w, \mathcal{I}_w > (\phi)\}. \tag{10.78}$$

Within $\mathcal{L}_{\mathbf{c}}$ we further introduce a domain of 'contexts', for which we admit variables and quantifiers, and the meaning within a context $\mathbf{u}$ of a statement "$\forall \mathbf{v}\, \phi$" is given by

$$\gamma_{\mathbf{u}}(\forall \mathbf{v}\, \phi) = \bigcap_{\mathbf{v}\in\Psi_{\mathbf{u}}} \gamma_{\mathbf{u}}(\phi). \tag{10.79}$$

The meaning of the Kripke necessity operator "$\triangle$", applied within the context $\mathbf{u}$, is now given by

$$\triangle\, \phi \equiv \forall \mathbf{v}\, [\mathbf{v}| : \phi]. \tag{10.80}$$

Note: The fixed shift operator[8] "$[\mathbf{v}| :$" must be used here because the floating shift operator is likely to lead to trivial results. $[\mathbf{v} : \phi]$ is trivially true within $\mathbf{u}$ if $\mathcal{J}_{\mathbf{v}|\mathbf{u}}(\Gamma^*_{\mathbf{u}}) = \emptyset$.[9]

The above construction provides a representation theorem for embedding any Kripke semantics into a contextual semantics. *Vice versa*, an interpretation of contextual semantics into Kripke semantics should be possible if the Kripke approach is suitably extended to deal with, in the first place, a network of languages, and, in the second place, compounds of several types of modalities at once, each with its own accessibility relation. For interpreting contextual logic, one would need a compound of two Kripke type modalities, one for representing the structure of the Ψs, the other one for representing the Γs.

[8]See (10.53)

[9]A floating shift operator *can* be used, however, if we apply the construction $[\bar{\mathbf{v}} : \Box(T\mathbf{v} \Rightarrow \phi)]$, where $\bar{\mathbf{v}}$ is the trivially complete extension of $\mathbf{v}$, see (10.18).

10.2.3 Epistemic logic

If we consider the problem of representing knowledge, or belief, or some other similar notion, about a system within a context $\mathbf{u}$ in which this system is described, then it seems natural to assume that knowlege can be formulated in a set of sentences $\phi_1 \cdots \phi_n$, interpreted within the state space $\Gamma_{\mathbf{u}}$ by means of the intersection of the states $\gamma_{\mathbf{u}}(\phi_1) \cdots \gamma_{\mathbf{u}}(\phi_n)$. Thus one might represent knowledge by an actualization $\mathbf{u}(\Gamma^*_{\mathbf{u}})$ of $\mathbf{u}$, with $\Gamma^*_{\mathbf{u}}$ equal to

$$\Gamma^*_{\mathbf{u}} = \gamma_{\mathbf{u}}(\phi_1) \cap \cdots \cap \gamma_{\mathbf{u}}(\phi_n). \tag{10.81}$$

In simple cases this may work quite well, but more generally it is a rather optimistic view on how much we can know and how well we are in fact able to handle knowledge. Suppose, for instance, that we know for a fact that a statement ϕ is true, then what does this tell us about the truth of another statement ψ? If the state space is sufficiently simple and the language sufficiently adequate, we may be able to compute an answer by calculating whether perhaps $\gamma(\phi \wedge \psi)$ or $\gamma(\phi \wedge \neg\psi)$ is empty. In a more complex case, however, this may be practically, or even fundamentally impossible. It may occur, for instance, that the two statements are contradictory without our being aware of it. This means that, even though we have in fact

$$\gamma(\phi \wedge \psi) = \emptyset, \tag{10.82}$$

an adequate representation of our knowledge would require

$$\Gamma^* \cap \gamma(\phi \wedge \psi) \supset \emptyset. \tag{10.83}$$

Evidently, these two conditions cannot be simultaneously satisfied and, accordingly, an epistemic logic, in which one tries to give an adequate representation of knowledge, must be based on a more general logical framework. In order to make it possible to satisfy eq.(10.83) states would have to be added to Γ in such a manner that eq.(10.82) is no longer satisfied. This would mean at the same time, however, that the added states would have to be very peculiar ones, violating a condition called *semantic compositionality*.

Semantic compositionality is the condition that the interpretation of a composite sentence is determined entirely by the interpretation of the atomic sentences of which it is composed. This condition is established by giving rules, such as $\gamma_{\mathbf{u}}(\phi \wedge \psi) = \gamma_{\mathbf{u}}(\phi) \cap \gamma_{\mathbf{u}}(\psi)$, saying *how* the interpretation the composite sentences depends on the interpretation of the atomic ones. If we want the state $\gamma_{\mathbf{u}}(\phi \wedge \psi)$ to give an adequate representation of our knowledge, however, then we must account for the possibility that we have not computed $\gamma_{\mathbf{u}}(\phi) \cap \gamma_{\mathbf{u}}(\psi)$, and that, in fact, it may not even be computable, and this means that we have to look for a logical system in which semantic compositionality can be violated.

In order to obtain a logical framework in which such a thing is possible without contradictions, we consider a representation of the state space as a set of valuations on $\mathcal{L}$. A valuation is an assignment of truth values to the formulas of a language.

Every state $\lambda \in \Gamma$ of $\mathcal{L}$ has an associated valuation $\mathcal{I}_\lambda : \mathcal{L} \rightarrow \{0, 1\}$ on the statements of $\mathcal{L}$, with

$$\begin{aligned} \mathcal{I}_\lambda(\phi) &= 1 \text{ if } \lambda \in \gamma(\phi) \\ &= 0 \text{ otherwise.} \end{aligned} \tag{10.84}$$

We may now consider Γ as merely a set of such valuations. For the proper states in Γ the associated valuations satisfy the standard requirements of semantic compositionality. It is easy, however, to imagine what it means to admit valuations that do not satisfy semantic compositionality. An example is a valuation that assigns truth value 1 to statements ϕ and ψ, but 0 to $\phi \wedge \psi$. Let us call such valuations *noncompositional valuations.* They are peculiar objects in the sense that, contrary to the proper valuations, they cannot be considered to represent mathematical structures.

Let us now imagine a mathematician who is working on some mathematical theory, formalized within a formal language $\mathcal{L}$. We may interpret this language within a context $< \mathcal{L}, \bar{\Gamma}, \bar{\gamma} >$, where $\bar{\Gamma}$ is a 'state space' that includes noncompositional valuations. As the mathematician sets out to prove more and more theorems, his increasing knowledge can be represented by a sequence of states $\bar{\Gamma}^*_1$, $\bar{\Gamma}^*_2$, $\bar{\Gamma}^*_3, \cdots$. If all his theorems are correct, this sequence will decrease monotonically towards the set Γ of proper states that do satisfy compositionality. In fact, his whole enterprise of proving theorems can be considered as an attempt to enforce semantic compositionality upon all states. If $\mathcal{L}$ is sufficiently rich, however, this may take a long time and there may even be formulas for which it cannot ever be decided whether they are theorems or not. In the meantime, knowledge is represented more accurately by a state $\bar{\Gamma}^* \subseteq \bar{\Gamma}$ than by a state $\Gamma^* \subseteq \Gamma$.

This approach can be developed into a proof–theoretical semantics by giving a set of proofs, $\mathcal{B}$, as an argument of $\bar{\Gamma}^*$:

$$\bar{\Gamma}^*(\mathcal{B}) = \{\lambda \in \bar{\Gamma} \,|\, \mathcal{I}_\lambda(\phi) = 1 \text{ if } \mathcal{B} \text{ contains a proof of } \phi \,\}. \tag{10.85}$$

This has indeed the effect that, as more and more proofs become available, the state $\bar{\Gamma}^*$ decreases monotonically towards Γ. Furthermore, if $\mathcal{C}$ is a set of proofs for *all* statements deducible in $\mathcal{L}$, then the logic $< \mathcal{L}, \Gamma, \gamma >$ is complete precisely iff $\bar{\Gamma}^*(\mathcal{C}) = \Gamma$.

I would suggest that an appropriate name for a logical theory concerning contexts of the type $< \mathcal{L}, \bar{\Gamma}, \bar{\Gamma}^*(\mathcal{B}), \bar{\gamma} >$ might be "constructive classical logic", as opposed to the intuitionistic approach which is standard in this field. The relation with intuitionistic logic can be established by taking a state space of the above type and imposing a specific interpretation of the connectives. We impose that $\mathcal{B}$ contains a proof of

$$\begin{aligned} \phi \vee \psi &: \text{ only if } \mathcal{B} \text{ contains a proof of } \phi \text{ or one of } \psi; \\ \phi \wedge \psi &: \text{ only if } \mathcal{B} \text{ contains a proof of } \phi \text{ and one of } \psi; \\ \phi \Rightarrow \psi &: \text{ only if } \mathcal{B} \text{ contains a proof of } \psi, \text{ with } \phi \text{ as a} \\ &\quad \text{premise;} \end{aligned} \tag{10.86}$$

and there are similar restrictions for the remaining connectives and the quantifiers. These rules characterize a Heyting semantics, and the structure so defined on the state space $\bar{\Gamma}^*$ imposes restrictions on the logical laws that are acceptable in a proof. The "law of the excluded middle", for instance,

$$\phi \vee \neg\phi, \tag{10.87}$$

is not acceptable because, by the rules stated above, it has a proof only if there is a proof of either ϕ or $\neg\phi$. By virtue of these restrictions, Heyting semantics provides an appropriate framework for interpreting Brouwer's intuitionism.

10.2.4 Tense logic

Tensed languages are languages containing special operators for dealing with temporal phenomena. The use of different tenses of a verb may be interpreted as being an instance of such an operator working on a sentence. For instance, the sentence "John is happy now but he wasn't yesterday" admits a formal representation of the form

$$\triangle_{\mathbf{now}} Hj \wedge \triangle_{\mathbf{yesterday}} \neg Hj, \tag{10.88}$$

including the two tense operators $\triangle_{\mathbf{now}}$ and $\triangle_{\mathbf{yesterday}}$.

Let us, in the first place, note that the use of tense operators for describing temporal phenomena is not strictly necessary; one may well do without them. Alternatively, one may simply take a state space consisting of a set of possible histories, and describe these as objects without introducing any special operators at all. Consider some set $\Gamma_{\mathbf{proc}}$ of alternative processes, i.e. possible histories. Each process may be described as a shape in some space including a time dimension. For instance, a particle may be described by its world line in spacetime, in more or less the same way as one might describe a string of spaghetti in 3–space. Neither of these cases strictly requires the introduction of special logical operators. Accordingly, tense operators apparently exist either for brevity and convenience, or for traditional reasons. Indeed, our procedure for introducing tense operators will be to start with a single context, with a semantics in terms of a state space of possible histories, i.e., processes, and a syntax without tense operators. Then we proceed by introducing contextual structure, in such a way that the corresponding contextual operators (connective and/or intensional) are tense operators.

In the special case that one is describing the evolution of a system with continuing existence, during some period $[0, T]$, the state space $\Gamma_{\mathbf{proc}}$ of possible histories may be generated by considering the possible trajectories through a system state space $\Gamma_{\mathbf{sys}}$. For instance, let $\Gamma_{\mathbf{sys}}$ be the phase space of some system of classical mechanics, then in order to obtain $\Gamma_{\mathbf{proc}}$ we take the set of continuous functions $[0, T] \rightarrow \Gamma_{\mathbf{sys}}$, and delete all trajectories which are considered impossible because of dynamical laws pertaining to the system. At the syntactic level, deleting impossible states may be represented by using the modal operator $\Box$ for formulating these dynamical laws. The use of $\Box$ is implicit in stating the dynamical laws as axioms in a theory, and can be made explicit by applying the logical inference rule of *necessitation*, saying that from ϕ one may infer $\Box\phi$ if the deduction of ϕ does not depend on any open premises. For instance, in classical mechanics one may thus choose to formulate the Hamilton equations with a prefixed modal operator, i.e.,

$$\Box(dq/dt = \partial H/\partial p), \quad \Box(dp/dt = -\partial H/\partial q), \tag{10.89}$$

and via (10.9) this implies that only trajectories satisfying these equations remain in $\Gamma_{\mathbf{proc}}$.

The introduction of tense operators

Consider a context $\mathbf{u}$ in which a class of temporal phenomena is described in a tenseless way, as explained above, with the state space $\Gamma_{\mathbf{u.proc}}$ containing the set of possible processes in a time interval $[0, T]$. We may now introduce contextual structure by considering subintervals, and by defining to such a subinterval $A \subseteq [0, T]$ a new context $\mathbf{u.per}(A)$, with $\Gamma_{\mathbf{u.per}(A)}$ containing the possible subprocesses for this period. For the language $\mathcal{L}_{\mathbf{u.per}(A)}$ some syntax may be assembled at convenience, requiring only the satisfaction of local criteria of adequacy.

Consider a set Ψ of contexts formed in this way. One may now construct a connective context $\mathbf{c}$ with $\Psi_{\mathbf{c}} = \Psi$, and in this context $\mathbf{c}$ tense operators arise immediately, in the form of the connective and intensional operators introduced earlier. For instance, the sentence "John is happy now but he wasn't yesterday" may be formulated in $\mathbf{c}$ either as

$$[Hj]\mathbf{u.per}(\text{now}) \;\wedge\; [\neg Hj]\mathbf{u.per}(\text{yesterday}), \text{ or as} \tag{10.90}$$

$$[\mathbf{u.per}(\text{now}) : Hj] \;\wedge\; [\mathbf{u.per}(\text{yesterday}) : \neg Hj]. \tag{10.91}$$

The choice between using intensional or connective operators may generally depend on the details of the syntax, but is for the rest largely immaterial.

> **Note:** This way of defining tense operators is very flexible. It is easy, for instance, to adapt the meaning of the operators to specific physical theories, such as special or general relativity, and entirely the same procedures may be applied for introducing operators referring to specific locations in space, or to regions of spacetime. Indeed, the emergence of tense operators within the framework of contextual logic is so natural and straightforward that one might question the convenience of developing tense logic as a special branch of logic at all. Moreover, discarding the very idea of tense logic now, may save us from the prospect of having special logics for left and right, for up and down, and for any other place or direction in spacetime. Accordingly, I do not agree, in this respect, with Van Benthem, who, in the preface of his book about the logic of time[10], seems rather happy with its development into a mature branch of logic.

[10] [22].

Identity relations

State spaces were introduced above as sets of possible worlds. Generally, these worlds will be full of structure. For instance, imagine a context in which we describe the solar system, then the state space may consist of a set of mental representations of the solar system, with the planets following different orbits in each of the states. This may lead us to question in what sense we can say that a planet, called Earth in one state, is the same object as one called Earth in another state. Suppose, for example, that we regard a state in which Earth, in the sense of the blue watery lukewarm planet, is moving in what we presently regard as the orbit of Mars, whereas Mars, the red arid cold planet, is following the current orbit of the Earth. Must we then say that Mars and Earth have changed position, or must we say that in this world Mars is a blue watery lukewarm planet and Earth is red, arid and cold?

Clearly, the most satisfactory answer to this question seems to be that we 'must' do nothing; we are simply free to do as we like. This also raises a problem, however. Apparently, a context is an object full of conventional identity relations, of the sort: object a in state λ = object a' in state λ'. The extra information contained in these conventions cannot simply be neglected, however. Some information is stored in the way in which we choose to assign names to objects. For instance, in the above case we may decide to stick to the rule that Earth is blue, watery and lukewarm. However, this information is incomplete if we do not, at the same time, remember to what extent, and in what way, the naming procedure is considered to be arbitrary. Therefore, the *ontology* of a context, containing the rules for the interpretation of the nonboolean terms, may have a much more complex structure than one might expect at first sight.

The same phenomenon also appears if we regard contextual structure. For instance, returning to the subject of tense logic, it may be questioned whether the John who is happy today is the same person as the unhappy John of yesterday. To some extent, it is implicit in our using the same name that we do regard this to be the case, but basically the matter remains conventional, and circumstances may occur in which this fact is put to use. For instance, his mother, upon seeing a him smile, might say something like: "I can see that you're a quite different person today", and John might react affirmatively. This shows that our use of such conventional identity relations is quite flexible. Even though, in practical cases, we are generally very fast in developing rules for standard application, we easily grasp the meaning of slightly nonstandard usage.

Within the framework of contextual logic we may analyze this kind of ambiguity in the following manner: associated with each possible pattern of identity conventions we may imagine a context in which rules are fixed in this particular way. Communication between people consists, to a certain, and presumably quite large extent, in choosing and fixing the contexts in which communication is to proceed. Most people are skilled 'context hoppers', however, and upon noticing a context shift they swiftly react, either by trying to follow the speaker to another context, or by conveying some utterance of disagreement. Looking at arbitrary fragments of daily conversations from this angle, may make it clear that we are, indeed, highly skilled in context shifting, but also that the process remains complex and that we are quite ready to take it to the limits of our capabilities, resulting in fair levels of miscommunication about which usually nobody cares.

The flow of time

Having set up a contextual framework for tense logic, we proceed by relating it to ideas about the 'flow of time'[11]. Our awareness of temporal phenomena is coupled to the circumstance that the future is unknown to us, in the sense that we have no direct recollection of it, whereas the past is to some extent known, and at all times we experience the 'now' as a moving point of culmination of conciousness, at which a transition from future to past is taking place. The now floods us each moment with with an awesome instream of information, part of which is picked up and processed, and lingers on in our memory for some time.

The subjective account

As our awareness of the flow of time may be regarded as referring primarily to our changing knowledge of the world, considerations about it may be based on a representation of what knowledge is. Making an adequate theory about the structure of knowledge is a very nontrivial business, but simplefied models are readily assembled. Within a given context, knowledge about an *actual world* may be represented by the set $\Gamma^* \subseteq \Gamma$ of possible worlds compatible with what we know about this actual world. The more we know about it the smaller Γ^* will be; and if, within the given context, our knowledge is complete, this means that only one possible world is left. Associated with the set Γ^* we may define a modal operator $\Box^*$, with $\Box^*\phi$ expressing that, according to our knowledge, ϕ must be the case, i.e., $\Gamma^* \subseteq \gamma(\phi)$. This may be arranged by defining

$$\begin{aligned} \gamma(\Box^*\phi) &= \gamma(\phi) \text{ if } \Gamma^* \subseteq \gamma(\phi) \\ &= \emptyset \text{ otherwise.} \end{aligned} \tag{10.92}$$

Evidently, we do not all know the same things, so a different $\Box^*$ should be assigned to each of us. Moreover, our knowledge changes in time, so, instead of a single $\Box^*$ per person, we should give everybody a family of operators $\Box^*(t)$, with t referring to time.

The time–dependence of $\Box^*$ gives a representation of the flow of time, in as far as the flow of time can indeed be regarded as a phenomenon relating to our changing knowledge of the world, and in as far as modeling knowledge by means of a subset $\Gamma^* \subseteq \Gamma$ is a more or less adequate way of doing it.

Clearly, in this analysis the flow of time is pictured as a subjective phenomenon[12]. As everybody has his own conciousness, everybody has his own flow of time. We

[11] Kroes [139] p. 113, Grünbaum [105] p. 197 ff, Dieks [65], Muller [164].

[12] In agreement with, e.g., Weyl [217] p. 116, see also Kroes [139] p. 63.

may, at this point, also quote Bertrand Russell, who wrote[13]:

> [...] asking ourselves why our feelings towards the past are so different from our feelings towards the future. The reason for this difference is wholly practical: our wishes can affect the future but not the past, the future is to some extent subject to our power, while the past is unalterably fixed. But every future will some day be past: if we see the past truly now, it must, when it was still future, have been just what we now see it to be, and what is now future must be just what we shall see it to be when it has become past. The felt difference of quality between past and future, therefore, is not an intrinsic difference, but only a difference in relation to us: to impartial contemplation, it ceases to exist.

This view is not generally approved of, however, and this is a touchy subject because of its links with debates about determinism and free will. The idea behind the use of state spaces and modalities, in the way in which these were introduced above, is that they relate to the way in which *we* conceptualize our knowledge. It is assumed that when a given context is used for the description of a given system there is a single actual state applying to the system. The other states in the state space can be thought of as representing a possible lack of knowledge. They do not have an objective relevance relating to *this* particular system, but merely provide a contrast class, containing information about how the system might have been if it were a different system of a similar type. Accordingly, state spaces and modalities refer to *classes* of possible systems, rater than singular ones, and our use of them is associated with a way of representing knowledge which is, in some respects, rather primitive but, nevertheless, quite effective. Applying this idea to the history of the world as such, the upshot is that there is only one actual history of the world; other possible histories we may conceive of exist only in our minds. For an ideal mind knowing everything, a state space containing only the single actual history of the world would be sufficient. This has the consequence that, for such an ideal mind, truth and necessary truth, as defined in terms of the natural necessity operator, would coincide. Everything which happens, happens necessarily. This might lead to the view that, objectively, the history of the world is in some sense already determined, and freedom of choice and free will are an illusion.

One may react to this line of reasoning in various different ways. The most sensible reaction, in my view, is that free will relates to the choice between possibilities conceived in our minds. We choose between *imagined* courses of action, but this always results in a single actual history, which may or may not coincide with what we imagined it to be. Accordingly, there is no incompatability as such between free will and the uniqueness of the actual course of history, and our use of temporal modalities originates in the fact that *we* are not the ideal minds who know everything. This does not tell us what free will is, or whether it exists at all, but one cannot expect answers to such questions, which are deeply related to the

[13]Russell [185] §III.

mind–body problem, to flow spontaneously from a logical framework or a theory of physics.

Further, this view by itself neither explains why we have recollection only of the past, but not of the future. An answer to this problem, however, *might* be produced by a physical theory, on the assumption that a memory is an ordinary physical system satisfying constraints such as the second law of thermodynamics. The observed time–asymmetry in physics may have a cosmological explanation[14]

The objective account

A quite different reaction to the problem of free will is to assume that recollection of the future is fundamentally impossible *because* the future is in some ontic sense *open.* In this case, physical reality is pictured as a continuously growing entity. The past exists and is fixed, the future is some kind of ocean of potentialities, and the *now* is the surface for a 'crystallization' process where new facts boom into existence. The transformation of potential into actual is, thus, pictured as an objectively real physical process. Associated with this physical actualization process, one may define a family of modal operators $\Box^*(t)$, as before, but these operators now have a fundamental physical meaning; they refer to *ontic modalities* and are conceived to have a meaning, entirely independent of our own knowledge and conceptualization of the world. This also implies that the associated semantics is much more complicated, involving, for each moment t, the physically real existence of an objectively fixed past and a set of possible futures.

Example: Consider a state space $\Gamma_{\mathbf{sys}}$ of some system evolving in time. Its dynamics, in a time interval $[0, T]$, is contained in the structure of a corresponding process space $\Gamma_{\mathbf{proc}}$, being some set of trajectories through $\Gamma_{\mathbf{sys}}$, i.e., each $\lambda_{\mathbf{proc}} \in \Gamma_{\mathbf{proc}}$ is a function $\lambda_{\mathbf{proc}} : [0, T] \to \Gamma_{\mathbf{sys}}$. At an intermediate moment $0 < t < T$, physical reality cannot be assumed to correspond to one particular, although perhaps unknown, state $\lambda_{\mathbf{proc}}$, because the upper part of the trajectory has not 'crystallized' into existence yet. At the moment t, physical reality contains the fixed past of the system and some set of potential futures. In terms of states, this may be analyzed in the following manner: the state of physical reality at t is described in a context $\mathbf{pr}(t)$, with a corresponding state space $\Gamma_{\mathbf{pr}(t)}$. Each state $\lambda_{\mathbf{pr}(t)} \in \Gamma_{\mathbf{pr}(t)}$ may now be conceived to consist of a set of trajectories $\lambda_{\mathbf{proc}}$, such that, for all times $t' \leq t$, these trajectories coincide with the fixed past of the system. In this way, a specific $\lambda_{\mathbf{pr}(t)}$ indeed provides us with a fixed past and a set of possible futures of the system.

Mathematically, $\Gamma_{\mathbf{pr}(t)}$ may be constructed by first defining a family of functions $f_t : \Gamma_{\mathbf{proc}} \to \wp(\Gamma_{\mathbf{proc}})$ by

$$f_t(\lambda_{\mathbf{proc}}) = \{\lambda'_{\mathbf{proc}} \mid \forall t' \leq t \;\; \lambda'_{\mathbf{proc}}(t') = \lambda_{\mathbf{proc}}(t'), \tag{10.93}$$

[14]My own favourite attempt at an answer is contained in § 7.6.1.

and then $\Gamma_{\mathbf{pr}(t)}$ itself by

$$\Gamma_{\mathbf{pr}(t)} = \{f_t(\lambda_{\mathbf{proc}}) \mid \lambda_{\mathbf{proc}} \in \Gamma_{\mathbf{proc}}\}. \tag{10.94}$$

Finally, the history of the system, in terms of its physically real state, may be described by constructing a state space $\Gamma_{\mathbf{pr.proc}}$, consisting of the set of all possible trajectories through the sequence of $\Gamma_{\mathbf{pr}(t)}$s.

Special relativity has been used as an argument against views of this type[15], because of the nonexistence of a preferred simultaneity hyperplane which can serve as an objective *now*. However, such arguments do not hold, because there is no reason to demand that the *now* must be the simultaneity hyperplane of any specific observer. Any timelike hypersurface would do[16]. In fact, one might even question whether it is, indeed, necessary for the hypersurface to be timelike. The assumption behind this is that, for causality reasons, no unfixed area can be located within the past lightcone of any fixed area, because, otherwise, a sentient being with free will might choose to perform, within the unfixed area, some action which would be incompatible with the already actual history in the fixed area. However, let us imagine the existence of an ideal sentient being, Johnnie, with free will and with total recollection of all fixed events, either in the past or in the future, with "past" and "future" here in the sense of referring to coordinate time. Suppose Johnnie has eaten porridge this morning, but he likes ham and eggs better. One might suggest that he should use his free will to choose what he has eaten this morning, but, presumably, he will dissmiss the idea as absurd, because this morning's history is already fixed and he can no longer have any influence on it. Now suppose that by some bend in the actualization surface this evening's history is also fixed, and, according to Johnnie's recollection, or rather his *pre*collection, he will eat spinach tonight. As Johnnie hates spinach, we might again suggest that he use his free will and choose a different meal. Johnnie, however, will once again dissmiss the idea as absurd, because he knows very well that it is quite impossible to undertake any actions that are incompatible with fixed history, and his prememory of it.

To us, Johnnie's attitude with respect to this evening's dinner may seem fatalistic. This is only because we ourselves do not have ideal minds, however. *We* remember only past events, and, presumably, there are good grounds for why our memories, as physical systems, behave in this way. *We* live with the, perhaps illusory, idea that the entire future is open, but we cannot know this. In fact, all we have is the rather weak compatibility requirement that, in as far as our memory is correct, all events we can recollect must refer to fixed history. This leaves a lot of room. Not only is it impossible for us to know that the future is entirely open, but we do not even know that the past is entirely fixed.

[15]E.g., Putnam [174], Rietdijk [180, 179].
[16]See Dieks [65].

Evaluation

Surveying the above analysis, I must say that I do not quite see the benefit of using heavy equipment, such as the introduction of ontic modalities, for expressing such vague notions as the idea that parts of history are in some deep sense 'fixed'. Further, it may be noted that time–asymmetry is imposed, in a sort of 'sneaky' way, via a construction in logic. This seems a rather suspect way of doing it. Accordingly, I would suggest that, if one wants ontic modalities at all, they should be initially introduced via a time–symmetric procedure. Time–asymmetry, if desired, may then be introduced more explicitly at a later stage, in close correspondence with a specific physical theory.

For the time being, however, I see little reason for going into all this trouble, just for imposing antropomorphic notions about the openness of the future on the structure of physical reality. I see the subjective account of the 'flow of time', and of the difference between past and future, as much more appealing. There seems nothing wrong with it, and, as the Americans say: "If it ain't broken, don't fix it." Such considerations may, however, depend to some extent on one's favourite theory of physics. In particular, if one has a preference for an interpretation of quantum mechanics in which the state reduction occurs as a physically real process, then one might think of it as providing just the kind of crystallization of potential into actual that is described above, by the ontic version of the operator $\Box^*(t)$. We shall not elaborate on this topic, but the example shows that we are dealing with an issue that is not strictly isolated but has at least some links with other fields in physics and philosophy[17], and which, in fact, is closely related to some of the main topics of the present work. There may well be an ocean of potential quantum theories containing ontic modalities.

A quite different point, which may, nevertheless, bear some responsibility for the fact that the present issue has risen at all, is that, in order to speak about future events or to speak about spacetime as in relativity theory, one must be working in a context in which these events are well–defined. Within such a context, the possible histories which are discussed must have a definite existence. If we are discussing future events now, this might lead to the confusing idea that these future events, too, must already exist now. This, however, is an unwarranted conclusion, because one may easily create contexts where totally imaginary events can be discussed apparently meaningfully, in spite of the fact that they do not exist, have never existed, and never will exist in physical reality. In essence, existence must be regarded as a context dependent concept, and there are limits to what concepts may be transported meaningfully from one context to another. In the present case, we may distinguish between two contexts, a context **st** relating to the totality of spacetime and a context **st.per**(now), relating to some small part of it. Existence of some event in **st.per**(now) should imply its existence in **st**, but the reverse is obviously false. Existence of a king of France somewhere in spacetime does not imply the existence of a present king of France.

[17] See also Stein [196] and Maxwell [159].

We may conclude that the existence of our discussion of future events in the now–context has no implications for the existence of these future events themselves in the now–context. Accordingly, the entire issue of history being already fixed may be regarded as originating from an attempt at transporting entities from one context to another, in a situation where this is illegal. This may serve to illustrate that, even though we generally seem to be very skilled in shifting contexts, it remains a difficult matter, and confusion arises easily. This indicates that, within science, a formalization of context shifting may be desirable, and I would suggest that contextual logic provides a relatively straightforward, and quite effective logical framework for doing this.

10.3 Relevance, truth, reality

Relevance, truth, and reality will be introduced here as three very closely related concepts which help to organize the structure and the flow of information within a labyrinth. Precisely how they are used to achieve this depends on the situation. We shall start by considering a fairly straightforward type of situation in which reality is supposed to be given and the other concepts can be defined in terms of it. Later on, we shall consider the much more complex case in which the roles are reversed, and truth and relevance serve as a basis for the construction of a reality.

Imagine a connective context $\mathbf{c}$, with $\Psi_{\mathbf{c}}$ containing a context $\mathbf{r}$ which is accepted as representing a reality for $\mathbf{c}$. $\Psi_{\mathbf{c}}$ may further contain any amount of other contexts, such as contexts in which observations are described or which contain simplified models of certain aspects of reality. Based on the given reality $\mathbf{r}$ we now introduce within $\mathcal{L}_{\mathbf{c}}$ a *truth predicate* T, satisfying the condition that $T(\phi)$ is true iff ϕ is meaningful in $\mathbf{r}$, i.e.[18], $M_{\mathbf{r}}(\phi)$, and true with respect to $\mathbf{r}$, i.e., $[\phi]\mathbf{r}$. Within $\mathcal{L}_{\mathbf{c}}$ this can be formulated by

Truth:

$$T(\phi) \equiv (M_{\mathbf{r}}(\phi) \wedge [\phi]\mathbf{r}). \tag{10.95}$$

We further introduce a binary relevance predicate R, with $R_{\mathbf{u}}(\phi)$ meaning something like: "What the context $\mathbf{u}$ has to say about ϕ is relevant to the subject, because this is how it really is." More specifically, the concept is defined by

Relevance:

$$R_{\mathbf{u}}(\phi) \equiv ((M_{\mathbf{u}}(\phi) \wedge M_{\mathbf{r}}(\phi)) \wedge ([\phi]\mathbf{u} \equiv [\phi]\mathbf{r})). \tag{10.96}$$

As an abbreviation we also introduce a unary predicate R by

$$R(\phi) \equiv \exists \mathbf{u}\ R_{\mathbf{u}}(\phi), \tag{10.97}$$

where the range of the quantifier is understood to be $\Psi_{\mathbf{c}}$.

[18]See (10.33).

Some theorems

We have now obtained truth and relevance in terms of reality. Before considering a construction the other way around we shall first prove some useful theorems.

$$T(\phi) \equiv \exists \mathbf{u}(R_{\mathbf{u}}(\phi) \wedge [\phi]\mathbf{u}) \tag{10.98}$$

Proof: From left to right: substitution of $\mathbf{u} = \mathbf{r}$ in (10.96) provides an example. From right to left: using (10.96) we obtain both $M_{\mathbf{r}}(\phi)$ and $[\phi]\mathbf{r}$, and, therefore, $T(\phi)$. □

$$R_{\mathbf{u}}(\neg\phi) \equiv R_{\mathbf{u}}(\phi) \tag{10.99}$$

Proof: From the definition (10.33) of M then follows

$$M_{\mathbf{u}}(\phi) \equiv M_{\mathbf{u}}(\neg\phi).$$

From $(M_{\mathbf{u}}(\phi) \wedge M_{\mathbf{r}}(\phi))$ follows

$$([\phi]\mathbf{u} \equiv [\phi]\mathbf{r}) \equiv ([\neg\phi]\mathbf{u} \equiv [\neg\phi]\mathbf{r}),$$

and the desired result follows directly from (10.96). □

$$(R_{\mathbf{u}}(\phi) \wedge R_{\mathbf{v}}(\phi)) \Rightarrow ([\phi]\mathbf{u} \equiv [\phi]\mathbf{v}) \tag{10.100}$$

Proof: Follows directly from (10.96). □

$$R(\phi) \equiv M_{\mathbf{r}}(\phi) \tag{10.101}$$

Proof: From left to right: $R(\phi)$ is defined to mean that there is a $\mathbf{u}$ such that $R_{\mathbf{u}}(\phi)$. $M_{\mathbf{r}}(\phi)$ then follows immediately from (10.96). From right to left: given $M_{\mathbf{r}}(\phi)$ it follows from (10.96) that $R_{\mathbf{r}}(\phi)$, and, accordingly, also $R(\phi)$ □

$$(R(\phi) \wedge R(\psi)) \Rightarrow R(\phi \vee \psi) \tag{10.102}$$

Proof: By virtue of the preceding theorem $R(\phi)$ is equivalent to $M_{\mathbf{r}}(\phi)$. From (10.34) follows

$$(M_{\mathbf{r}}(\phi) \wedge M_{\mathbf{r}}(\psi)) \Rightarrow M_{\mathbf{r}}(\phi \vee \psi).$$

□

$$R(\phi) \equiv (T(\phi) \vee T(\neg\phi)) \tag{10.103}$$

Proof: From left to right: $R(\phi)$ is equivalent to $M_{\mathbf{r}}(\phi)$. From $M_{\mathbf{r}}(\phi)$ follows

$$M_{\mathbf{r}}(\phi) \wedge ([\phi]\mathbf{r} \vee \neg[\phi]\mathbf{r}),$$

because the second part is a tautology, and

$$(M_{\mathbf{r}}(\phi) \wedge [\phi]\mathbf{r}) \vee (M_{\mathbf{r}}(\phi) \wedge \neg[\phi]\mathbf{r}),$$

because of distributivity. $T(\phi) \vee T(\neg\phi)$ then follows directly from (10.95). From right to left is trivial. □

10.3.1 Building a reality

With reality given, the definition of truth and relevance is fairly straightforward. Let us now consider the case that, *vice versa*, reality is the thing to be constructed. Imagine the kind of messy situation which may be typical for the conditions in which we usually have to operate. We have some set of observations, some of which may be good, and some unreliable, and we have some set of theoretical contexts, each one also with its more and its less reliable sides. Can we build a reality out of this?

Clearly the present situation is much more complicated than the preceding one. Although in specific cases a natural and straightforward construction may exist, I do not see that there would be a general recipe leading to guaranteed success, and it is not even clear how to generally define "success" here. It rather depends on what kinds of criteria one has for being satisfied with whatever one is able to achieve. In any case, however, the project would have to include some way of constructing a suitable language $\mathcal{L}_{\mathbf{c}}$, and checking whether ideas one has about the reliability of the various member contexts can indeed be developed into a coherent relevance concept, satisfying conditions like (10.99), (10.100), and (10.102). As a next step, if one lacks more ingenuous ways to proceed, a state space may be constructed by considering valuations on the set of relevant sentences $\mathcal{L}_{\mathbf{c}}^R$. More in particular, let $\mathcal{G}_{\mathbf{c}}^R$ be a generating set for $\mathcal{L}_{\mathbf{c}}^R$, preferably the smallest possible one, in the sense that every sentence in $\mathcal{L}_{\mathbf{c}}^R$ is compounded from sentences in $\mathcal{G}_{\mathbf{c}}^R$.

Note: $\mathcal{G}_{\mathbf{c}}^R$ is not necessarily the set of atomic sentences of $\mathcal{L}_{\mathbf{c}}^R$, because one may, for instance, have $R(\phi \wedge \psi)$ without also having $R(\phi)$ or $R(\psi)$.

Disregarding problems of the type discussed in § 10.2.3, the valuations on $\mathcal{L}_{\mathbf{c}}^R$ are determined by those on $\mathcal{G}_{\mathbf{c}}^R$, and one can, therefore, take this latter set as a starting point for finding a representation of the state space $\Gamma_{\mathbf{r}}$ of the reality to be. The next step may then consist of removing those states considered to be impossible, i.e., one then obtains some set

$$\Gamma_{\mathbf{r}} \subseteq \{0,1\}^{\mathcal{G}_{\mathbf{c}}^R}. \tag{10.104}$$

Example: Consider two sentences $[\phi]\mathbf{u}$ and $[\psi]\mathbf{v}$, expressing observable phenomena, and assume that $R_{\mathbf{u}}(\phi)$ and $R_{\mathbf{v}}(\psi)$. Further, suppose that "$\phi \Rightarrow \psi$" is a relevant physical law in some theoretical model $\mathbf{t} \in \Psi_{\mathbf{c}}$, i.e., one has $[\Box(\phi \Rightarrow \psi)]\mathbf{t}$ and $R_{\mathbf{t}}(\Box(\phi \Rightarrow \psi))$. In this case, all valuations assigning 1 to ϕ but 0 to ψ can be removed from $\Gamma_{\mathbf{r}}$, because they are presumed to be impossible.

Following these steps one may succeed in producing something which, at the level of a semiinterpreted language, can be consistently accepted as a reality, but whether the result will satisfy us is a quite different matter. In the first place, building a reality from many fragmentary pieces will generally result in a state space with a very complicated structure, unless one has very specific ideas about the relations between the fragments. Further, one would generally require the elements of $\Gamma_{\mathbf{r}}$ to have more structure than a valuation on a language. $\Gamma_{\mathbf{r}}$ should be a set of 'worlds', structures in which not only sentences but also other terms of $\mathcal{L}_{\mathbf{c}}$ can be interpreted, by reference to objects in these worlds, i.e., one must still define an ontology for **r**. This part of the project may be difficult, and, in particular, it may involve much freedom of choice. Generally, one will try to import ontological structure from relevant parts of other contexts, but this may not always be straightforward, because there may be dramatic ontological differences between different theoretical models, even if they are empirically equivalent. The difference between various interpretations of QM is clearly a point in case.

We may conclude that, from a theoretical point of view, building a reality the 'bottom–up' way cannot be expected to be logically straightforward, in spite of the attractive simplicity of the 'top–down' picture, in which the relations between a number of fragmentary contexts are organized by means of a *given* reality . Looking at scientific practice, and at the rather opaque discussions in philosophy of science about subjects like realism and rationality, may further convince us that there is indeed little reason to believe in the existence of failsafe standard procedures, and that, if we find something that pleases us, we are by no means justified in assuming that it is unique and everlasting. This, of course, leads us back to topics discussed in the earlier chapters of this book. Briefly said, building a total and uncontroversial scientific reality for the entire body scientific knowledge is way beyond our capabilities, so we proceed step by step and in small fragments. We may sometimes like to believe that a step by step and piece by piece procedure automatically leads to a consistent and coherent overall picture. Especially for realists it would be comforting to know that the scientific process would be simply additive in some comprehensible way and that it would, perhaps slowly but irrevocably, lead us ever closer to the unique literal truth about nature. In this case science should proceed in a linear way, without great conceptual shocks which would make it necessary to reconsider earlier results, but at this point there is considerable room for disagreement, as one may always argue about when to call an earthquake a shock.

10.3.2 Shadows

The circumstance that we have no direct knowledge of a given, definitive reality compels us to find things out the hard way and remain uncertain about the true relevance of our findings. One might say that we are forever wandering through the shadows, and, indeed, one may construct for this fine sounding platonic metaphor a quite literal meaning within the present framework. Let us call a context **v** a shadow of **u**, $\mathbf{v}S\mathbf{u}$, if its class of relevant features is a subset of those of **u**, i.e.,

$$\mathbf{v}S\mathbf{u} \equiv \forall\phi\,(R_{\mathbf{u}}(\phi) \Rightarrow R_{\mathbf{v}}(\phi)) \tag{10.105}$$

Shadows have great practical importance, because a shadow of a context can be a much more manageable system than the original context itself. Sometimes it is much easier to show $[\phi]\mathbf{v}$ and $[\phi]\mathbf{v} \equiv [\phi]\mathbf{u}$ than to demonstrate $[\phi]\mathbf{u}$ directly within **u** itself. In this case the route via the shadow **v** can be used as a shortcut for an argument within **u**.

How do we make a shadow? Imagine, for example, that we are working on a dynamical model of the solar system. Evidently, because the solar system is part of a larger system of stars, it is subjected to external perturbative influences. Our first step may be to make a rough estimate of the influence of these perturbations and further neglect them. Subsequently, we can then regard the solar system as an isolated system and disregard the rest entirely.

Two different kinds of steps are involved in this procedure; the first step is one of *approximation*: we neglect small perturbations. The second step that of *isolation*: we confine our considerations to a subsystem of some larger system. Both steps produce a shadow, and both steps tend to reduce the relevant content of a context. Another shadow producing procedure, this time leaving the relevant contents invariant, consists in merely adding irrelevant structure, generally with the purpose of getting a mathematically more manageable structure. We may call this *formal extension*, and a much applied example is that of analytic continuation of a real function in the complex plane.

10.3.3 Realistic, empiristic, and pluralistic labyrinths

Without going into discussion again about the realism debate, let us note that contextual logic at least seems to provide the proper framework for investigating the structure of the arguments in the debate. Let us slightly elaborate upon this point by making a distinction between some different types of labyrinths.

Let us define a *realistic context* as one containing a single reality for everything considered relevant within that context. Further, an *empiristic context* will be defined as a realistic context satisfying some additional requirements. In the first place, the set of contexts $\Psi_{\mathbf{e}}$ of an empiristic context $\mathbf{e}$ must contain a subset Ψ_{obs} of contexts which serve to describe, what are considered within $\mathbf{e}$ as, direct observations. In the second place, everything considered relevant within $\mathbf{e}$ must be relevant with respect to one of the contexts in Ψ_{obs}, i.e.

$$R(\phi) \Rightarrow \exists \mathbf{u} \in \Psi_{obs} R_{\mathbf{u}}(\phi). \tag{10.106}$$

Finally, a *pluralistic context* is a context in which there are several different realities $\mathbf{r}_1 \cdots \mathbf{r}_n$, instead of merely a single one, each with its own associated relevance and truth concepts. In the way in which reality was introduced above, there is no objection against making a lot of realities at once, as long as they are not confused and all associated concepts are properly indexed.

Continuing with labyrinths, a *realistic labyrinth* may be defined as one in which all realities, of all contexts within the labyrinth, are pure shadows of a single global reality, where a *pure shadow* is one in which no additional irrelevant features are introduced, relative to the source context. An *empiristic labyrinth* can now be defined as one containing a sublabyrinth Ψ_{obs} of 'direct observation contexts', and with a global relevance concept satisfying (10.106). Finally, a *pluralistic labyrinth* is one containing either a number of different and mutually incompatible realistic contexts, or one or more pluralistic contexts, or both.

With respect to the actual structure of science, let us now make two observations. In the first place, disagreement among scientists, even over prolonged periods of time, is an entirely common affair in science. In the second place, the relations between different scientific theories and disciplines are generally so complex that we cannot tell with any degree of reliability to what extent theories are mutually compatible or not. On the basis of just these facts one may already conclude that the realistic labyrinth is entirely unrealistic as a model for the current structure of actual science; and how science might ideally be in some remote future, nobody knows. The pluralistic labyrinth, on the other hand, can capture much more of what is going on. It may help us to perceive, appreciate, and develop science as a rich and varied cultural environment, and I may seize the occasion to strongly recommend it as a tool for investigating the logic of science.

10.3.4 More about truth

Within a realistic context one may consistently, and conveniently, impose the rule that ϕ is true whenever $T(\phi)$ is true, i.e.,

$$T(\phi) \Rightarrow \phi. \tag{10.107}$$

It now becomes interesting to see to what extent the predicate T can express semantic truth, as given by $I_\lambda(\phi) = 1$ or $\lambda \in \gamma_c(\phi)$. T would correctly represent semantic truth precisely iff the condition

$$T(\phi) \equiv \phi \tag{10.108}$$

would hold, conform Tarski's definition of truth[19]. It is readily inferred, however, that in fact we have

$$T(\phi) \equiv (\phi \wedge R(\phi)), \tag{10.109}$$

on the basis of our earlier definitions.

> **Proof:** From left to right: from $T(\phi)$ follows ϕ by virtue of (10.107), and $R(\phi)$ by virtue of (10.95) and (10.101). From right to left: from $R(\phi)$ follows $(T(\phi) \vee T(\neg\phi))$ by virtue of (10.103). From $T(\neg\phi)$ follows $\neg\phi$; therefore, only $T(\phi)$ remains. □

We thus arrive at a concept which is slightly weaker than Tarskian truth: (10.108) is contained as a special case under the condition that $\phi \Rightarrow R(\phi)$, but, alternatively, there may be sentences ϕ for which $\neg R(\phi)$, such that neither $T(\phi)$ nor $T(\neg\phi)$ is true. In this sense, the truth predicate T is admitted to be partial, and it is natural to look for a link with Kripke's theory of truth, in which truth is partial in a similar sense.

Tarski's and Kripke's theory of truth differ crucially in the way in which they manage to evade the liar paradox, so let us briefly consider this topic. Consider

[19] [204].

a realistic connective context $\mathbf{c}$, with a language $\mathcal{L}_\mathbf{c}$ in which the liar sentence L, defined by

$$L = \neg T(L), \tag{10.110}$$

can be formulated as a grammatically correct sentence. Assume, furthermore, that the syntax of $\mathcal{L}_\mathbf{c}$ includes some standard set of rules of logic, such as natural deduction. In this case, one readily infers that $T(L)$ leads to a contradiction, because from $T(L)$ follows L and therefore $\neg T(L)$. Accordingly, $\neg T(L)$, and therefore L itself, is deducible, and because $L \wedge \neg T(L)$ implies $\neg R(L)$, one has proof of the existence of at least one irrelevant sentence. Assuming soundness of the calculus, one has to conclude that $\lambda \in \gamma_\mathbf{c}(L)$ for arbitrary λ, and therefore $\gamma_\mathbf{c}(L) = \Gamma_\mathbf{c}$. It is clear, then, that T cannot faithfully represent semantic truth, because it can apparently occur that

$$\lambda \in \gamma_\mathbf{c}(\phi) \text{ but } \lambda \notin \gamma_\mathbf{c}(T(\phi)). \tag{10.111}$$

Suppose that we want to stick to languages with a standard logical calculus, then there are apparently two alternatives: either one finds a way to deny grammatical correctness to the liar sentence, or one accepts that the truth predicate is partial. Tarski's theory of truth elaborates the former alternative, Kripke's the latter.

Tarski

In Tarski's theory the sentence $T(L)$ is not well-formed, and this is achieved by introducing a strict hierarchy of languages. In terms of contextual logic, his approach amounts to the following: let $\mathbf{r}$ be the reality of $\mathbf{c}$. It is now assumed that the language of $\mathbf{r}$ is contained in the language of $\mathbf{c}$, i.e.,

$$\mathcal{L}_\mathbf{r} \subseteq \mathcal{L}_\mathbf{c}, \tag{10.112}$$

and the domain of the truth predicate is restricted to sentences of $\mathcal{L}_\mathbf{r}$ only. In other words, $T(\phi)$ is well-formed only for $\phi \in \mathcal{L}_\mathbf{r}$, and the liar paradox is evaded by not admitting $T(\phi) \in \mathcal{L}_\mathbf{r}$. Relative to each other, $\mathcal{L}_\mathbf{r}$ and $\mathcal{L}_\mathbf{c}$ are respectively called the object language and the metalanguage. As a consequence of this division, nested truth sentences, such as $T(T(\phi))$ can be admitted only if $\mathbf{r}$, in its turn, is a connective context, with a reality $\mathbf{r}_1$, such that $\mathcal{L}_\mathbf{r}$, in its turn, is a metalanguage of $\mathcal{L}_{\mathbf{r}_1}$. With $\phi \in \mathcal{L}_{\mathbf{r}_1}$, $T(T(\phi))$ can then be admitted as well-formed in $\mathcal{L}_\mathbf{c}$. This procedure only works, however, if the set of languages $\mathbf{c}$, $\mathbf{r}_1$, $\mathbf{r}_1$, ..., is strictly ordered into a hierarchy, i.e., loops lead to problems.

> **Example:** Suppose, for example, that $\mathbf{c} \in \Psi_\mathbf{c}$. Assuming $R_\mathbf{c}(\phi)$ for all $\phi \in \mathcal{L}_\mathbf{c}$ then leads to the liar paradox, because the liar sentence $L = \neg T(L)$ is well-formed, or at least it is not automatically excluded by virtue of Tarski's distinction between object and metalanguages. Longer loops still lead to similar problems. For instance, with $\mathbf{r} \in \Psi_\mathbf{c}$ and $\mathbf{c} \in \Psi_\mathbf{r}$ assume that neither $\mathcal{L}_\mathbf{c}$ nor $\mathcal{L}_\mathbf{r}$ contains irrelevant sentences, i.e., in $\mathbf{u}$ we have $R_\mathbf{v}(\phi)$ for all $\phi \in \mathcal{L}_\mathbf{v}$ and in $\mathbf{v}$ we have $R_\mathbf{u}(\psi)$ for all $\psi \in \mathcal{L}_\mathbf{u}$. The sentence $L' = \neg T(T(L'))$ then produces a liar paradox.

Accordingly, Tarski's theory easily runs into difficulties in parts of a labyrinth containing loops, and Kripke's theory seems more natural here.

Kripke

Kripke's theory of truth is based on a fixed point construction involving a sequence of domains $\mathcal{K}_0, \mathcal{K}_1, \cdots \subset \mathcal{L}_\mathbf{c}$ on which a sequence of truth predicates T is defined. In our present terminology, the main argument may be summarized in the following manner: assume that $\mathbf{r}$ is $\mathbf{c}$'s reality and that, again, $\mathcal{L}_\mathbf{r} \subseteq \mathcal{L}_\mathbf{c}$. For simplicity we here assume $\mathcal{L}_\mathbf{c}$ to be a countable language with connectives $\neg$ and $\vee$, and without quantifiers. A truth predicate T for the sentences in $\mathcal{K}_0 := \mathcal{L}_\mathbf{r}$ is defined by $T(\phi) \equiv [\phi]\mathbf{r}$, and on this domain it is also safe to assume $T(\phi) \equiv \phi$. Then, as a next step, the domain of T is extended ever further by defining

$$\begin{aligned} \mathcal{K}_{i+1} &= \mathcal{K}_i \cup \{T(\phi), \neg\phi, \phi \vee \psi | \, \phi, \psi \in \mathcal{K}_i\} \\ T(\phi) &\equiv \phi \text{ for all } \phi \in \mathcal{K}_i. \end{aligned} \tag{10.113}$$

By repeating the procedure an infinite number of times one eventually arrives at a fixed point in the sequence of $\mathcal{K}_i$s, being a set $\mathcal{K}_\sigma$ such that $\mathcal{K}_{\sigma+1} = \mathcal{K}_\sigma$. Let $\mathcal{K} \subset \mathcal{L}_\mathbf{c}$ be the fixed point obtained by the above procedure, then for all $\phi, \psi \in \mathcal{K}$ one also has

$$T(\phi), \neg\phi, \phi \vee \psi \in \mathcal{K}. \tag{10.114}$$

Apart from $\mathcal{K}$ there are many other fixed points satisfying this condition, but $\mathcal{K}$ is minimal in the sense that it is contained in all the other ones. Sentences in the minimal fixed point $\mathcal{K}$ are called *grounded.* It is possible to extend the domain of the truth predicate to some of the sentences outside $\mathcal{K}$, for instance, to the ungrounded sentence A defined by $A = T(A)$. Including this sentence, and adding all compounds such as $T(T(A))$ etc. that can be formed from it, one arrives at a nonminimal fixed point $\mathcal{K}'$. For all sentences in such a $\mathcal{K}'$ we have $T(\phi) \equiv \phi$, and every such $\mathcal{K}'$ is a language containing its own truth predicate, contrary to the languages in Tarski's hierarchy. Apart from sentences like $A = T(A)$, or $B = T(B) \vee \neg T(B)$, which can be consistently included in a fixed point of this type, there are also sentences, and the liar sentence L is one of them, which cannot be included in any such fixed point.

A match with the contextual truth predicate

Consider one of Kripke's fixed point languages, for instance the minimal fixed point $\mathcal{K}$ itself, because it seems intuitively the most attractive one. Sentences outside $\mathcal{K}$ can be treated in different ways. A first option is to leave T undefined outside $\mathcal{K}$. A second possibility, however, is to *close off* T, by defining $\neg T(\phi)$ for every ϕ outside $\mathcal{K}$[20]. For such sentences one now has both $\neg T(\phi)$ *and* $\neg T(\neg\phi)$. In contextual terminology this means that the condition "$\phi \in \mathcal{K}$" precisely corresponds to $R(\phi)$, and the construction of $\mathcal{K}$ amounts to the construction of a relevance concept.

We may now note that the same relevance concept can be obtained by creating a loop in the labyrinth. Let us take $\mathbf{c} \in \Psi_\mathbf{c}$, and impose the condition that

$$R(\phi) \Rightarrow R_\mathbf{c}(R(\phi)). \tag{10.115}$$

[20] See also Kripke [137] p.715

By virtue of (10.97), $R_c(\phi)$ yields $R(\phi)$. Therefore, we can infer from (10.115) that

$$R(\phi) \Rightarrow R(R(\phi)). \tag{10.116}$$

From $R(\phi) \wedge R(\psi)$ one may infer $R(\phi \vee \psi)$ and $R(\phi \wedge \psi)$, by virtue of (10.99), (10.102), and the definition of $\wedge$. Therefore, from $R(\phi) \wedge R(R(\phi))$ we conclude to $R(\phi \wedge R(\phi))$, and to $R(T(\phi))$, because of (10.109). The upshot is that for all $\phi, \psi \in \mathcal{L}_c$ such that $R(\phi)$ and $R(\psi)$ one also has

$$R(T(\phi)), R(\neg\phi), \text{ and } R(\phi \vee \psi), \tag{10.117}$$

which provides the link with (10.114).

Conclusion

It may be concluded that contextual truth theory admits a wide variety of constructions for the concept of truth, and that, depending on the details of the construction, a natural relation can be established both with Tarski's and with Kripke's approach.

At an earlier occasion we considered the question to what extent a truth predicate T could faithfully represent semantic truth, as given by $\mathcal{I}_\lambda(\phi) = 1$, or $\lambda \in \gamma_c(\phi)$. Clearly, in Tarski's theory the fit is perfect, but the domain of T is restricted to the object language $\mathcal{L}_r$. Outside this domain $T(\phi)$ is not even considered as grammatically well–formed. In Kripke's theory $T(\phi)$ can be considered as grammatically well–formed for all $\phi \in \mathcal{L}_c$, but the match with semantic truth is perfect only within the domain $\mathcal{K}$ of relevant sentences. In fact, the liar paradox is evaded by denying truth, in the sense of $T(L)$, to the liar sentence in spite of its being deducible. One may react to this in two alternative ways. The first option is to admit that T can represent semantic truth only for relevant sentences. This makes it possible to have $\mathcal{I}_\lambda(L) = 1$, but $\mathcal{I}_\lambda(T(L)) = 0$. The alternative is to observe that assigning truth to something like the liar sentence is a murky business, even in spite of its being deducible, and one may prefer to make semantic truth too into a partial function, restricted itself to the domain of relevant sentences. This has the consequence that the soundness of the calculus is violated, but because the violation is restricted to irrelevant sentences only this may be considered as not a serious objection.

10.4 End

Conceiving reality as a self–made construction, and realizing the freedom inherent in this, can be both a liberating and a frightning experience. Further, realizing the difficulties involved in the bottom–up construction process, especially as opposed to the simplicity of the top–down picture, can be extremely annoying. As people are not particularly fond of fear and annoyance, they have been quite inventive in finding ways of overcoming them. Divine inspiration, mystic insight, rational thought, objective observation; the roads to absolute, true, and final knowledge are plenty. Besides a good deal of irony, the subject deserves a touch of care in dealing with it. We do have religious, mystic, and other sorts of feelings, we do have thoughts in many varieties, and we do have our senses. A fruitful way of looking at these mental faculties may be to regard them as being comparable to arms and legs. It is quite a job to have all of them cooperate in some sort of harmony, and there is little sense in crippling ourselves by playing them against each other, especially because, just like our muscular power, our mental abilities are rather limited, and none of the above mentioned faculties can bear the weight of being taken fully serious. Whenever possible they should be used in a playful mode, straining but not breaking them.

Finding a proper balance seems difficult, however, perhaps because we are too frantically trying to achieve the impossible, searching answers to questions that have no solution and longing for certainties that do not exist. Another factor which may be relevant in this respect is that we are very vulnerable to committing, what may be called, the fallacy of obsessive ordering. People seem to have a bad intuition for dealing with partial orderings. Merely the hint that *some* sort of ordering applies seems sufficient to induce illusions of a *strict* hierarchy, preferably even with a top element. We are interested in 'the best' rather than in variation. This tendency has played a role in many obsessions and perversities which have pervaded our culture in the course of the centuries, ranging from major crimes, such as genocide in the name of Christian civilization, to a multitude of minor oddities, such as the idea of monotheism itself, the obsession with absolute truth, and the search for absolute foundations for ethics and aesthetics. Within the area of philosophy of science the rather extensive attention received over the past few decades by subjects like rationality and realism may be viewed from this angle.

During the present century, empirical positivism has been important in defending freedom of thought by liberating science from unwarranted metaphysical constraints. The downfall of empirical positivism has further liberated us from the empiricist obsession with direct observation, thus providing the possibility of an even fuller view on how much freedom there may actually be, and how important 'design' is in the architectonic of science, even though individual scientists may not always be in a position to be aware of this. Rather than welcoming the new situation, exploring its possibilities, and studying the complications that come with it, many scientists and philosophers have chosen to feel offended, and perhaps frightened by the threatening prospect of conceptual anarchy. This has resulted in a revival of realism and an extensive search for some historically tenable form of rationality. I regard this as regrettable and unfruitful. With respect to both the theory of rationality and the idea of realism, the situation is still as diffuse as ever, and theoretically there is little reason to assume that this will ever improve. Science as such, however, has little need for either realism or a theory of rationality. Both of them mainly serve to feed the complacency of scientists, and perhaps to some extent as rhetorical instruments for directing cash flows, but no doubt adequate substitutes can be found for both functions.

Meanwhile, I would say that too extensive research into realism and rationality is not only a waste of time, but actually works counterproductive by blurring the view on more important problems, concerning the complications associated with the presence of a great deal of conceptual freedom in science, which is there *irrespective of whether we wish to see it or not.* Problems of *design* cannot be solved by denying them and suggesting that one should look for *the* right way to do it. Further, from a realistic point of view the possibility of communication between different people and different communities may seem evident, because after all they are all talking about the same reality. From a pluralistic point of view, however, communication is made possible by the *construction* of a communication channel, i.e., one needs some sort of interface between the different views, but such interfaces are not in the habit of coming out of the blue. The existence of adequate communication channels between different views, different disciplines, and different research groups is of crucial importance to the future of science, not only for trying to save it as some sort of cultural unity, in whatever vague sense, but also for organizing and financing it. With science evolving to ever higher levels of complexity, however, one can expect this to become increasingly difficult.

Contextual logic may be able to contribute to the construction of interfaces between different views by providing some general conceptual background for analyzing the relations between them. Irrespective of how useful this particular approach will turn out to be, however, the issue of retaining effective communication channels in science will presumably be of increasing importance, and philosophy of science would seem an appropriate discipline for dealing with at least some of the theoretical aspects of this. It may sound somewhat more prosaic than dealing with sweeping topics like "reality" and "rationality", but eventually it should be much more rewarding.

Literature: Barwise and Etchemendy [10], Benthem [23, 22], Bergstra *et al.* [25], Beth [27, 26], Bowen [31], Cooke [46], Dieks [65], Van Fraassen [92, 89, 93, 96], Fourman and Grayson [87], Gregerson *et al.* [104], Grünbaum [105], Haack [106], Hansson [110], Hempel [115], Hintikka [119, 118, 117], Kripke [138, 137, 136], Kroes [139], Lewis [150], Maxwell [159], Moulines [163], Muller [164], Putnam [174], Rietdijk [180, 179], Russell [185], Scheibe [186], Stein [196], Sneed [194], Suppe [201], Suppes [202], Tarski [204, 205], Weyl [217].

Bibliography

[1] L. Accardi. Foundations of quantum mechanics: a quantum probabilistic approach. In G. Tarrozzi and A. Van der Merwe, editors, *Quantum Paradoxes*, pages 119–147, Kluwer, Dordrecht, 1987.

[2] L. Accardi. The probabilistic roots of the quantum mechanical paradoxes. In Diner et al., editors, *The Wave–Particle Dualism*, Reidel, Dordrecht, 1984.

[3] L. Accardi. Quantum probability and the foundations of quantum theory. In R. Cooke and D. Constantini, editors, *Statistics in Science. The Foundations of Statistical Methods in Biology, Physics and Economics*, Kluwer, Dordrecht, 1990.

[4] L. Accardi. Topics in quantum probability. *Physics Reports*, 77:169–172, 1981.

[5] R.J. Adler, M.J. Bazin, and M. Schiffer. *Introduction to General Relativity.* McGraw–Hill, New York, 1975.

[6] Y. Aharonov, P.G. Bergmann, and J.L. Lebowitz. Time asymmetry in the quantum process of measurement. *Physical Review B*, 134:1410–1416, 1964. reprinted in Wheeler and Zurek eds. [219].

[7] D.Z. Albert, Y. Aharonov, and S. D'Amato. Curious new statistical predictions of quantum mechanics. *Physical Review Letters*, 54:5–7, 1985.

[8] R. Alicki and K. Lendi. *Quantum Dynamical Semigroups and Applications.* Springer, Berlin, 1987.

[9] M. Barr. Book review of Penrose: The Emperor's New Mind. *American Mathemathical Monthly*, 938–943, 1990.

[10] J. Barwise and J. Etchemendy. *The Liar: An Essay in Truth and Circularity.* Oxford University Press, Oxford, 1987.

[11] W. Bechtel. *Philosophy of Mind, An Overview for Cognitive Science.* Lawrence Erlbaum Ass., Hillsdale, New Jersey, 1988.

[12] F.J. Belinfante. *A Survey of Hidden Variable Theories.* Pergamon, Oxford, 1973.

[13] J.S. Bell. The measurement theory of Everett and De Broglie's pilot wave. In M. Flato, Z. Maric, A. Milojevic, D. Sternheimer, and J.P. Vigier, editors, *Quantum Mechanics, Determinism, Causality and Particles*, Reidel, Dordrecht, 1976.

[14] J.S. Bell. On the problem of hidden variables in quantum mechanics. *Reviews of Modern Physics*, 38:447–452, 1966.

[15] E.G. Beltrametti and G. Cassinelli. *The Logic of Quantum Mechanics.* Addison–Wesley, Reading Mass., 1981.

[16] Y. Ben–Dov. Everett's theory and the "many–worlds" interpretation. *American Journal of Physics*, 58:829–832, 1990.

[17] Y. Ben–Dov. Noncollapse versions of quantum mechanics. In M. Cini and J.M. Lévy–Leblond, editors, *Quantum Theory without Reduction*, pages 140–150, Adam Hilger, Bristol, 1990.

[18] Y. Ben–Dov. An "observer decomposition" for Everett's theory. *Foundations of Physics Letters*, 3:383–387, 1990.

[19] P. Benioff. Finite and infinite measurement sequences in quantum mechanics and randomness: the Everett interpretation. *Journal of Mathematical Physics*, 18:2298–2295, 1977.

[20] P. Benioff. A note on the Everett interpretation of quantum mechanics. *Foundations of Physics*, 8:709–720, 1978.

[21] Y. Ben–Menahem. Equivalent descriptions. *British Journal for the Philosophy of Science*, 41:261–279, 1990.

[22] J. van Benthem. *The Logic of Time.* Kluwer Academic Publishers, Dordrecht, 1983. second edition 1991.

[23] J. van Benthem. Four paradoxes. *Journal of Philosophical Logic*, 7:49–72, 1978.

[24] H. van den Berg, D.J. Hoekzema, and H. Radder. Accardi on quantum theory and the "fifth axiom" of probability. *Philosophy of Science*, 57:149–157, 1990.

[25] J.A. Bergstra, J. Heering, and P. Klint. *Module Algebra.* Technical Report Report P8823, University of Amsterdam, Programming Research Group, 1988.

[26] E.W. Beth. *Inleiding tot de Wijsbegeerte der Exacte Natuurwetenschappen.* Standaard N.V, Antwerpen, 1953.

[27] E.W. Beth. *Natuurphilosophie.* Noorduyn, Gorinchem, 1948.

[28] G. Birkhoff and J. von Neumann. The logic of quantum mechanics. *Annals of Mathematics*, 37:823–843, 1936.

[29] D. Bohm. A suggested interpretation of the quantum theory in terms of "hidden variables". *Physical Review*, 85:166–179 and 180–193, 1952.

[30] H. Bondi. *Cosmology.* Cambridge University Press, Cambridge, 1952.

[31] K.A. Bowen. *Model Theory for Modal Logic.* Volume 127 of *Synthese Library*, Reidel, Dordrecht, 1979.

[32] G.L. Bowie. Lucas' number is finally up. *Journal of Philosophical Logic*, 11:279–285, 1982.

[33] R. Boyd. Realism, approximate truth, and philosophical method. In C. Wade Savage, editor, *Scientific Theories*, page , University of Minneapolis Press, Minneapolis, 1990.

[34] R. Boyd. What realism implies and what it does not. *Dialectica*, 43:5–30, 1989.

[35] V.B. Braginsky, Y.I. Vorontsov, and K.S. Thorne. Quantum nondemolition measurements. *Science*, 209:547–557, 1980. Reprinted in Wheeler and Zurek [219]:749–768.

[36] A.A. Broyles. Space–time position operators. *Physical Review*, D1:979–988, 1970.

[37] J. Bub and H. Brown. Curious properties of quantum ensembles which have been both preselected and postselected. *Physical Review Letters*, 56:2337–2340, 1986.

[38] M. Čapek. Particles or events. In R.S. Cohen and M.W. Wartofsky, editors, *Physical Sciences and History of Physics*, pages 1–28, Reidel, Dordrecht, 1983.

[39] N. Cartwright. *How the Laws of Physics Lie.* Clarendon Press, Oxford, 1983.

[40] C. Castaneda. *Tales of Power.* Simon and Schuster, New York, 1974.

[41] C. Castaneda. *The Teachings of Don Juan.* University of California Press, Berkeley, 1968.

[42] P.M. Churchland. *Scientific Realism and the Plasticity of the Mind.* Cambridge University Press, Cambridge, 1979.

[43] P.M. Churchland. Semantic content, in defence of a network approach. *The Behavioural and Brain Sciences*, 9:139–140, 1986.

[44] P.M. Churchland and C.A. Hooker, editors. *Images of Science.* The University of Chicago Press, Chicago, 1985.

[45] R. Cooke, M. Keane, and W. Moran. An elementary proof of Gleason's theorem. *Mathematical Proceedings of the Cambridge Philosophical Society*, 98:117–128, 1985.

[46] R.M. Cooke. Mystical realism. In A. Hartkämper and H.J. Schmidt, editors, *Structure and Approximation in Physical Theories*, pages 165–169, Plenum Press, New York, 1981.

[47] R.M. Cooke and J. Hilgevoord. The algebra of physical magnitudes. *Foundations of Physics*, 10:363–373, 1979.

[48] L.N. Cooper. Wavefunction and observer in the quantum theory. In J. Mehra, editor, *The Physicist's Conception of Nature*, pages 668–683, Reidel, Dordrecht, 1973.

[49] O. Costa de Beauregard. Causality as identified with conditional probability and the quantal nonseparability. In D. Greenberger, editor, *New Techniques and Ideas in Quantum Measurement Theory*, pages 317–325, Annals of the New York Academy of Sciences 480, 1986.

[50] O. Costa de Beauregard. On the zigzagging causality EPR model. *Foundations of Physics*, 17:775–75, 1987.

[51] J.G. Cramer. Generalized absorber theory and the Einstein, Podolsky, Rosen paradox. *Physical Review*, D 22:362–376, 1980.

[52] J.G. Cramer. The transactional interpretation of quantum mechanics. *Reviews of Modern Physics*, 58:647–688, 1983.

[53] J.T. Cushing. *Theory Construction and Selection in Modern Physics, The S–Matrix.* Cambridge University Press, Cambridge, 1990.

[54] J.T. Cushing, C.F. Delaney, and G.M. Gutting, editors. *Science and Reality.* University of Notre Dame Press, Notre Dame, 1984.

[55] A. Daneri, A. Loinger, and G.M. Prosperi. Quantum theory of measurements and ergodicity conditions. *Nuclear Physics*, 33:297–319, 1962.

[56] E.B. Davies. *Quantum Theory of Open Systems.* Academic Press, London, 1976.

[57] E.B. Davies and J.T. Lewis. An operational approach to quantum probability. *Communications in Mathematical Physics*, 17:239–260, 1970.

[58] P.C.W. Davies. Extension of the Wheeler–Feynman quantum theory to the relativistic domain I. scattering processes. *Journal of Physics*, A 4:836–845, 1971.

[59] P.C.W. Davies. Extension of the Wheeler–Feynman quantum theory to the relativistic domain II. emission processes. *Journal of Physics*, A 5:1025–1036, 1972.

[60] P.C.W. Davies. *The physics of Time Asymmetry.* University of California Press, Berkeley, 1977.

[61] B. De Witt and N. Graham, editors. *The Many–Worlds Interpretation of Quantum Mechanics.* Princeton University Press, Princeton, 1973.

[62] W. Demopoulos. Locality and the algebraic structure of quantum mechanics. In P. Suppes, editor, *Studies in the Foundations of Quantum Mechanics*, pages 119–144, PSA, East Lansing, Michigan, 1980.

[63] D. Dieks. Quantum mechanics without the projection postulate and its realistic interpretation. *Foundations of Physics*, 19:1397–1423, 1989.

[64] D. Dieks. Resolution of the measurement problem through decoherence of the quantum state. *Physics Letters A*, 142:439–446, 1989.

[65] D. Dieks. Special relativity and the flow of time. *Philosophy of Science*, 55:456–460, 1988.

[66] D. Dieks. On the covariant description of wavefunction collapse *Physics Letters*, 108A:379–383, 1985.

[67] J. Dorling. Schrödinger's original interpretation of the Schrödinger equation: a rescue attempt. In C.W. Kilmister, editor, *Schrödinger, Centenary Celebration of a Polymath*, pages 16–40, Cambridge University Press, Cambridge, 1987.

[68] U. Eco. *The Name of the Rose.* Pan Books Ltd., London, 1984.

[69] U. Eco. *Reflections on the Name of the Rose.* Martin Secker and Warburg Ltd., London, 1985.

[70] A. Einstein. *Ideas and Opinions.* Bonanza Books, New York, 1954.

[71] A. Einstein. On the method of theoretical physics. The Herbert Spencer Lecture, June 10 1933. published in *Mein Weltbild*, Amsterdam, Querido,1934; reprinted in *Ideas and Opinions*, New York, Bonanza Books, 1954, pages 270–276.

[72] A. Einstein. Physics and reality. *The Journal of the Franklin Institute*, 221, 1936. reprinted in *Ideas and Opinions*, New York, Bonanza Books, 1954, pages 290–323.

[73] B. d' Espagnat. *Conceptual Foundations of Quantum Mechanics.* W.A. Benjamin, Inc., Reading, Mass., second edition, 1976.

[74] H. Everett. "Relative state" formulation of quantum mechanics. *Reviews of Modern Physics*, 29:454–462, 1957.

[75] P.K. Feyerabend. *Against Method.* NLB, London, 1975.

[76] P.K. Feyerabend. On the critique of scientific reason. In C. Howson, editor, *Method and Appraisal in the Physical Sciences*, pages 309–339, Cambridge University Press, Cambridge, 1976.

[77] P.K. Feyerabend. *Problems of Empiricism.* Volume 2 of *Philosophical papers*, Cambridge University Press, Cambridge, 1981.

[78] P.K. Feyerabend. *Realism, Rationalism and Scientific Method.* Volume 1 of *Philosophical papers*, Cambridge University Press, Cambridge, 1981.

[79] A. Fine. Einstein's realism. In J.T. Cushing, C.F. Delaney, and G.M. Gutting, editors, *Science and Reality*, University of Notre Dame Press, Notre Dame, 1984.

[80] A. Fine. The natural ontological attitude. In J. Leplin, editor, *Essays on Scientific Realism*, University of California Press, Berkeley, 1984.

[81] J.A. Fodor. *The Modularity of the Mind.* The MIT Press, Cambridge Mass., 1983.

[82] J.A. Fodor. Observation reconsidered. *Philosophy of Science*, 51:23–43, 1984.

[83] J.A. Fodor. *Psychosemantics.* The MIT Press, Cambridge Mass., 1987.

[84] A.D. Fokker. *Zeitschrift für Physik*, 58:386, 1929.

[85] H.J. Folse. *The Framework of Complementarity.* North Holland, 1985.

[86] H.J. Folse. Niels Bohr's concept of reality. In P. Lahti and P. Mittelsteadt, editors, *Symposium on the Foundations of Modern Physics*, pages 161–179, World Scientific Publishing Co., Singapore, 1987.

[87] M.P. Fourman and R.J. Grayson. Formal spaces. In A.S. Troelstra and Dalen D. van, editors, *The L.E.J. Brouwer Centenary Symposium*, North Holland, Amsterdam, 1982.

[88] B.C. van Fraassen. A formal approach to the philosophy of science. In R.G. Colodny, editor, *Paradigms and Paradoxes*, University of Pittsburg Press, Pittsburgh, 1972.

[89] B.C. van Fraassen. *Formal Semantics and Logic.* Macmillan, New York, 1971.

[90] B.C. van Fraassen. Hidden variables and the modal interpretation of quantum theory. *Synthese*, 42:155–165, 1979.

[91] B.C. van Fraassen. The labyrinth of quantum logics. In R.S. Cohen and M.W. Wartofsky, editors, *Logical and Epistemological Studies in Contemporary Physics*, pages 224–254, Reidel, Dordrecht, 1974.

[92] B.C. van Fraassen. On the extension of Beth's semantics of physical theories. *Philosophy of Science*, 37:325–339, 1970.

[93] B.C. van Fraassen. *The Scientific Image.* Clarendon Press, Oxford, 1980.

[94] B.C. van Fraassen. *Quantum Mechanics, an Empiricist View.* Clarendon Press, Oxford, 1991.

[95] B.C. van Fraassen. Semantic analysis of quantum logic. In C.A. Hooker, editor, *Contemporary Research in the Foundations and Philosophy of Quantum Theory*, Reidel, Dordrecht, 1973.

[96] B.C. van Fraassen. The semantic approach to scientific theories. In N. Nersessian, editor, *The Process of Science*, Nijhoff, Dordrecht, 1987.

[97] G.C. Ghirardi, A. Rimini, and T. Weber. Disentanglement of quantum wave functions: answer to "comment on 'unified dynamics for microscopic and macroscopic systems' ". *Physical Review*, D 36:3287–3289, 1987.

[98] G.C. Ghirardi, A. Rimini, and T. Weber. The puzzling entanglement of Schrödingers wave function. *Foundations of Physics*, 18:1–27, 1988.

[99] G.C. Ghirardi, A. Rimini, and T. Weber. Unified dynamics for microscopic and macroscopic systems. *Physical Review*, D 34:470–491, 1986.

[100] A.M. Gleason. Measures on the closed subspaces of a Hilbert space. *Journal of Mathematics and Mechanics*, 6:885–893, 1957.

[101] C. Glymour. Theoretical realism and theoretical equivalence. In R.S. Cohen and M.W. Wartofsky, editors, *Boston Studies in the Philosophy of Science 8*, pages 274–288, Reidel, Dordrecht, 1971.

[102] T. Gotō, K. Yamaguchi, and Sudō. On the time operator in quantum mechanics. *Progress of Theoretical Physics*, 66:1525–1538, 1981.

[103] T. Gotō, K. Yamaguchi, and Sudō. The time as an observable in quantum mechanics. *Progress of Theoretical Physics*, 64:1–17, 1980.

[104] F. Gregersen and S. Koppe. Against epistemological relativism. *Studies in the History and Philosophy of Science*, 19:447–489, 1988.

[105] A. Grünbaum. The meaning of time. In E. Freeman and W. Sellars, editors, *Basic Issues in the Philosophy of Time*, Open Court, Lasalle, 1971.

[106] S. Haack. *Philosophy of Logic.* Cambridge University Press, Cambridge, 1980.

[107] I. Hacking. *Representing and Intervening.* Cambridge University Press, Cambridge, 1983.

[108] K. Hannabus. Dilations of a quantum measurement. *Helvetica Physica Acta*, 57:610–620, 1984.

[109] N.R. Hanson. *Patterns of Discovery.* Cambridge University Press, Cambridge, 1961.

[110] B. Hansson. Paradoxes in a semantic perspective. In J. Hintikka, I. Niiniluoto, and E. Saarinen, editors, *Essays on Mathematical and Philosophical Logic*, pages 371–385, Reidel, Dordrecht, 1978.

[111] E. Harrison. *Darkness at Night.* Harvard University Press, Harvard, 1987.

[112] S. Hawking. *A Brief History of Time: from the Big Bang to Black Holes.* Bantam Books, Toronto, 1988.

[113] S.W. Hawking. The quantum state of the universe. *Nuclear Physics*, B239:257–276, 1984.

[114] W. Heisenberg. *The Principles of the Quantum Theory.* University of Chicago Press, Chicago, 1930. Translated into English by C. Eckart and F.C. Hoyt; reprinted in 1949 by Dover Publications Inc., New York.

[115] C.G. Hempel. *Aspects of Scientific Explanation and Other Essays in the Philosophy of Science.* Basic Books, New York, 1965.

[116] J. Hilgevoord and J. Uffink. Uncertainty in prediction and in inference. *Foundations of Physics*, 21:323–341, 1990.

[117] K.J.J. Hintikka. The logic of science as a model– oriented logic. In P.D. Asquith and P. Kitcher, editors, *Proceedings of the 1984 Biennal Meeting of the Philosophy of Science Association*, pages 177–185, Philosophy of Science Association, East Lansing, Michigan, 1984.

[118] K.J.J. Hintikka. Individuals, possible worlds, and epistemic logic. *Noûs*, 1:33–62, 1961.

[119] K.J.J. Hintikka. Modality and quantification. *Theoria*, 27:110–128, 1961.

[120] D.J. Hoekzema. Contextual quantum process theory. *Foundations of Physics*, 22:467–486, 1992.

[121] D.J. Hoekzema. Locality and realism in contextual theories. *Foundations of Physics*, 17:787–797, 1987.

[122] D.J. Hoekzema. A note on contextual quantum formalisms. In J. Mizerski, A. Posiewnik, J. Pykacz, and M. Żukowski, editors, *Problems in Quantum Physics II; Gdańsk '89*, World Scientific, Singapore, 1989.

[123] D.J. Hoekzema. Quantum event theory, a Fokker–Tetrode version of quantum field theory. *Foundations of Physics*, 22:487–506, 1992.

[124] A. Holevo. *Probabilistic and Statistical Aspects of Quantum Theory.* North Holland, Amsterdam, 1982.

[125] C.A. Hooker. The nature of quantum mechanical reality: Einstein versus Bohr. In R.G. Colodny, editor, *Paradigms and Paradoxes*, University of Pittsburg Press, Pittsburgh, 1972.

[126] L.P. Horwitz and Y. Lavie. Scattering theory in relativistic quantum mechanics. *Physical Review*, D 26:819–838, 1982.

[127] F. Hoyle and J.V. Narlikar. *Action at a Distance in Physics and Cosmology.* W.H. Freeman and Co., San Fransisco, 1974.

[128] F. Hoyle and J.V. Narlikar. Electrodynamics of direct interparticle interaction II. relativistic treatment of radiative processes. *Annals of Physics*, 62:44–97, 1971.

[129] M. Jammer. *The Conceptual Development of Quantum Mechanics.* Mc Graw–Hill, New York, 1966.

[130] M. Jammer. *The Philosophy of Quantum Mechanics.* Wiley, New York, 1974.

[131] J. Jauch and C. Piron. *Helvetia Physica Acta*, 36:829, 1963.

[132] J.E. Johnson. Position operators in relativistic quantum mechanics. *Physical Review*, 181:1755–1764, 1969.

[133] E. Joos. Comment on "unified dynamics for microscopic and macroscopic systems". *Physical Review*, D 36:3285–3286, 1987.

[134] S. Kochen and E.P. Specker. The problem of hidden variables in quantum mechanics. *Journal of Mathematics and Mechanics*, 17:59–87, 1967.

[135] K. Kraus. General state changes in quantum mechanics. *Annals of Physics*, 64:311–335, 1971.

[136] S. Kripke. *Naming and Necessity.* Basil Blackwell, Oxford, 2nd edition, 1980.

[137] S. Kripke. Outline of a theory of truth. *Journal of Philosophy*, 72:690–716, 1975.

[138] S. Kripke. Semantical analysis of modal logic I: normal modal propositional calculi. *Zeitschrift für mathematische Logic und Grundlagen der Mathematik*, 9:113–116, 1963.

[139] P.A. Kroes. *An Inquiry into the Structure of Physical Time.* PhD thesis, University of Nijmegen, 1982.

[140] I. Lakatos. Falsification and the method of scientific research programmes. In I. Lakatos and A. Musgrave, editors, *Criticism and the Growth of Knowledge*, Cambridge University Press, Cambridge, 1970.

[141] I. Lakatos. *Proofs and Refutations.* Cambridge University Press, Cambridge, 1976.

[142] I. Lakatos and A. Musgrave, editors. *Criticism and the Growth of Knowledge.* Cambridge University Press, Cambridge, 1970.

[143] L.D. Landau and E. Lifshitz. *The Classical Theory of Fields.* Pergamon Press, Oxford, 4–th english edition edition, 1975.

[144] L. Laudan. *Progress and its Problems.* Routledge and Kegan, London, 1977.

[145] L. Laudan. A refutation of convergent realism. *Philosophy of Science*, 48:218–249, 1981.

[146] L. Laudan. *Science and Hypothesis.* Reidel, Dordrecht, 1981.

[147] L. Laudan. *Science and Values. Pittsburgh Series in the Philosophy and History of Science*, University of California Press, Berkeley, 1984.

[148] J. Leplin, editor. *Essays on Scientific Realism.* University of California Press, Berkeley, 1984.

[149] J.M. Levy–Leblond. Quantum mechanics, a half century later. In J. Leite Lopes and M. Paty, editors, *Towards a Proper Quantum Theory*, pages 171–206, Reidel, Dordrecht, 1977.

[150] D. Lewis. *On the Plurality of Worlds.* Basil Blackwell, 1986.

[151] J. Lighthill. *Proceedings Royal Society*, A407:35, 1986.

[152] G. Lindblad. On the generators of quantum dynamical semigroups. *Communications in Mathematical Physics*, 48:119–130, 1976.

[153] F.W. London and E. Bauer. *La Theorie de l'Observation en Mecanique Quantique.* Volume 775 of *Actualités scientifiques et industrielles: Exposés de physique generale, publiés sous la direction de Paul Langevin*, Hermann, Paris, 1939. English translation in Wheeler an Zurek [219].

[154] J.R. Lucas. Minds, machines and Gödel. *Philosophy*, 36:120–124, 1961.

[155] G. Ludwig. *Deutung des Begriffs "Physikalische Theorie" und axiomatische Grundlegung der Hilbertraumstruktur der Quantenmechanik durch Hauptsätze des Messens.* Springer, Berlin, 1970.

[156] M.J. Maczynski. Boolean properties of observables in axiomatic quantum mechanics. *Reports on Mathematical Physics*, 2:135–150, 1971.

[157] B.B. Mandelbrot. *The Fractal Geometry of Nature.* W.H. Freeman and Co., San Fransisco, 1983.

[158] G. Maxwell. The ontological status of theoretical entities. In H. Feigl and G. Maxwell, editors, *Scientific Explanation, Space, and Time*, pages 3–27, University of Minnesota Press, Minneapolis, 1962.

[159] N. Maxwell. Are probabilism and special relativity compatible? *Philosophy of Science*, 52:23–43, 1985.

[160] J. McCarthy. Book review of Penrose: The Emperor's New Mind. *Bulletin of the American Mathemathical Society*, 23:606–616, 1990.

[161] B. Mielnik. Generalized quantum mechanics. *Communications in Mathematical Physics*, 37:221–256, 1974.

[162] C.W. Misner, K.S. Thorne, and J.A. Wheeler. *Gravitation.* Freeman, San Fransisco, 1973.

[163] C.U. Moulines. A general scheme for intertheoretic approximation. In A. Hartkämper and H.J. Schmidt, editors, *Structure and Approximation in Physical Theories*, pages 123–146, Plenum Press, New York, 1981.

[164] F.A. Muller. Worldlines are growing. Preprint. Utrecht State University, 1992.

[165] J. von Neumann. *Mathematische Grundlagen der Quanten Mechanik.* Springer, Berlin, 1932. Translation into English by R.T. Beyer [1955]: Mathematical Foundations of Quantum Mechanics, Princeton: Princeton University Press.

[166] R.B. Partridge. *Nature*, 244:263, 1973.

[167] R. Penrose. *The Emperor's New Mind.* Oxford University Press, Oxford, 1989.

[168] S. Pokorsky. *Gauge Field Theories.* Cambridge University Press, Cambridge, 1987.

[169] K.R. Popper. *Conjectures and Refutations.* Routledge and Kegan Paul, London, 1963.

[170] K.R. Popper. *The Logic of Scientific Discovery.* Harper and Row, New York, 1959.

[171] H. Primas. The measurement process in the individual interpretation of quantum mechanics. In M. Cini and J.M. Lévy–Leblond, editors, *Quantum Theory without Reduction*, pages 140–150, Adam Hilger, Bristol, 1990.

[172] H. Putnam. *Reason, Truth and History.* Cambridge University Press, Cambridge, 1975.

[173] H. Putnam. *The Refutation of Conventionalism*, pages 153–191. Volume Philosophical Papers II, Cambridge University Press, Cambridge, 1975.

[174] H. Putnam. Time and physical geometry. *Journal of Philosophy*, 64:240–247, 1967.

[175] W.V. Quine. *From a Logical Point of View.* Harvard University Press, Cambridge Mass., 1953.

[176] B.S. Rajput. *Advanced Quantum Mechanics.* Pragati Prakashan, Meerut (India), 1990.

[177] M. Redhead. *Incompleteness, Nonlocality and Realism.* Clarendon Press, Oxford, 1987.

[178] L.E. Reichl. *The Transition to Chaos in Conservative Classical Systems: Quantum Manifestations.* Springer, Berlin, 1992.

[179] C.W. Rietdijk. On the reality of now at a distance. *Annales de la Fondation Louis de Broglie*, 15:233–236, 1990.

[180] C.W. Rietdijk. A rigorous proof of determinism derived from the special theory of relativity. *Philosophy of Science*, 33:341–344, 1966.

[181] K.V. Roberts. An objective interpretation of Lagrangian quantum mechanics. *Proceedings of the Royal Society*, A360:135–155, 1978.

[182] F. Rohrlich. Quantum electrodynamics of charged particles without spin. *Physical Review*, 80:666–687, 1950.

[183] N. Rosen. General relativity with a background metric. *Foundations of Physics*, 10:673–703, 1980.

[184] D.M. Rosenbaum. Super Hilbert space and the quantum mechanical time operator. *Journal of Mathematical Physics*, 10:1127–1144, 1969.

[185] B. Russell. Mysticism and logic. *Hibbert Journal*, July 1914. Reprinted as ch. I in *Mysticism and Logic*, Unwin Books, London, 1963.

[186] E. Scheibe. A comparison of two recent views on theories. In A. Hartkämper and H.J. Schmidt, editors, *Structure and Approximation in Physical Theories*, pages 197–215, Plenum Press, New York, 1981.

[187] P. Schilpp, editor. *Albert Einstein, Philosopher–Scientist. Library of Living Philosophers*, Open Court, Evanston Illinois, 1949.

[188] L.S. Schulman. Remote two–time boundary conditions and special states in quantum mechanics. *Foundations of Physics Letters*, 2:515, 1989.

[189] L.S. Schulman. Two time localization. *Annals of Physics*, 183:320, 1988.

[190] J. Schwinger. On gauge invariance and vacuum polarization. *Physical Review*, 82:664–679, 1951.

[191] F. Selleri, editor. *Quantum Mechanics versus Local Realism, The Einstein–Podolsky–Rosen Paradox.* Plenum Press, New York, 1988.

[192] A. Shimony. The status of hidden variable theories. In P. Suppes, L. Henkin, G.R.C. Moisil, and A. Joja, editors, *Logic, Methodology, and Philosophy of Science*, North Holland, 1973.

[193] H. Siegel. *Relativism Refuted.* Reidel, Dordrecht, 1987.

[194] J. Sneed. *The Logical Structure of Mathematical Physics.* Reidel, Dordrecht, 1977.

[195] F. Staal. *Over Zin en Onzin in Filosofie, Religie en Wetenschap.* Meulenhoff, Amsterdam, 1986.

[196] H. Stein. On relativity theory and the openness of the future. *Philosophy of Science*, 58:147–167, 1991.

[197] H. Stein. Yes but...—some skeptical remarks on realism and anti–realism. *Dialectica*, 43:47–65, 1989.

[198] W.F. Stinespring. *Proceedings of the American Mathematical Society*, 6:211–216, 1955.

[199] E.C.G. Stueckelberg. La significaton du temps propre en mécanique ondulatoire. *Helvetia Physica Acta*, 14:322–323, 1941.

[200] E.C.G. Stueckelberg. Remarque à propos de la création de paires de particules en théorie de la rélativité ondulatoire. *Helvetia Physica Acta*, 14:588–594, 1941.

[201] F. Suppe, editor. *The Structure of Scientific Theories.* University of Illinois Press, Urbana, 1974.

[202] P. Suppes. *Probabilistic Metaphysics.* Basil Blackwell, Oxford, 1984.

[203] P. Suppes, L. Henkin, GR.C. Moisil, and A. Joja, editors. *Logic, Methodology, and Philosophy of Science.* Volume 74 of *Studies in Logic and the Foundations of Mathematics*, North Holland, 1973.

[204] A. Tarski. Der Wahrheitsbegriff in den formalisierten Sprachen. *Studia Philosophica*, 1:, 1935. English translation in Tarski [205].

[205] A. Tarski. *Logic, Semantics, Metamathematics.* Oxford University Press, Oxford, 1956.

[206] H. Tetrode. *Zeitschrift für Physik*, 10:317, 1922.

[207] F.J. Tipler. Interpreting the wave function of the universe. *Physics Reports*, 137:231–275, 1986.

[208] S. Toulmin. *Physical Reality.* Harper and Row, New York, 1970.

[209] T.S. Kuhn *The Structure of Scientific Revolutions.* The University of Chicago Press, Chicago, 1962.

[210] W. Unruh. Quantum nondemolition and gravitational detection. *Physical Review*, D19:2888–2896, 1979.

[211] F. Waisman. Verifiability. In A.G.N. Flew, editor, *Logic and Language*, pages 117–144, Basil Blackwell, Oxford, 1963.

[212] K.W. Wan. Superselection rules, quantum measurement, and Schrodinger's cat. *Canadian Journal of Physics*, 58:976–982, 1980.

[213] P. Watzlawick. *How real is 'real': confusion, disinformation, communication.* Random House, New York, 1976.

[214] R.M. Wald *General Relativity.* University of Chicago Press, Chicago, 1984.

[215] S. Weinberg. *Gravitation and Cosmology.* Wiley, New York, 1972.

[216] P. Wesson, K. Valle, and R. Stabell. The extragalactic background light and a definitive resolution of Olbers's paradox. *The Astrophysical Journal*, 317:601–606, 1991.

[217] H. Weyl. *Philosophy of Mathematics and Natural Science.* Atheneum, New York, 1963 (1927).

[218] J.A. Wheeler. The "past" and the "delayed–choice" double slit experiment. In A.R. Marlow, editor, *Mathematical Foundations of Quantum Theory*, Academic Press, New York, 1978.

[219] J.A. Wheeler and W.H. Zurek, editors. *Quantum Theory and Measurement.* Princeton University Press, Princeton, 1983.

[220] J.A. Wheeler and R.P. Feynman. Classical electrodynamics in terms of direct interparticle action. *Reviews of Modern Physics*, 21:425–433, 1949.

[221] J.A. Wheeler and R.P. Feynman. Interaction with the absorber as the mechanism of radiation. *Reviews of Modern Physics*, 17:157–181, 1945.

[222] E.P. Wigner. Interpretation of quantum mechanics. In J.A. Wheeler and W.H. Zurek, editors, *Quantum Theory and Measurement*, pages 260–314, Princeton University Press, Princeton, 1983.

[223] E.P. Wigner. *Symmetries and Reflections.* Indiana University Press, Bloomington, 1967.

[224] J. Worrall. Structural realism: the best of both worlds? *Dialectica*, 43:99–124, 1989.

[225] H.D. Zeh. Measurement in Bohm's versus Everett's quantum theory. *Foundations of Physics*, 18:723–730, 1988.

[226] H.D. Zeh. Towards a quantum theory of observation. *Foundations of Physics*, 3:109–116, 1973.

[227] W.H. Zurek. Pointer bases of quantum apparatus: into what mixture does the wave packet collapse? *Physical Review*, D 24:1516–1525, 1981.

Index

Fundamental Theories of Physics

Series Editor: Alwyn van der Merwe, *University of Denver, USA*

1. M. Sachs: *General Relativity and Matter*. A Spinor Field Theory from Fermis to Light-Years. With a Foreword by C. Kilmister. 1982 ISBN 90-277-1381-2
2. G.H. Duffey: *A Development of Quantum Mechanics*. Based on Symmetry Considerations. 1985 ISBN 90-277-1587-4
3. S. Diner, D. Fargue, G. Lochak and F. Selleri (eds.): *The Wave-Particle Dualism*. A Tribute to Louis de Broglie on his 90th Birthday. 1984 ISBN 90-277-1664-1
4. E. Prugovečki: *Stochastic Quantum Mechanics and Quantum Spacetime*. A Consistent Unification of Relativity and Quantum Theory based on Stochastic Spaces. 1984; 2nd printing 1986 ISBN 90-277-1617-X
5. D. Hestenes and G. Sobczyk: *Clifford Algebra to Geometric Calculus*. A Unified Language for Mathematics and Physics. 1984 ISBN 90-277-1673-0; Pb (1987) 90-277-2561-6
6. P. Exner: *Open Quantum Systems and Feynman Integrals*. 1985 ISBN 90-277-1678-1
7. L. Mayants: *The Enigma of Probability and Physics*. 1984 ISBN 90-277-1674-9
8. E. Tocaci: *Relativistic Mechanics, Time and Inertia*. Translated from Romanian. Edited and with a Foreword by C.W. Kilmister. 1985 ISBN 90-277-1769-9
9. B. Bertotti, F. de Felice and A. Pascolini (eds.): *General Relativity and Gravitation*. Proceedings of the 10th International Conference (Padova, Italy, 1983). 1984 ISBN 90-277-1819-9
10. G. Tarozzi and A. van der Merwe (eds.): *Open Questions in Quantum Physics*. 1985 ISBN 90-277-1853-9
11. J.V. Narlikar and T. Padmanabhan: *Gravity, Gauge Theories and Quantum Cosmology*. 1986 ISBN 90-277-1948-9
12. G.S. Asanov: *Finsler Geometry, Relativity and Gauge Theories*. 1985 ISBN 90-277-1960-8
13. K. Namsrai: *Nonlocal Quantum Field Theory and Stochastic Quantum Mechanics*. 1986 ISBN 90-277-2001-0
14. C. Ray Smith and W.T. Grandy, Jr. (eds.): *Maximum-Entropy and Bayesian Methods in Inverse Problems*. Proceedings of the 1st and 2nd International Workshop (Laramie, Wyoming, USA). 1985 ISBN 90-277-2074-6
15. D. Hestenes: *New Foundations for Classical Mechanics*. 1986 ISBN 90-277-2090-8; Pb (1987) 90-277-2526-8
16. S.J. Prokhovnik: *Light in Einstein's Universe*. The Role of Energy in Cosmology and Relativity. 1985 ISBN 90-277-2093-2
17. Y.S. Kim and M.E. Noz: *Theory and Applications of the Poincaré Group*. 1986 ISBN 90-277-2141-6
18. M. Sachs: *Quantum Mechanics from General Relativity*. An Approximation for a Theory of Inertia. 1986 ISBN 90-277-2247-1
19. W.T. Grandy, Jr.: *Foundations of Statistical Mechanics*. Vol. I: *Equilibrium Theory*. 1987 ISBN 90-277-2489-X
20. H.-H von Borzeszkowski and H.-J. Treder: *The Meaning of Quantum Gravity*. 1988 ISBN 90-277-2518-7
21. C. Ray Smith and G.J. Erickson (eds.): *Maximum-Entropy and Bayesian Spectral Analysis and Estimation Problems*. Proceedings of the 3rd International Workshop (Laramie, Wyoming, USA, 1983). 1987 ISBN 90-277-2579-9

Fundamental Theories of Physics

22. A.O. Barut and A. van der Merwe (eds.): *Selected Scientific Papers of Alfred Landé*. [*1888-1975*]. 1988 ISBN 90-277-2594-2
23. W.T. Grandy, Jr.: *Foundations of Statistical Mechanics*. Vol. II: *Nonequilibrium Phenomena*. 1988 ISBN 90-277-2649-3
24. E.I. Bitsakis and C.A. Nicolaides (eds.): *The Concept of Probability*. Proceedings of the Delphi Conference (Delphi, Greece, 1987). 1989 ISBN 90-277-2679-5
25. A. van der Merwe, F. Selleri and G. Tarozzi (eds.): *Microphysical Reality and Quantum Formalism, Vol. 1*. Proceedings of the International Conference (Urbino, Italy, 1985). 1988 ISBN 90-277-2683-3
26. A. van der Merwe, F. Selleri and G. Tarozzi (eds.): *Microphysical Reality and Quantum Formalism, Vol. 2*. Proceedings of the International Conference (Urbino, Italy, 1985). 1988 ISBN 90-277-2684-1
27. I.D. Novikov and V.P. Frolov: *Physics of Black Holes*. 1989 ISBN 90-277-2685-X
28. G. Tarozzi and A. van der Merwe (eds.): *The Nature of Quantum Paradoxes*. Italian Studies in the Foundations and Philosophy of Modern Physics. 1988 ISBN 90-277-2703-1
29. B.R. Iyer, N. Mukunda and C.V. Vishveshwara (eds.): *Gravitation, Gauge Theories and the Early Universe*. 1989 ISBN 90-277-2710-4
30. H. Mark and L. Wood (eds.): *Energy in Physics, War and Peace*. A Festschrift celebrating Edward Teller's 80th Birthday. 1988 ISBN 90-277-2775-9
31. G.J. Erickson and C.R. Smith (eds.): *Maximum-Entropy and Bayesian Methods in Science and Engineering*. Vol. I: *Foundations*. 1988 ISBN 90-277-2793-7
32. G.J. Erickson and C.R. Smith (eds.): *Maximum-Entropy and Bayesian Methods in Science and Engineering*. Vol. II: *Applications*. 1988 ISBN 90-277-2794-5
33. M.E. Noz and Y.S. Kim (eds.): *Special Relativity and Quantum Theory*. A Collection of Papers on the Poincaré Group. 1988 ISBN 90-277-2799-6
34. I.Yu. Kobzarev and Yu.I. Manin: *Elementary Particles. Mathematics, Physics and Philosophy*. 1989 ISBN 0-7923-0098-X
35. F. Selleri: *Quantum Paradoxes and Physical Reality*. 1990 ISBN 0-7923-0253-2
36. J. Skilling (ed.): *Maximum-Entropy and Bayesian Methods*. Proceedings of the 8th International Workshop (Cambridge, UK, 1988). 1989 ISBN 0-7923-0224-9
37. M. Kafatos (ed.): *Bell's Theorem, Quantum Theory and Conceptions of the Universe*. 1989 ISBN 0-7923-0496-9
38. Yu.A. Izyumov and V.N. Syromyatnikov: *Phase Transitions and Crystal Symmetry*. 1990 ISBN 0-7923-0542-6
39. P.F. Fougère (ed.): *Maximum-Entropy and Bayesian Methods*. Proceedings of the 9th International Workshop (Dartmouth, Massachusetts, USA, 1989). 1990 ISBN 0-7923-0928-6
40. L. de Broglie: *Heisenberg's Uncertainties and the Probabilistic Interpretation of Wave Mechanics*. With Critical Notes of the Author. 1990 ISBN 0-7923-0929-4
41. W.T. Grandy, Jr.: *Relativistic Quantum Mechanics of Leptons and Fields*. 1991 ISBN 0-7923-1049-7
42. Yu.L. Klimontovich: *Turbulent Motion and the Structure of Chaos*. A New Approach to the Statistical Theory of Open Systems. 1991 ISBN 0-7923-1114-0

Fundamental Theories of Physics

43. W.T. Grandy, Jr. and L.H. Schick (eds.): *Maximum-Entropy and Bayesian Methods.* Proceedings of the 10th International Workshop (Laramie, Wyoming, USA, 1990). 1991 ISBN 0-7923-1140-X
44. P.Pták and S. Pulmannová: *Orthomodular Structures as Quantum Logics.* Intrinsic Properties, State Space and Probabilistic Topics. 1991 ISBN 0-7923-1207-4
45. D. Hestenes and A. Weingartshofer (eds.): *The Electron.* New Theory and Experiment. 1991 ISBN 0-7923-1356-9
46. P.P.J.M. Schram: *Kinetic Theory of Gases and Plasmas.* 1991 ISBN 0-7923-1392-5
47. A. Micali, R. Boudet and J. Helmstetter (eds.): *Clifford Algebras and their Applications in Mathematical Physics.* 1992 ISBN 0-7923-1623-1
48. E. Prugovečki: *Quantum Geometry.* A Framework for Quantum General Relativity. 1992 ISBN 0-7923-1640-1
49. M.H. Mac Gregor: *The Enigmatic Electron.* 1992 ISBN 0-7923-1982-6
50. C.R. Smith, G.J. Erickson and P.O. Neudorfer (eds.): *Maximum Entropy and Bayesian Methods.* Proceedings of the 11th International Workshop (Seattle, 1991). 1993 ISBN 0-7923-2031-X
51. D.J. Hoekzema: *The Quantum Labyrinth.* 1993 ISBN 0-7923-2066-2

KLUWER ACADEMIC PUBLISHERS – DORDRECHT / BOSTON / LONDON